INSOLVENT

INSOLVENT

INSOLVENT

HOW TO REORIENT COMPUTING
FOR JUST SUSTAINABILITY

CHRISTOPH BECKER

THE MIT PRESS   CAMBRIDGE, MASSACHUSETTS   LONDON, ENGLAND

The MIT Press would like to thank the anonymous peer reviewers who provided comments on drafts of this book. The generous work of academic experts is essential for establishing the authority and quality of our publications. We acknowledge with gratitude the contributions of these otherwise uncredited readers.

This book was set in Stone Serif by Westchester Publishing Services. Printed and bound in the United States of America.

Library of Congress Cataloging-in-Publication Data

Names: Becker, Christoph (Director of the Digital Curation Institute), author.
Title: Insolvent : how to reorient computing for just sustainability / Christoph Becker.
Description: Cambridge, Massachusetts : The MIT Press, [2023] | Includes bibliographical references and index.
Identifiers: LCCN 2022038283 (print) | LCCN 2022038284 (ebook) | ISBN 9780262545600 | ISBN 9780262374651 (epub) | ISBN 9780262374668 (pdf)
Subjects: LCSH: Electronic data processing—Social aspects. | Computer systems—Environmental aspects. | Information technology—Social aspects. | Sustainable development.
Classification: LCC QA76.9.C66 B435 2023 (print) | LCC QA76.9.C66 (ebook) | DDC 303.48/34—dc23/eng/20221121
LC record available at https://lccn.loc.gov/2022038283
LC ebook record available at https://lccn.loc.gov/2022038284

10   9   8   7   6   5   4   3   2   1

CONTENTS

ACKNOWLEDGMENTS

I want to first acknowledge that colonialism is not in the past. It is present, and it continues to shape our lives. The land on which the University of Toronto operates, and on which I work and live, is the traditional land of the Huron-Wendat, the Seneca, and most recently, the Mississaugas of the Credit River. They have been here for thousands of years. Today, this land is still the home to many Indigenous people. I am grateful that I can live and work on it.

Coming to realize that I am a settler was an uncomfortable grappling, nudged by encounters with others who helped me see beyond my horizon and simultaneously understand better how the past shapes us and continues. This book talks about a concept I call the critical friends of computing: fields that offer a productive, constructive relationship with computing, which needs their insights. Critical friends tell us what others cannot or will not say, and we listen to them because we respect them and trust they have our best interest at heart. In this sense, I regard many of the people who have helped make this book a reality as critical friends.

First, this book really started because of my dear colleagues and friends in the Karlskrona Initiative for Sustainability Design, even though it is in many ways a departure from and even critique of our joint work. It is in no small part because of their enthusiasm, collegiality, and friendship that I was able to shift the central focus of my research to sustainability design and, later, just sustainability design. I hope the book offers them some critical friendships.

Since then, kind and wise people have been invaluable in providing ideas, advice, critique, and suggestions. Andrew Feenberg graciously agreed to co-organize the Dagstuhl Seminar on Values in Computing in 2019, in which his presence and thoughts were crucial to not only ensure participation of key participants outside of computer science, but also to provide a framework of thought that was instrumental in shaping the discussion at this meeting point between computing and some of its critical friends. His advice on what I should read were keenly followed by this student and very helpful. Elizabeth Patitsas and Andrea Thomer gave me early opportunities to present some of the ideas here to a wider audience at McGill University and the University of Michigan. Bill Tomlinson graciously provided spontaneous support during my sabbatical stay at UC Irvine, cut untimely short by the sudden emergence of a deadly virus. Geoff Bowker took the better part of a day to learn about my struggle to reconcile science and technology studies and critical systems thinking and offered useful suggestions.

Laura Portwood-Stacer's book proposal accelerator program in 2019 helped me immensely in shaping the argument of the book. Being nudged to articulate what the book was saying in more detail than I was prepared for eventually resulted in a much clearer arc. Her encouragement and constructive feedback were as helpful as her inside-outside view on how books are made. Brian Cantwell Smith commented on the book proposal and, later, on early chapters with his usual precision and nuance. I am very grateful to Gita Manaktala at MIT Press for believing in this project and finding such excellent reviewers in times where those seem impossible to find, to Suraiya Jetha for her support in navigating the complex process, to Molly Seamans for designing the striking cover, and to Patsy Baudoin, whose developmental editing helped to sharpen, tighten, and iron out my English. And of course, I greatly appreciate the constructive and encouraging feedback received by my manuscript readers.

A group of colleagues and friends took it on them to read draft chapters as I was writing, and I am deeply grateful for the significant amount of time they spent on my imperfect writing and the helpful critique and suggestions they offered—true critical friends, as I like to think. Ola Leifler read the whole book, chapter by chapter, encouraging me to go further and be clearer. Stefanie Betz read the first half, offered very helpful reflections and suggestions, and similarly spent hours debating these ideas. Doris Allhutter made sure I was not making a fool of myself in chapter 5 when talking about STS. Elizabeth Patitsas did the same for several earlier chapters, and our conversations have helped me make sense of my own disability and gain new perspectives on crip technoscience.

This book's flaws are most certainly my own, but the argument emerges from and draws on the collaborative research I have been fortunate to be involved in over the past decade. This includes the Karlskrona group who I have coauthored with: Sedef Akinli Kocak, Stefanie Betz, Ruzanna Chitchyan, Leticia Duboc, Steve Easterbrook, Birgit Penzenstadler, Jari Porras, Norbert Seyff, and Colin Venters. I want to highlight some whose involvement was particularly generous and insightful. Leticia Duboc was the first to embark on a critical friendship adventure between requirements engineering and critical systems heuristics with me, and her interest, ethical commitment, and friendship means a lot to me. Chapter 10 would have never happened were it not for our collaboration and the participants of the CREW workshops in 2020 and 2021, including Francien Dechesne, Antonio Vetrò, Matti Nelimarkka, Christoph Schneider, and others mentioned here. Doris Allhutter's critical friendship, also featured in chapter 10, is offering a lot more than the book can capture, and I look forward to exploring it further. That Fabian Fagerholm decided to apply for a postdoctoral fellowship I advertised was my incredible luck. The hours we spent discussing intertemporal choice and conducting our studies were endlessly enriching, and his many insights resonate all across chapters 7 and 11. The studies on intertemporal choice would have never taken place were it not for the collaborations Fabian and I developed with Alexander Chatzigeorgiou, Stefanie Betz, Leticia Duboc, Birgit Penzenstadler, Rahul Mohanani, Colin Venters, Carol Cardenas Castro, and Jenny Gil. Some of the arguments made in later chapters have benefited from my collaborative work and interactions with critical friends in science and technology studies, participatory design, HCI, and adjacent fields, including Lucy Suchman, Andrew Clement, Ann Light, Chris Frauenberger, Victoria Palacin, Wolmet Barendregt, Douglas Schuler, Syed Ishtiaque Ahmed, Rachel Charlotte Smith, Pedro Reynolds Cuéllar, David

Nemer, Erhardt Graeff, Shion Guha, EunJeong Cheon, Devansh Saxena, Kenneth R. Fleischmann, and Maria Angela Ferrario. And I would like to thank the Fellows I had the honor of hosting at the Digital Curation Institute during the pandemic, whose research, collegiality and discussions have enriched the pandemic isolation and made it feel a little less lonely: Victoria Palacin, Vanessa Thomas, Ren Loren Britton, Isabel Paehr, Doris Allhutter, and David Wachsmuth.

Like many academics, I learn a lot from my students. I am grateful to the students at the just sustainability design course I have taught since Fall 2020, whose responses, discussions, and insights have helped me shape the argument presented here. I have been fortunate to advise PhD students Dawn Walker and Curtis McCord, who I continue to learn from and who have been involved in some of the projects discussed here too. Curtis has explored and pushed forward the use of critical systems thinking in systems design in his own PhD research as well as our joint work. And my graduate research assistants at the Faculty of Information at the University of Toronto, especially Andres De Los Rios, Tara Tsang, Enning Zhang, and Rachel Booth, have been wonderful to work with.

I wrote most of this book during two years of pandemic isolation. I am grateful to those who supported me during this time, especially Zippy, Andres, Elana, and Sarah. Nox and Lux were the best supervisors one can hope for.

Some research covered here was supported by the Canadian National Science and Engineering Research Council through RGPIN-2016–06640, the Canada Foundation for Innovation, the Ontario Research Foundation, and the Connaught Fund. Early work on this manuscript was supported by a research stay at the Leibniz Center for Informatics at Dagstuhl, sponsored by the German Academic Exchange Service. And some of the groundwork was funded by the Vienna Science and Technology Fund, the best-run funding agency I know.

INTRODUCTION
JUST SUSTAINABILITY DESIGN

A systems approach begins when first we see the world through the eyes of another.
—Churchman (1979a, 231)

It took a plague to make some of the people realize that things could change.
—Butler (1993)

NOT ALL IS WELL IN COMPUTING

In April 2022, a single bitcoin transaction produced over a thousand kilograms of CO_2 emissions and 350 grams of electronic waste. These numbers are enormous when compared to conventional payment systems like credit cards: the carbon footprint of a single purchase paid in bitcoin equals that of over 2.6 *million* credit card transactions (Digiconomist 2022). This is not coincidence: it is a systemic consequence of a key design decision in the bitcoin system that leverages the relative strength of compute power as a mechanism to decide what is valid (Nakamoto 2008). The largest pool of processing power decides, and the power consumption of the network as a whole is bound to rise together with the economic valuation of the currency: in 2021 alone, the carbon footprint of bitcoin transactions tripled. If its use continues, its emissions *alone* could cause global warming in excess of 2°C (Mora et al. 2018). The impact of this warming, and of other human-driven decline of nature, falls disproportionately on the world's poorest, who do not use bitcoin at all, and on nonhuman nature. One million animal and plant species are now on the fast track to extinction (Díaz et al. 2019).

The deep entanglement of information technology (IT) with our societies has raised high hopes for a transition to more sustainable and just societies. In principle, computing can be key to environmental sustainability and social justice. It can enable open information access, support sustainable lifestyles, drive the dematerialization of goods and services, and support the partial decoupling of value creation from material resource consumption. But in practice, the opposite is more common. Everywhere on this planet, computational systems drive environmental damage, increase the demand for resource extraction, reinforce inequality and injustice, and enable the erosion of privacy and democratic governance. This is especially true for those systems that are touted as success stories.

The computational systems we live with today often advance economic objectives at the expense of social and environmental considerations. Those we design tomorrow must advance social and environmental values simultaneously with technical and economic objectives. This fact has prompted many in computing to work on designing for sustainability and to tackle questions of justice and fairness in computing. But their good intentions have not changed the trajectory of computing. On the contrary: the rising tide of reports about the harmful implications of computing indicate that its debts are mounting. But these debts are treated as what economists call *externalities*—they are outsourced, offshored, foisted on those distant from the design processes.[1] From designed obsolescence to excessive energy consumption, from racist algorithms to the indirect implications of platform economies and surveillance capitalism, the choices made in the design and development of computational systems have implications at farther distances than ever before.

These choices come in many forms. I use the term *systems design* to refer to the broad set of intentional activities that directly or indirectly shape computational systems.[2] This encompasses activities such as software development that are often seen as "technical," and activities such as the facilitation of workshops with stakeholders that are often seen as "social." This simplistic distinction between technical and social hides that most systems design activities are simultaneously social and technical because they explicitly or implicitly reconcile social aspirations, expectations, needs, and concerns with technical constraints and material conditions. This is openly acknowledged in areas such as user interface design, architectural design, requirements engineering, or project management. Because of its wide sociotechnical spectrum, systems design is taught and practiced in a range of disciplines including human–computer interaction (HCI) and software engineering by many stakeholders, not all of whom consider themselves computing professionals.

The social and temporal distance between design choices and their outcomes entails uncertainty, ambiguity, and what moral philosopher Gardiner (2014) terms *asymmetric vulnerability*: those affected by technology development and design decisions have little means of influencing the outcomes—in the case of future stakeholders, none at all. One might say that these stakeholders carry the debts of computing: the wide-ranging ways in which the outcomes of systems design weigh on our planet and its societies.

Once we account for these debts, the question is whether computing can pay them back or whether we should consider it *insolvent*. As the title suggests, I will argue that in its current form, computing is indeed insolvent: It is incapable of paying back the debts it owes to this planet and its societies. It is so stuck in its ways of thinking that, to make progress, we need to rethink and restructure it.

This book will start from the premise that the rise of computing's debts is shaped by the ways of thinking that dominate it. This book's central argument is that computing can only pay them back and become a force for sustainability and justice once we rewrite central narratives of systems design. Three ideas in particular have shaped the discourse of computing like a river's undercurrents—that algorithms and computational systems are politically neutral tools; that problems and requirements are objectively given facts; and that the individuals and teams who design and develop computational systems are rational agents whose reasoning processes are admittedly flawed and biased and incomplete but, nevertheless, best described and approximated using the computational metaphor of information processing. The modern ideas of technological neutrality, scientific objectivity, and rational decision-making have played central roles in shaping the self-understanding of computing throughout the past seven decades, even though their validity and footing vary across the range of fields and perspectives in computing. In systems design practice and research, these ideas manifest *as myths*—"foundational narratives that are ritualistically circulated within groups to reinforce collective beliefs" (Ames 2019, 18)—when their limited scope is over-extended into a context in which their validity is not empirically or theoretically supported. Together, these myths drive a narrow focus that views design as using computational tools to solve problems described by scientific reasoning, based on the assumption that the resulting technology is value neutral. This narrow ideology is stuck. To overcome its impasse, we need to rewrite the narratives it is built on.

This book examines the role and validity of the myths of computing in order to reorient systems design practice and theory in computing toward the values of sustainability and justice. My aim is to illustrate *how* these ideas manifest in systems design *as myths*; to demonstrate how the myths interact to produce misleading beliefs about the nature and implications of systems design practice; to show how these misleading beliefs prevent meaningful engagement with the challenging questions raised by social justice and sustainability; and on that basis, to develop concrete steps to reorient systems design practice, research, and education.

In other words, to genuinely progress toward sustainability and justice, we need to combine the conversation about how computing shapes our lives with a close and critical look at how it comes to life—how we design computational systems and how we should design them. Doing so shows that despite many promising efforts, computing in its current orthodox form is not paying the dues it owes to our planet and its societies because its dominant mode of engagement through computational problem solving cannot recognize and reckon with the politics and values that shape technology design. The politics of technology design, the social dynamics of participation in design, and the cognitive factors of engineering and design decisions reveal how the myths of computing distort the discourse and shape what we can talk about when we talk about sustainability and justice.

WHAT IS JUST SUSTAINABILITY DESIGN?

In this book, I introduce just sustainability design (JSD) as a framework for systems design practice, research, and pedagogy that privileges sustainability and justice and, therefore, the asymmetric and uneven effects of systems design choices at a distance. The purpose of JSD is to bring about improvement, not just avoid damage. Just sustainability design is sensitive to the discursive nature of systems design and the dangers presented by misleading narratives that keep systems design practice captive to a false consciousness. Because it centers the concerns of sustainability and justice, JSD must attend to five important factors they raise:

First, most profound effects that we recognize as sustainability or justice do not happen right away. Instead, they take place later, over longer time periods, removed from the moment in time and place where the decisions took place that caused them. So JSD must account for the uneven *dispersal* of design effects and their ripple effects across spatial and temporal scales.

Second, because the dispersed effects are distant, fragmented and scattered across these scales, they are *uncertain as well as ambiguous*. Uncertainty captures the understanding that there is a probability associated with these effects: few are certain, but some are more likely than others. Ambiguity captures the uncertainty about uncertainty: for many potential outcomes, we will not have probability distributions, and for many, their meaning to those affected is not immediately clear to the designers. Ambiguous outcomes can be interpreted in more than one way. These two aspects are related but distinct because addressing uncertainty is very different from addressing ambiguity.

Third, the many stakeholders—people involved in or affected by systems design activities, and non-human life affected by it—are not all able to participate equally, and future stakeholders are unable to participate at all. This causes a *fragmentation* of perspective and agency across the many stakeholders involved in or affected by design decisions.

Fourth, some stakeholders will have the ability and interest to influence systems design to various degrees; others will be affected indirectly and have much less influence, even if they are interested. This causes *power dynamics* across direct and indirect stakeholders with unevenly distributed influence over these decisions in systems design.

Finally, the nature of these complex ripple effects, and the fragmentation of perspectives, means that multiple interpretations of the same situation or outcome may be articulated based on assumptions that are mutually incompatible. Similarly, two stakeholders may evaluate outcomes in ways that cannot be reconciled easily, because there is no common scale of measurement. JSD must account for this *incommensurability* across different views of the complex design space it aims to address.

These five factors—dispersal, uncertainty and ambiguity, fragmentation, power dynamics, and incommensurability—raise questions that are social, cognitive, and political as well as technical. To address them, we need a theoretical framework for systems design in the twenty-first century guided by the criteria outlined in broad strokes below.

1. In a world that is deeply unsustainable and unjust, the uncritical pursuit of technology design as instrument in the unreflective service of dominant interests cannot possibly be a force for sustainability and justice. But we still need to build technology that makes sustainable and just societies a reality. Critique alone, as important as it is, will not bring about the constructive approach still required to do that. So JSD must be simultaneously *constructive and critical*: that is, it needs to build positive change while constantly questioning hegemonic norms and assumptions.
2. Because technology is social and society is technological, JSD must prioritize the integrated understanding of sociotechnical systems and their environments over the isolated analysis of individual components. In one word, it must be *systemic*.
3. Because sustainability and justice involve such a broad range of dispersed, fragmented, uneven, and sometimes incommensurable perspectives, JSD must represent a *pluralist* ground on which divergent worldviews can meet without requiring full consensus.
4. Because just sustainability is about the future, JSD must consider the *temporal* scale and dynamics of design in a historical context.
5. Technology never happens in a social or cultural vacuum. Far from being "independent of culture," JSD must be aware that its *perspective* will always be partial, built on unspoken assumptions, and it must be equipped to reflect on its already given context, boundaries, and assumptions.
6. In situations of ambiguity and incommensurability, goals of optimization are only appropriate or meaningful in a very limited way, as we will

see. The primary concern of JSD must be whether its proposed interventions are *legitimate*, rather than whether they are optimal.

These directions for JSD emerged through collaborative research in which I explored the role of computational systems in sustainability and developed tools that support the collaborative exploration of distant effects in systems design among heterogenous stakeholders (Penzenstadler et al. 2018; Becker et al. 2016). This work, discussed in chapter 1, allowed me to recognize the limitations of conventional systems approaches applied to sustainability and justice, as explored throughout part I. The principles of JSD are expanded and refined in chapter 9.

In its commitments, JSD is consistent with critical approaches to design such as design justice (Costanza-Chock 2020) and data feminism (D'Ignazio and Klein 2020). It shares its commitment to be constructive and critical with critical making (DiSalvo 2014; Ratto 2011), feminist HCI (Bardzell 2010), and crip technoscience (Hamraie and Fritsch 2019). It shares the aim to find paths off the rationalist highway with Rosner (2018) and its embrace of epistemic pluralism with Escobar (2018). What distinguishes it in orientation from these allies is an agenda of social change that aims to orient systems design practice toward the values of sustainability and justice by introducing a *critically systemic* perspective and approach to reflection, emancipation, and critique into the engineering methods of systems design in computer science and cognate fields.

REORIENTING SYSTEMS DESIGN TO SUSTAINABILITY AND JUSTICE

There is no need to belabor the urgency of environmental sustainability in the age of the climate crisis. Others before me have mapped out some of the roles that IT can play in addressing sustainability (Tomlinson 2010) and highlighted the problematic role that IT also plays in fostering environmental destruction (e.g., Hilty and Aebischer 2015b; Crawford 2021). It is important to add the long-standing argument that environmental sustainability should never be treated in isolation from social justice (Agyeman, Bullard, and Evans 2003; Agyeman et al. 2016; Agyeman 2013), but the role of IT in social justice has been explored more competently by others

(Benjamin 2019; Costanza-Chock 2020; Eubanks 2018; S. Noble 2018). What can a book on sustainability, justice, and IT add to the above?

To my knowledge, this is the first book that explicitly focuses on the intersection of sustainability and justice in the context of systems design and explores the consequences. In doing so, I am less interested in adding substantively to the existing forceful critique of computing's material implications in systemic racism and inequity, or to the substantive exploration of the environmental effects of IT production and consumption. Instead, I focus on how issues of sustainability and justice get discursively recognized and addressed in systems design practice, research, and education. I am especially interested in how the myths of computing cause these concerns to be recognized, framed, and interpreted in particular ways, and I want to examine how conducive these framings are to genuine improvement.

Three examples will illustrate how the myths distort the discussions in subtle ways. A few years ago, I attended a week-long workshop on modeling for sustainability at Dagstuhl, the prestigious informatics center in Germany. Three dozen carefully selected experts, leading representatives of the fields of model engineering and sustainability in computer science, came together on invitation of the workshop chairs to discuss the role that modeling research should play in supporting environmental sustainability. The seminar room was full of critical thinkers, highly educated computer scientists who want computing to help the world transition to a sustainable way of life. At one point in the middle of the week, we agreed that we needed to discuss the intricate role that human values play *in modeling*. A breakout group formed to map out and discuss the thorny questions that arise. Or so I thought. After half an hour, the conversation took a turn, I took a backseat, and a mere twenty minutes later, the group felt they had completed the task. They had *modeled human values*. A neat chalk figure on a blackboard documented a decomposition of values into constituent elements and outlined a number of relations that would have to be represented in a model of human values. Our subject matter had been thoroughly defeated. Modeling had once again remained victorious. I must have had an incredulous look on my face when I asked if no one else thought that we had gotten off track? I was certainly met with bewildered looks—there was a palpable sense of disconnection in the room. What was I talking about? Wasn't this what we set out to do?

I struggled to explain that I was hoping to discuss how human values operate *in* modeling. What I and one other colleague had argued for was different in nature—we wanted to examine how deeply held beliefs and commitments (values) play out in social processes where human beings create models to reason about environmental sustainability.[3] We wanted a conversation that combined the constructive view, how model engineering can help to orient our life toward sustainability, with a close and critical look at how these models themselves come to life. That week is when I resolved to write what became this book.

Many in computing want our field to be a force for good—a force for sustainability, equity, diversity, justice, and similar values. Many are involved in projects with lofty aims along these lines, and many of these projects are worth pursuing. But more often than not, their work does little to make the world a more sustainable or just place. On an aggregate level, computing's environmental impact continues to worsen, and the reports of harmful technology and its implications in the erosion of our democratic governance systems continue unabated. At the time of the workshop, I had been working on sustainability in and through software systems for over a decade. I had shifted from a focus on the sustainability of digital software-dependent resources and information systems to a broader concern with the longer-term social and environmental implications of software technology design. Four years before the workshop, I had cofounded the Karlskrona Initiative for Sustainability Design. Its manifesto, discussed in chapter 1, calls attention to the need to consider sustainability holistically in every software project. It had received widespread attention and support in software engineering and requirements engineering. But a nagging suspicion had grown in me. I worried that the energy, momentum, and support for sustainability in computing were getting misdirected, seeing that the consequences and actions drawn from our arguments did not appear to enact lasting, transformative, sustainable change—merely business-as-usual under a new name.

The incident at the workshop brought home one central reason why. My colleagues did not consider how models embed values. Instead, they placed the arguments for the importance of human values in modeling into the conceptual and narrative structures arising from their disciplinary ways of thinking—above all, computational thinking. Inside that frame, the issue of human values *in* modeling becomes a challenge *of* modeling. Our question was meant to be a critical one: How are the resulting models shaped by the values of those who model? But this question is inaccessible to computational thinking and off limits to a reasoning mode of deductive logic in which the correspondence of modeled artifacts to "the real world" is taken as the only validity criterion of epistemology and then bracketed off from debate as an a priori condition not worth debating (B. C. Smith 1993). The question didn't even have to be explicitly reformulated—it was directly processed as a question of modeling. Why? Because the discursive worlds of the computing professions make it appear that way. They continue to lay out narratives that tell a story of computing and software technology as politically neutral tools built to address objectively given problems in processes best described as "problem solving" by groups of individuals whose cognitive processes are viewed as "information processing."

This was not the first time I had run against this brick wall. A year earlier, some of the same experts met at a different venue to discuss the same topic. One of the first presentations, presented by a civil engineer with environmental sustainability expertise, argued for *sustainability engineering* as an approach to "the *wicked problem* of sustainability." Early on, a slide with the ten properties of wicked problems. On the next slide: *How to solve this wicked problem.* Sustainability engineering was meant to be "objective; repeatable; sensitive to, but independent of culture; universal; and complete." I was baffled because, in my mind, the exact opposite was required: An approach that was not objective but dialectic; not repeatable but replicable; not independent of culture but embedded *with* culture, acknowledging the inevitable entanglement as a founding condition; not universalist but situated and contingent; and certainly not complete, but proudly incomplete and evolving. What was presented in this session as "systems thinking" really wasn't. It was deeply reductionist and failed to appreciate the nature of wicked problems (Rittel and Webber 1973). I scribbled furiously into my notebook.

Nor was it the last. A year after the values modeling exercise, I was back at Dagstuhl on a retreat. For the duration of one week, during lunchtime,

I was seated with the participants of a Dagstuhl seminar on blockchains. By that time, I had started work on this book, and when asked about it, I spoke about it in terms of social justice and sustainability. Inevitably, some colleagues from formal areas of computer science would struggle to see how that was "computer science." Others would nod instantly and say that they also worked on this topic—usually, they had large amounts of grant money for some variation of "artificial intelligence (AI) for good." In one case, a grant on AI and blockchain for justice resulted in the worst lunch conversation I had in a very long time. It turned out that "justice," in the eyes of this eminent member of the distributed databases research community, meant an equal distribution of measurable risks and benefits among a set of individuals out of an identified population. I suggested that this narrow version of "fairness" is not the same as justice and that social scientists and others had something substantive to say about that topic. For example, we can hardly bring about justice merely by *defining* supposedly ideal institutions for justice and assuming that people's behavior will be compliant with those (Sen 2009), and it is crucial to distinguish between equal distribution of goods, equitable treatment of people, and justice. Algorithmic justice goes far beyond debiased fairness to include a historical understanding of oppression and injustice (cf. Costanza-Chock 2020, 63–65). But my comments were brushed off. As Jacobs and Wallach (2021) wrote later, the operationalization of essentially contested concepts of fairness and justice into narrow observable attributes is "appealing to computer scientists because they operate within the boundaries of a single computational system, without reference to the broader societal context in which the system is situated . . . these operationalizations necessarily lack aspects of the substantive nature of fairness" (382). This was not long after I had written the commentary for a brilliant polemic demonstrating just how inadequate computational reasoning mechanisms are when it comes to fairness, justice, and ethics (Keyes, Hutson, and Durbin 2019). I left the lunch a bit shaken. It was just too real.

These are not isolated incidents. I single them out here to illustrate how the undercurrent of the computing field shapes its discourse about sustainability and justice concerns. In hindsight, we can recognize in these stories the contours of the myths of computing. In the first instance, the meaning and texture of "human values in computing" were lost in a modeling view that considered only the problem of how to model values and failed to see that modeling itself is subject to values. Because the technology of model engineering itself was seen as a neutral tool, the idea that values might play a role in the construction of its artifacts was not even considered a possibility. Instead, the only question was how to use this tool to solve the problem at hand. In the second, the nature of sustainability as a wicked problem was misunderstood by an engineering view that interpreted it as a particularly difficult problem to be solved. And in the last, the nature of justice was interpreted through a computational lens that sees only a computational problem, unable to grasp what lies beyond that narrow field of vision. The individuals I encountered were not bad people, and they are brilliant thinkers in their fields. Like fish in water, my colleagues did not pay close attention to how their views were socially shaped and limited by the conceptual structures of their fields. Most fish in water have little appetite for learning how to walk, but fortunately, humans aren't fish, and many carry an enormous appetite for new grounds.

I examine four undercurrents and their interactions. In their naïve form, these are:

1. **The myth of value-neutral technology** tells a story of technology as neutral: that is, impartial and value-free. In this story, technology bears only facts. Values arise only in its interpretation.
2. **The myth of objective problems** tells a story of problems as objective entities that are the objects of problem solving and design. In this story, problems need to be represented correctly so that they can be solved using technology design.
3. **The myth of rational decision-making** tells a story of the human brain as an information processor making decisions. In this story, rationality is defined by reference to a normative ideal of decision-making embodied by a computer, and deviations from this ideal are treated as biases or mistakes. The story admits that the brain is imperfect but retains allegiance to the idea that rationality is the appropriate ideal in terms of which to understand it as an approximation.
4. **The myth of solvency** tells an optimistic story of technology as the savior solving problems for our world. In this story, the central activity of technology design is problem solving, and computer science is a problem-solving discipline. To design is to collaboratively solve

objective problems using value-neutral technology through a series of rational decisions. The story is headed to a happy ending. Unintended side effects are lamentable collateral damage, but they are exceeded by the benefit—it is "worth it." In using the terms *in/solvency*, I connect their common meaning—(in)solvency as the (in)ability to pay one's dues—to the notion of problem solving. A *solvent* in chemistry is also something that provides a solution.

To do justice to their texture and influence, this book will give space to each story, recognizing its merit, its scope of validity, and its historical evolution. It will develop a framework drawing from a range of fields I call the *critical friends of computing* to examine the limitations of each story and the distorting influence that it has on systems design for sustainability and justice. In more humorous terms, it will also sometimes add entries to a Devil's Dictionary of Computing.[4] Here are two:

> **Problem**, n.: something that can be <u>fixed</u> or solved.
> **Fix**, n. & v.: the source of tomorrow's <u>problems</u>.

MY STANDPOINT

Two of these critical friends, feminism and critical systems thinking, emphasize the importance of clarifying the position from which knowledge is produced. Feminist standpoint theory speaks of embodied, *situated knowledges* (Haraway 1988). Since every perspective is partial, we need to understand in which way. Disclosing my own standpoint makes visible the partiality of my own perspective. Critical systems thinking speaks of implicit *reference systems* (Ulrich 1983), emphasizes the importance of making visible the boundaries and membership of what Midgley (2000) calls the "knowledge producing system," and to reflect on its positionality in the world by critiquing those boundaries. These concepts will return in chapter 5. For now, here is my standpoint.

I am a white cis man, born in Austria, of central European descent, with an invisible disability. I was born without that disability into relative privilege—a household of teachers and practicing musicians in Salzburg, a city defined by classical music and baroque history. I attended a conservative "humanist" high school where I learned Latin, Ancient Greek, and some Western philosophy. I read a book a day for most of my childhood; I played competitive chess and devoured chess books; and I learned to play the piano. My visible orientation as a nerd, combined with classroom dynamics and bad luck, resulted in permanent bullying throughout my school years. Chess opened a window to what I would later recognize as computational thinking, and I discovered, by way of chess, that I was really good at programming.

I was diagnosed with Type I diabetes at age seventeen, but interestingly, I never considered myself as living with a disability until much later, when I had lived through years of experiences of oppression through the North American sickness industry (pardon: healthcare system).[5] It took the lived experience of what Bowker and Star call the *torque* of classification systems (Bowker and Star 1999); of seeing my most intimate life signals monetized as data by "healthcare" startups; of lacking meaningful agency or control over the repurposing of my body's minute-by-minute statistics as data assets ("the new oil," remember?); of agonizing over endlessly repeated attempts to receive refunds for officially covered treatments from for-profit insurance providers; and of realizing just how much of my time is spent on managing my condition in ways able-bodied persons never have to see,[6] to recognize my situation as different. I am well aware that this experience is still vastly helped and buffered by my privilege as a tenured academic with private healthcare, and as a settler in Canada. Recognizing myself as such took me a long time (Lowman and Barker 2015).

My educational path to this position was supported by luck and a free public education system. This being central Europe, I never had a student loan or a credit score. Education was free, and healthcare was a given—a position of safety, freedom, and agency much more common in Europe than in the Global South or in North America. How privileged I was in comparison to others only became clear to me much later. My choice to study computer science opened ample opportunities to gain income and experience. Throughout my student years, I worked as freelance software developer, then software architect, then project manager, in projects of all sizes and various domains, from solo assignments on geodesic software to a contract as a software architect in the IT department of one of the world's largest financial corporations. This allowed me to develop a professional competence of delivering working technology that turned out to be invaluable later on. Only in part due to my work focus, I had terrible grades in the first few years of my bachelor's degree. But work experience taught me why I wanted the degree, and by the end of my first master's degree I had excellent grades. Yet I ended up in academia by serendipity. I hadn't applied for a doctoral program at a prestigious university, and I wasn't even thinking about it when I graduated in spring 2006. Instead, my thesis supervisor had a large European grant incoming and asked me to join it as a doctoral student and salaried full-time research assistant. I was mildly uninterested at first about the obscure subject of "digital preservation." I thought I belonged in the IT industry. I was convinced otherwise by a week-long summer school in a monastery in Tuscany, sharing breakout sessions in the garden with archivists and academics at the intersection of humanities and computing. Suddenly, I found myself in a group of people who genuinely cared about books. My doctoral program was free too, entirely unstructured, and formally detached from my grant-funded full-time employment as "project assistant."

I happened to thrive in this environment because I found myself able to deploy my project management and software engineering skills to deliver systems for decision support that helped shape a roadmap for digital preservation. This allowed me to win a European grant while graduating and a more foundational national research grant a bit later as a postdoctoral principal investigator. I realized that I loved the intellectual freedom of academic research. At that point, I had expertise in software technology, project management, and digital preservation, and I understood that "freedom" primarily as a way to find interesting problems to work on, get funding to do so, and pursue my intellectual curiosity in the process of solving those problems. Social responsibility or privilege were not yet a central part of understanding that freedom—nor was the possibility that some of these problems may have been "wicked." But I had also traveled widely; publishing articles in conference proceedings in various disciplines during my doctorate, I had the privilege of access to substantial travel funding. In the absence of care duties, I was able to append time for travel after many conferences. A formative side effect was a heightened appreciation of colonialism, inequality, social justice, and environmental destruction, as well as a growing awareness of the partiality of perspective, as I began to notice the blind spots of the European colonial tradition I inherited in education and through my social life in central Europe.

I undoubtedly had also subscribed to every myth identified in this book. The ideals of enlightenment rationality, computational thinking, problem solving, and IT as a savior were—implicitly—very prominent in my thinking. My Austrian-funded project included a peculiar research objective: "Longevity engineering" was motivated by the recognized lack of long-term thinking in systems design. The aim was to deliver metrics, measures, and evidence-based explanatory models for the lack of longevity in some systems but not others in order to integrate considerations of longevity into the lifecycle of information systems from the beginning (Proenca et al. 2013). A curious shift took place during this project. The failing attempt to quantify, measure, and model the "longevity" of systems so that it could be predicted and increased led me to recognize the framing as misguided. (From some different standpoints, of course, this would have been obvious all along.)

What began as an engineering-focused project led to *sustainability design*. It emphasizes a set of misperceptions about sustainability in systems design and proposes, in response, a set of commitments that make up the sustainability design approach (Karlskrona Initiative 2015; Becker et al. 2014). The double switch in terminology was significant: Instead of the engineering focus on measurement, prediction, and optimization, the manifesto places a focus on design—what Winograd and Flores (1986) call "the interaction of creation and understanding" (4); and instead of a *long life*, which can only be verified ex-post, it places a focus on the prospective "capacity to endure": sustainability (Fowler and Fowler 1995). The first small step toward the position in this book was taken.

In 2013, I joined the Faculty of Information at the University of Toronto on the tenure track. Thrown into a space where roughly a third of my colleagues hold a computer science doctorate, a third a social science doctorate, and a third a humanities doctorate, I began to read up on what seems like a self-replenishing Borgesian library department of all the books one must have read, and to appreciate the situated nature of knowledges. I faced the challenges of intercultural communication and some of the invisible systemic barriers that immigrants face in North America. As one of a few without a degree from what Toronto considers a "peer institution," the lack of a high-profile support network gave me some first-hand experience of the dubious mechanisms behind academic "meritocracy" (Labaree 2019; Iwen 2019; Appiah 2018) and the reactionary disciplinary politics of the Canadian peer-review system (Semeniuk 2017).

The unevenly distributed toll of the COVID-19 pandemic reminds me that despite the challenges I face as an immunocompromised person, I am incredibly lucky to be in a secure and privileged position. The vulnerability that comes with my disability has created a strange juxtaposition of privilege and oppression that has taught me to reflect and develop a certain sensitivity to privilege. I believe it also helps me to appreciate the nuance afforded by an intersectional approach that considers the *matrix of domination* (Hill Collins 1990).

As we learn simultaneously that the pandemic has dampened the accelerating rise of CO_2 emissions significantly, but only temporarily (Le Quéré et al. 2020), that the carbon footprint of Jeff Bezos's eleven-minute joyride in space exceeds the individual lifetime footprint of the planet's poorest billion people (Chancel et al. 2021), and that the shareholders of large tech companies are benefiting enormously from COVID-19 (N. Klein 2020), while poor communities, including those employed by Big Tech, were hit disproportionately hard in yet another illustration of systemic racism (K.-Y. Taylor 2020), it is clear that sustainability and justice are the challenges and opportunities of our lifetimes. Our societies *can* reshape significant parts of the way they operate. Each of us *has* "room for maneuver," to paraphrase Feenberg (2002)[7]—some have more, some have less.

Our future is in the hand of the collective actions of our societies. The design of computational systems plays one relatively central role in the reshaping of our ways of life. How we understand this role, and how we play our various parts, is our responsibility. This book aims to help us carry that responsibility through a systematic reflection on how we can think about systems design for sustainability and justice. It aims to help delineate the room for maneuver that exists for those participating in systems design, and writing this book is an expression of the room for maneuver I see for myself, from my standpoint.

Hindsight allows me to recognize the factors that stimulated my own appetite to recognize and transcend the water I was swimming in. My move from computer science to an information faculty entailed constant encounters with other disciplines, other norms, other values, and other literatures; in my research, the project of longevity engineering was going nowhere; and when the Karlskrona group committed to taking *systems thinking* as the transdisciplinary starting point for sustainability design in 2014 following compelling advocacy by Steve Easterbrook (2014a), I embarked on a chronological reading of systems thinking literature throughout the twentieth century.

THE CRITICAL TURN IN SYSTEMS THINKING

This book draws significant inspiration from the work of critical systems thinkers. It uses their arguments to advance its own, and it shows some direct applications of their work to systems design practice. But I have learned that the broad term "systems thinking" is open to significant misinterpretations because it is used to reference an enormous variety of work that is typically classified in a wide range of disciplines. It is beyond the scope of this book to provide a comprehensive historical overview, but I want to retrace a few core concepts and developments.[8]

The core idea of Systems Thinking is the realization that some aspects of interest of the world cannot be decomposed into constituent segments without disappearing or losing their meaning: "to make sense of the complexity of the world, we need to look at it in terms of wholes and relationships rather than splitting it down into its parts and looking at each in isolation" (Ramage and Shipp 2009, 1). This idea, now commonplace, arose in many different fields—from biology to family therapy, from psychology to climate science, from ecology to pedagogy, from economics to organizational theory. Systems Thinking therefore comes in countless forms. Biologists Maturana and Varela explored the biological roots of cognition and knowledge and defined the concept of *autopoiesis* (Maturana 1980; Maturana and Varela 1992), while cultural anthropologist Margaret Mead thought about the dynamics of social change (Ramage and Shipp 2009). Economist Kenneth Boulding placed an understanding of economics into a comprehensive set of hierarchies and "considered the world variously as a physical, biological, social, economic, political, communication and evaluative system" (Ramage and Shipp 2009, 70), while professional practice scholar Donald Schön wrote about reflective practice in professions such as music and design. These approaches to systems thinking, to name just a few, have little in common in terms of their *content*. What they have in common is a way of thinking that pays primary attention to thinking about wholes and how they are constituted and organized in given contexts.

The proliferation of the "systems" term in colloquial language—the healthcare system, the education system, the transportation system, the information system—indicates that many fields are inherently looking at "systems" of education, management, ecology, public health, and so on. The century-long history of Western systems thinking has also resulted in systemic thought dissipating deep into the way we reason about the world. Few disciplines have *not* seen some form of systems thinking. Because of the diversity inherent in systems thinking, it can be difficult to understand what makes an approach *systemic*. Systems thinking often sits uneasy with disciplinary categories because disciplinary science "is essentially isolated by its disciplinary politics" (Churchman 1979b, 12) and therefore unable to engage with the full meaning of the world. In contrast to the "vertical" deep dive of investigations based on disciplinary theories, systems thinking approaches tend to transcend or even disregard disciplinary perspectives. The main reason is that disciplinary boundaries are rarely aligned with the phenomena of interest. Systems thinking therefore often favors a grounding in the empirical complexity and structure of the world. Churchman's systems approach to real-world planning—today, we would call it design—is motivated by the *environmental fallacy* (1979a). This fallacy describes an approach to solving an identified problem without regard for the environment it arises in and therefore without understanding the wider effects the intervention will have. The result, more often than not, is a situation that is worse than before. Instead of narrowing the perspective, a systems approach begins by expanding it. To understand what a bicycle is, for example, a systems approach will begin not by taking it apart but by asking what other systems it forms a part of. This may lead to an understanding of urban transport, commuting, and road safety.

Because systems thinking approaches have been built across such diverse fields out of heterogeneous ontological and epistemological commitments, they ultimately have very little in common other than the commitment to struggle with the aim for a "holistic" understanding of the issues they face.

Whatever the scale a particular systems approach emerges in—whether it is a living unit such as a cell as in biology, a cognitive unit such as a human person as in psychology, a classroom as in pedagogy, or a natural habitat as in ecology—the *system* concept "embodies the idea of a set of elements connected together which form a whole, this [whole] showing properties which are properties of the whole, rather than properties of its component parts" (Checkland 1981, 3). These properties are called *emergent properties*. The concept of emergence is relevant to all systems approaches, but they diverge wildly on how they conceptualize it.

Systems approaches also differ in how they organize the elements they are interested in. Some provide specific structural models of systems. For example, system dynamics considers measurable variables that represent structural properties of the world, such as the number of people currently alive or the amount of carbon dioxide emitted yearly, and organizes these variables with quantitative causal relationships that allow it to predict overall behavior as the emergent property. Soft systems methodology (SSM) considers purposeful human activities as its *element* and the *contingency* of each activity on other activities its main organizational relationship. Where system dynamics represents structural relationships between quantifiable variables as causal loops that can be expressed as equations (D. H. Meadows 2008), the purposeful activity models in SSM are graphs used to represent the contingency of goal-directed activities relevant to a social situation (Checkland 1981). Other approaches, such as the critical systems heuristics we will meet later (Ulrich 1983), do nothing of the sort.

Even within the approaches listed here, not all models are in fact models of reality. Many approaches do assume that their models represent elements or structural properties of the real world. They are referred to as "hard" systems approaches not because their systems are hard but because their epistemology is. For others, such as Checkland, systems are *ideas*: that is, mental or discursive constructs. SSM rejects the idea that the social world can be meaningfully represented in models. The purpose of activity models in SSM is not to model real activities but to develop a structure for a conversation about a social situation. And yet other approaches are agnostic to this distinction. Instead of debating whether claims about systems correspond to reality, they focus on how claims are discursively established. For example, critical systems heuristics focuses all its attention on boundary judgments as the central discursive act of design. These three kinds of epistemic assumptions are commonly used to distinguish between "hard," "soft," and "critical" systems thinking. A central concept that unites and simultaneously distinguishes all three groups of systems thinking approaches is whether a given system's *environment* is treated as an unproblematic given, a central design choice, or a never-ending question.

JSD primarily relates to two types of systems approaches: system dynamics and critical systems thinking. System dynamics is a hard systems approach that underpinned the influential report *Limits to Growth* (D. L. Meadows and Club of Rome 1972) and which is central to various efforts to understanding, modeling, and predicting the human influence on the climate—so much so that its language has seeped into popular culture through concepts such as feedback loops, tipping points, carrying capacity, and leverage points. Within its conceptual framework, we can distinguish direct, indirect, and large-scale structural impact of systems design choices and explore nonlinear counterintuitive effects. It is within system dynamics that the concept of *leverage points*—places where small actions can cause large consequences—becomes operationalized into a very useful set of twelve categories (D. H. Meadows 1999) discussed in chapter 9. Its epistemology is structuralist—the assumption is that by modeling structural properties of the world, its future evolution can be predicted given a set of starting conditions. Because of its "hard" orientation, system dynamics has little to say about where its structures come from, where its assumptions come from, how to justify itself, and who gets to decide. Its direction is decidedly *analytic*—it proceeds by breaking down the whole system into constituent variables until they are considered sufficient to explain the observable behavior of interest, without being reflective or critical about the boundaries of the system of interest. In its ontological assumptions, system dynamics continues early systems thinking approaches that used terms like "systems theory," "systems analysis," or "systems engineering." These early flavors of systems thinking have left a bitter taste: functionalist, imperialist, positivist, managerial, controlling, reductionist, instrumentalist. This is one reason to place these approaches in the context of the broader family of thought known as systems *thinking*. It is important to be aware that their shortcomings have been forcefully critiqued *within systems thinking.*[9] This realization is complicated somewhat by the coopting of the term "systems thinking" by prominent system dynamics proponents (Senge 1990; D. H. Meadows 2008). In systems approaches based on a positivist or structuralist epistemology, the systems idea often translates into a misleading belief that the systems approach is *comprehensive.* That belief can only be maintained in an analytic, inward-looking approach.

Critical systems thinkers maintain that a well-understood systems approach must begin by recognizing the tension between the desire to be comprehensive and the realization that it is impossible to be comprehensive (Ulrich 1983). As Churchman put it in the epigraph of this introduction, "a systems approach begins when first we see the world through the eyes of another." Shortly afterward, "the systems approach goes on to discover . . . that every world-view is terribly restricted" (Churchman 1979a, 4). *Critical Systems Thinking* opposes the purely analytic attitude that is prevalent among hard systems thinkers. It emphasizes a "synthetic" attitude, which maintains that things can only be truly understood once they are seen as embedded into their environment. Context matters in systems thinking, as it does in feminist thought (D'Ignazio and Klein 2020). It is this critical approach to systems thinking—reflective, emancipatory, and pluralist (Midgley 1996)—that will be of significant interest in this book.

In the 1960s, in rationalistic systems approaches such as operations research and organizational cybernetics, the scientific apparatus was brought to bear on social questions to shape societies at an unprecedented scale. Together with computing, these approaches had developed mechanisms of large-scale prediction and control (Churchman 1979a; Erickson et al. 2013). Critical systems thinkers grappled with the hubris of that project, with its epistemological challenges, and with its ethical implications. The critical turn they took, and the insights they developed on the way, remain invaluable today. Computing now is about to implement the project that the rationalistic systems approaches dreamt of in the 1960s, and it does so largely on the same epistemological foundations that underpinned rationalism then, despite well-articulated critiques (e.g., Winograd and Flores 1986). Whether we consider this utopian or dystopian, we have a lot to learn from this history that can help us orient systems design in the twenty-first century for sustainability and justice.

THE AIM AND STRUCTURE OF THIS BOOK

This book makes the case for privileging the intersecting issues and values of sustainability and justice in systems design in computing and beyond, and it outlines an approach to do so. It argues that we must replace central narratives that have shaped the discourse of computational thought with more accurate narratives that are historically informed and provide a more nuanced set of metaphors to reorient the discussion about systems design. The ideas of value-neutral technology, objective problems, rational decision-making, and solvency form widely held but false beliefs underneath the discourse and will be treated as myths. While they are rarely taught as "facts," and many in computing will laugh about them outright, they remain central to the orthodox rationalistic discourse of computer science. Just sustainability design provides methodological principles and commitments that help researchers, practitioners, and educators avoid the distorting influence of these myths and design for sustainability and justice. This book is designed to enable computing professionals and researchers to identify these myths, point to them, surface them in comparable situations, and thus chart paths around their gravity wells in research, education, and practice.

Part I poses and responds to the question: "Is computing able to pay back its debts to societies?" Chapter 1 sets the stage by laying out the role of computing in environmental sustainability. I survey the recognition of the lifecycle impacts of computing technology and the role of computing in improving environmental sustainability. I compare different paradigms of sustainability and sustainable development to highlight key challenges and developments in ICT for sustainability and sustainable HCI, and I describe the emerging paradigm of sustainability design. Chapter 2 expands this view to address social justice. It introduces the concept of *just sustainabilities* and the connections between sustainability and social justice. On this basis, it explores what I call the *debts of computing*—the widespread, often hidden, and usually indirect effects of software systems on their social, economic, personal, and natural environment. Chapter 3 examines the adequacy of computing's primary forms of reasoning—computational thinking—to these challenges. It concludes that the current discourse of computing is shaped by the myths of objective problems, rational decision-making, and value-neutral technology. It outlines how these beliefs manifest in systems design *as myths*, and it demonstrates how these myths interact to produce misleading arguments about the nature and implications of systems design practice.

Chapter 4 shows how these three myths (value-neutral technology, rational decision-making, and objective problems) interact to form the fourth (solvency, or the idea that computational problem-solving makes the world a better place). Solvency is central to what I call *problemism*—a rationalist preoccupation with framing and solving problems. In computing, problemism occurs when the lens of computational thinking turns data-driven problem-solving into a tunnel vision, unaware of the social construction of problems and data, the varied forms of individual and social cognition and decision-making, and the politics of stakeholder engagement. A set of examples illustrates how these myths substantively constrain the direction of systems design activities and reinforce the unjust, unsustainable modes of current mainstream practice. The conclusion: the dominant discourse in computing is currently unable to address its debts to societies. Rather than to open bankruptcy proceedings, I suggest a restructuring of the narratives of systems design with help from neighboring and distant disciplines.

Part II introduces the critical friends of computing. With the help of their insights, it restructures the central narratives of systems design to facilitate the reorientation toward sustainability and justice. The conceptual basis for the central arguments is drawn from two areas that have a lot to offer computing: science and technology studies and critical systems thinking. These are introduced in chapter 5. The subsequent chapters address each myth in turn to examine where it came from, how it prevents meaningful engagement with the challenging questions raised by sustainability and justice, and how to replace or sidestep each myth.

Chapter 6 explores the myth of value-neutral technology. It contrasts it with insights from HCI, science and technology studies, and the philosophy of technology to show how values become facts in systems design. Chapter 7 explores the myth of rational decision-making in light of research from cognitive psychology and related areas. The conclusions suggest that a significant portion of behavioral research in systems design commits what I call the "normative fallacy"—it relies on normative frameworks to describe what people do, and thereby fails to provide reliable insights. This has significant implications on how we understand systems design practice, particularly how practitioners make decisions that have uncertain and ambiguous effects at a distance. Chapter 8 explores the myth of objective problems, drawing from a range of areas interpreted through critical systems thinking frameworks. Because problems are inevitably framings of particular aspects of a situation created by someone for a purpose, the politics of stakeholder engagement need to be made visible. That requires a critically systemic view on what it means to design in *wicked problem situations*.

The implications of this restructuring lie partly in practice, partly in education, and partly in research, and the chapters in part III explore each of these domains. Chapter 9 describes JSD as a coherent framework to systems design that is critical but engaged in productive design and engineering work. It then briefly introduces the idea of *leverage points* to organize possible avenues for change. The subsequent chapters present concrete methods and research projects that translate this methodological foundation into a research and design practice reoriented toward the values of sustainability and justice. Chapter 10 describes *critical requirements engineering*, an emerging approach to requirements engineering that combines its normative frameworks with critical systems heuristics, and it illustrates its application in a project. Chapter 11 develops new research directions in the cognitive dimensions of decision-making in systems design. It explores the implications of the *more-than-rational* view of judgment and decision-making in light of the challenge of psychologically distant effects, and it illustrates this new research direction with recent empirical studies on intertemporal choice. Chapter 12, finally, asks how this reorientation relates to professional competence. With a focus on social responsibility and collective organizing, it surveys the recent tide of initiatives for social change in computing and discusses the limitations of, and alternatives to, professional ethics codes in computing. The conclusion places the proposed reorientation of computing into a broader context of societal reorganization.

AUDIENCE

This book is addressed to two audiences. First, I hope that it can help those within computing to orient their thinking, assess the role of implicit false narratives in their work, and position the angle of their work to avoid the torque of false narratives. For those who already focus on environmental sustainability and social justice, such (re)positioning will help them identify those leverage points they can act on to lift computing out of its unjust and unsustainable ways. Others may find new angles on their existing focus that help them to address issues with social import that currently resist transformative change. This applies to graduate and undergraduate students as well as researchers and professionals for whom JSD should provide useful guidance. For tech workers keen on social change, I hope the book provides arguments that can help them disarm what Cathy O'Neil called the "weapons of math destruction" and instead build "algorithms of liberation" (Roberts et al. 2018).

If you are a computer scientist who normally reads technically oriented papers, I thank you for your curiosity and I ask for your patience. This book will build a vocabulary beyond the terms common in fields like software engineering, and it will use that vocabulary to speak to and with these fields. In doing so, it will uproot some of the terms commonly used in computing to question what they really (should) mean. This will take some time.

Second, I write it for those in neighboring fields that are not focused on design or engineering—information studies, communication, media studies, social justice, environmental sustainability, science and technology studies (STS)—who collaborate with the more design- and engineering-focused computer scientists in the space of sustainability and justice in IT. My hope is that it may help them understand how exactly computational modes of reasoning can limit the discourse and perspectives of computer science, show how the perspectives of *their* disciplines can be brought to bear on the blind spots of computing to effectively offer a helping hand as a critical friend, and perhaps help them to identify and make visible other myths of systems design that have not been recognized as such.

I

IS COMPUTING INSOLVENT?

===

We face a looming global environmental tragedy. Given that we see it coming, why has our response been so limited? . . . the metaphor of the perfect moral storm . . . has three dimensions; global, intergenerational, and theoretical. At its core is the asymmetric power of the rich, the current generation, and humanity as such over the future of the planet, and the corresponding vulnerability of the poor, future generations, and the rest of nature. These asymmetries make it tempting for the powerful to externalize the costs (including the serious harms) of their activities over space, time, and species. The possibility of such buck-passing threatens to undermine ethical action, and even moral discourse itself.

—Gardiner (2014, 439)

1

THE DESIGN OF SUSTAINABILITY

The dominant discourses about the nature of the climate threat are scientific and economic. But the deepest challenge is ethical. What matters most is what we do to protect those vulnerable to our actions and unable to hold us accountable, especially the global poor, future generations, and nonhuman nature.
—Gardiner (2014, xii)

SOFTWARE HAS BECOME PART OF THE FABRIC OF OUR SOCIETIES

Technologies are social, and modern societies are technological. It is now widely acknowledged that there is a mutually constitutive relationship between the social construction of scientific and technological artifacts and the technologically mediated social relationships that make up what we call "society." But we are still struggling to grasp its implications. In computing, the question *how computing shapes societies* has been welcomed: It resonates with the widely accepted role of computer science as a source of innovation, and it aligns with the demand for impact.[1] The reverse question *how societies shape computing* has been largely left to other disciplines. Those who responded to it often come from critical and feminist perspectives in sociology, science and technology studies, information, communications, media studies, geography, and the humanities.[2] I call these researchers *critical friends* of computing.

Within computer science, the development of algorithms and especially of software systems is the primary purview of the discipline of *software engineering*. The field sees its focus as the development of technical systems with clear boundaries and identifiable parts and connections, modules, and dependencies. But fifty years after the founding of software engineering as a field,[3] the boundaries between software and its social and environmental contexts are rapidly dissolving. Software systems now have become part of our societies' fabrics and shape the relationships that constitute them, through their information storage, collection, aggregation, and routing; their algorithmic sorting and filtering; their communication and control capacities; and their ability to learn and predict patterns. Communication systems, dating platforms, travel booking services, and procurement systems influence the private, social, economic, and natural environment through far-reaching effects on how we communicate and form relationships, how we travel, and what we buy. The indirect effects of these systems generally remain invisible in the software development process, and developers routinely believe that the effects of their designs are not their responsibility. Their responsibility, as per their job descriptions, is to create "software systems." But the boundaries that matter for understanding *what system has been designed* in each case become increasingly difficult to ascertain. Since every relevant software system is deeply interlinked with people, economic processes, social relationships, and other elements, what we need to consider *as system* inevitably transcends the boundaries of the software.

Software development then always creates sociotechnical, socioeconomic, sociocultural systems.[4] When this book refers, for the sake of readability, to "software systems," it will be always based on that understanding. But when software became social, software engineering research did not make the leap of fully incorporating social theory into its foundational body of literature, and neither did software engineering education (Dittrich, Klischewski, and Floyd 2002; Ralph, Chiasson, and Kelley 2016). As a consequence, both lack the theory, conceptual tools, and people to ask the critical questions needed to understand software technology's role in our societies (Leticia Duboc, McCord, et al. 2020).

Now that this role has become formative, we are beginning to recognize the need for change. For the sake of future generations, the software systems we design in the next fifty years must advance social, economic, and environmental causes simultaneously with technical ones. The ability to meet urgent needs of the present without compromising future generations' ability to flourish requires that all stakeholders must jointly understand and address the social, human, and technical sides of systems design and must explore systemic effects across the physical, societal, economic, and human environments that software is entangled in. But as we will see, they are currently unlikely and ill-equipped to do so.

Two implications of the pervasive role of software and information technology are central to the argument of this book: the long-range concerns of sustainability and the wide-ranging concerns of justice. Both have become urgent as computing now pervades our societies. The unprecedented breadth, depth, and length of its entanglement means that the choices made in systems design have farther-reaching implications than ever before. These distant effects are often uncertain and ambiguous. This complicates the political role of technology design, the social dynamics of participation in design, and the cognitive factors of design decisions. The book grapples with each of these implications. To establish its domain, this chapter will trace the evolving perspectives on the relationship between computing and sustainability, loosely following its historical evolution.

COMPUTING AND SUSTAINABILITY HAVE
A CONFLICTED RELATIONSHIP

Sustainability at its heart is the "capacity to endure" (Fowler and Fowler 1995) or, sometimes, the "ability to *be maintained* at a certain level" (*Oxford English Dictionary* 2021). Its primary prominence arose through the concept of *sustainable development* as defined in the United Nations' "Brundtland report"—"development that meets the needs of the present without compromising the ability of future generations to meet their needs" (Brundtland 1987). The use of the concept has evolved considerably: The initial view focused on avoiding the depletion of a stock of natural resources held in identified ecosystems. It drew on a global view of the planet at large and its limited resources. As famously outlined and demonstrated in the classic *Limits to Growth* report in 1972, human civilizations were already then on course to exceed the *carrying capacity* of our planet (D. L. Meadows and Club of Rome 1972). Today, the focus lies on a systemic view of the interlinked dimensions of social, environmental, and economic processes that are now seen to constitute sustainable development. For example, the UN's Sustainable Development Goals (SDGs), which now provide the framework for most large-scale initiatives and policies (International Council for Science 2017), contain "public participation" as a core goal. In North America, many businesses now use a version of the "triple-bottom-line," a model that asks companies to balance economic, social, and environmental accounts (Elkington 2004).

Sustainable development may still be the hegemonic paradigm, but it is a problematic framing, far from innocent. We will revisit it in chapter 8. In this book, I focus on the role that systems design plays in sustainability, rather than sustainable development, while acknowledging that design always implies a form of "development": it aims to shape the world according to the designers' intentions. The crucial issue is: whose intentions?

Two competing views of sustainability are illustrated in figure 1.1.[5] At the bottom left, the strict hierarchical "strong" view of sustainability places the economy inside a society, which is placed inside the natural

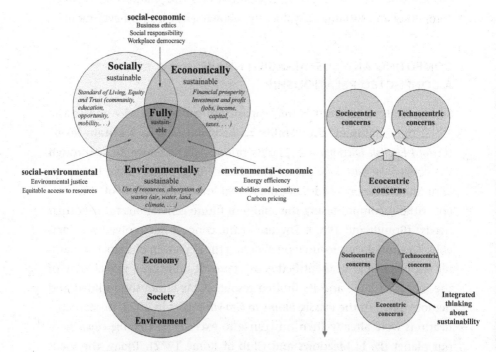

1.1 Sustainability visualized as hierarchy, interacting domains, and spheres of concerns.

environment. This view maintains that real limits of each of these strictly hierarchically contained systems must be respected for the continued viability of the whole. Each level of the hierarchy must remain within established limits, or else it damages its surroundings. This approach distances itself from what its proponents call "weak" sustainability, shown on the top left. This paradigm assumes that some resources within one domain may be substitutable by resources from another domain. For example, fossil fuel energy has substituted human labor, and economic profits could be used to plant trees. In this interpretation, some resources may be depleted as long as other resources continue to fulfill their functions. While this may appear reasonable, there are serious objections to this view.[6]

It is important to recognize the contrasting epistemological basis of different models. The perspective on the top left focuses on the *domains* to be addressed and organizes these according to their intersections and overlaps. For the purpose of systems design, this book follows the argument that Dodds and Venables make in the context of civil engineering (2005): The key focus is to integrate the *thinking about* these relevant spheres within a process that invariably involves a multitude of stakeholder perspectives and knowledge areas. The crucial emphasis then is one of *concerns*, as illustrated on the right (Becker et al. 2015). This shift sidesteps the debate between the views shown on the left and provides a more encompassing perspective than a resource-based view.

Integrating these concerns presents at least three types of challenges to individual and collective action. First, the effects of human choices on the environment are dispersed in space and time. Each individual action may make only a small, uncertain, opaque contribution to climate change, but their effects interact and accumulate over time. Future generations will be disproportionately affected, but they cannot speak up about their expectations of inheriting a livable planet. This creates what moral philosopher Stephen Gardiner (2014) calls *asymmetric vulnerability*. We will later argue that it is present not just in the global challenge of climate change, but in many seemingly "smaller" issues of systems design, where dispersed causality, fragmented agency, and asymmetric vulnerability combine to create a dangerous undertow.

Second, the dispersion of cause and effect across spatial and temporal timescales also raises difficult questions about the nature of human cognition and reasoning. As judgment and decision-making researcher

Elke Weber has argued, climate change "doesn't scare" the people making choices yet because it happens, for many in the developed world, at a distance, so our knowledge of it lacks the viscerality of personal experience (Weber 2006). This *psychological distance* affects our choices, but little is understood of how this plays out in systems design (see chapter 11).

Finally, the integrated consideration of social, environmental, economic, and technological notions of sustainability raises serious political and epistemic challenges, since it is difficult to reconcile the diverse assumptions and worldviews from which the concerns arise. Specific trade-offs across dimensions can be agreed on and have been formalized—for example, it requires an economic investment to decarbonize specific industries, and decarbonization will also result in direct and indirect economic benefits. But the stakeholders will also bring their mutually incompatible worldviews on what constitutes such foundational things as social life, individual prospering, economic activity, and nature. These epistemic differences frame what constitutes a valid measure within each dimension and how the dimensions should relate to each other. As a result of this epistemological conundrum of incommensurability, sustainability becomes "immeasurable" precisely when the aim is to evaluate it.[7] Bell and Morse (2008) respond with a shift to a subjectively conceived and intersubjectively negotiated system of sustainability indicators developed through a participatory approach: "Systemic Sustainability Analysis [is] the participatory deconstruction and negotiation of what sustainability means to a group of people, along with the identification . . . of indicators to assess that vision of sustainability" (147).[8]

Computing researchers recognized early that IT has a role to play in sustainable development.[9] Several areas took an interest. Each sees the topic through its own disciplinary lens, which shapes how researchers frame and approach it.[10]

GREEN IT: ADDRESSING THE ENVIRONMENTAL IMPACT OF COMPUTING

Computerization has led to a rise in mineral extraction, energy consumption, and electrical and electronic equipment waste. A range of fields tackle these issues, driven by environmental impact assessment research (see Widmer et al. 2005; Robinson 2009). Research partnerships with IT were soon labeled "green IT" and "green computing" (Jenkin, Webster, and McShane 2011; Murugesan 2008).

On the production side, the extraction of rare minerals and other ingredients for IT products has turned into a supply chain challenge[11] and, more importantly, a human and nonhuman rights question.[12] The public's attention rarely focuses on where the material and labor come from that make possible the global IT infrastructure, and where those materials go after they cease to be useful. But the material roots that link modern IT devices to the material and social conditions of their production span massive spatial and temporal scales. The brilliant visualization *Anatomy of an AI*[13] drew attention to just how far one has to go to identify the sources and inputs to one "smart home" device. The media's reception of this piece demonstrated vividly how surprising this realization was to a mainstream audience, even though scholars have for many years emphasized the invisible labor behind IT infrastructure and its uneven global distribution.

At the other end of the life cycle stands obsolescence. E-waste is the general term for "all items of electrical and electronic equipment (EEE) and its parts that have been discarded by its owner as waste without the intent of re-use" (Kuehr 2014). In 2017, the amount of e-waste generated globally reached 44.7 million metric tons, as about half of the planet's population has started using internet services. Only a fifth gets recycled, while the rest remains unaccounted for and is most likely not recycled (Balde et al. 2017). While developed countries have reached high levels of saturation in terms of technology use, consumption in countries with lower purchase power is growing by around 20 percent annually (Balde et al. 2017). Because obsolescence is not a law of nature but a sociotechnical phenomenon, we can design to avoid obsolescence or to generate it (Slade 2009). Unfortunately, the latter is the norm. Despite technological advances, the lifespans of electronic equipment continue to be extremely short. In 2005, a study estimated an annual disposal rate of 11 percent for personal computers (Widmer et al. 2005); in 2017, a similar level was estimated for laptops (Balde et al. 2017). Regulation efforts such as Extended Producer Responsibilities have made progress in placing some of the burden on the production side, but most global e-waste is still moved from rich countries to poor countries, where it is recycled under extremely unsafe conditions (e.g., Rifat, Prottoy, and Ahmed 2019). The manual extraction of the residual valuables inside electronic equipment (such as copper and gold) exposes laborers to a wide range of toxic chemicals and metals, including lead and mercury, with significant damage to their physical and mental health.[14] Unsurprisingly, these effects are very unevenly distributed, since production, consumption and disposal happen in distinct parts of the world separated by their socioeconomic development status.

Computing has responded in varied ways. Industry and research have continued miniaturization and cloud computing, based on the claim that efficiency gains would reduce waste accumulation. With the growth of mobile device use and cloud computing, the share of consumption has shifted from desktop PCs to mobile devices and data centers, while overall consumption continues to increase (Belkhir and Elmeligi 2018). Researchers and advocates have collaborated with policy makers to measure and make visible the burden of e-waste. The methods, often built on established environmental impact assessment frameworks such as life-cycle assessment, faced challenges of evidence collection, indicator definitions, and measurement (Guinée et al. 2011; Balde et al. 2017). The evidence base has led to a slow change in policy and regulation, which has not changed the direction of the overall trend: e-waste numbers continue to grow. Outside the mainstream, advocates, activists, community organizers, and some researchers have argued for extended technology lifespans through maintenance, repair, reuse, and modularity (S. Jackson 2014). These interventions certainly reduce waste, but they remain marginal efforts.

Between production and disposal stands usage, and with it the steadily increasing energy use of computing. IT accounts for a growing fraction of the world's rising electricity demand (Van Heddeghem et al. 2014; International Energy Agency 2013). The training of one single machine learning model emits as much CO_2 as five average cars in their entire life cycle (Strubell, Ganesh, and McCallum 2019). A significant portion of research in green computing, green IT, and green software engineering has thus focused on "the problem of energy use," with the declared aim of reducing the environmental impact of software systems by increasing efficiency in software production and use (Calero and Piattini 2015). Energy-efficient algorithm research is based on the claim that "algorithmic solutions can help reduce energy consumption in computing environments" (Albers 2010). At first glance, this seems direly needed. Advances in the energy efficiency of algorithms have not however reduced aggregate energy consumption. On the one hand, the share of data centers in global energy use has remained steady while their scale has rapidly expanded (IEA 2019). This is attributed largely to the efficiency gains made by extremely large-scale hardware and facility design (Shehabi et al. 2016), including the algorithmic control enabled by deploying sensors in large-scale data centers (Jones 2018; Hölzle 2020).

The energy efficiency of algorithms themselves may seem a promising research subject, since on the level of processing, there is a clear link between algorithmic complexity, memory operations, processing cycles, and energy use. But evidence is scarce that improvements on algorithmic efficiency have tangible effects. Numerous papers published in computer science conferences each year promote their algorithms as more energy efficient than previous algorithms. The strongest business driver for this trend is cost (Albers 2010), but many researchers in energy efficiency motivate their work by environmental sustainability. Few consider that their work may push an important lever in the wrong direction. The reason lies in a paradox already observed in the nineteenth century: increased technological efficiency often *increases* overall resource use (Alcott 2005). Why? Aggregate energy consumption is a product of average efficiency multiplied by total use. The expectation may be that efficiency gains result in lowered consumption, but use is not independent of efficiency. An increase in technology efficiency or capacity often induces increased usage. As a result, aggregate consumption *rebounds* against these expectations. For example, historically, despite a *one hundredfold* efficiency increase from the first light bulb to a contemporary LED bulb, increase in electricity demand *for light bulbs* has entirely offset these gains. A landmark study concluded that "global energy use for lighting has experienced 100% rebound over 300 years, six continents, and five technologies" (Saunders and Tsao 2012; Tsao et al. 2010). Just take a look around your home and try to count the bulbs.

The rebound effect is relevant for any comparable situation and often exceeds the efficiency gains, as has been observed for coal use in railways, for highway capacity increments, and for countless other cases (Alcott

2005; Sorrell 2009; Freeman, Yearworth, and Preist 2016). The compound outcome depends on economic, social, and cultural factors. It is not possible then to make a direct claim from individual-level efficiency to aggregate contribution without assessing the change of aggregate energy use. But despite the fact that rebound effects have been well studied and published, few studies in energy efficiency in computing even mention them (Knowles 2013, 4). This severely limits the relevance and value of energy efficiency work in particular (Coroama and Mattern 2019) and green computing research in general (Knowles 2013, 89–91).

Consider a bitcoin miner using a software running an algorithm that, on their hardware, produces $1 worth of bitcoin per hour with an energy cost of 90 cents, for a return on investment (ROI) of 11 percent (ignoring for simplicity the hardware costs). Decreasing the amount of energy required to produce 1$ worth of bit coin by 20 percent will increase the ROI to 27 percent. Will the miner mine the same amount of bitcoin (reduced energy), or expend the same amount of energy to mine more bitcoin (even energy balance), or invest in additional hardware as a response to the changed incentive structure (increased energy and environmental waste)? Considering the scalability of their enterprise and the fact that their ROI almost tripled, the answer seems clear. In fact, the increased attractiveness to investors may well spur new entrants into the market.[15]

If research and development into energy efficiency aims to improve environmental sustainability, the "system in question" to consider is not the bitcoin mining algorithm in isolation but the bitcoin mining algorithm *in use by a person*. Neglecting to do so commits what Churchman (1979a) calls the *environmental fallacy*: It takes too narrow a perspective to understand the effects of an intervention.

In the system of interest, increased efficiency *increases energy demand*. The outcome is often an increase in aggregate consumption. Just as with lighting, the real effect of increased efficiency in bitcoin mining is not reduced usage but increased productivity.[16] Resolving this paradox thus requires a reframing of the system of interest, and the extension of system boundaries comes with an expansion of the knowledge domain required to perform this type of research, from a mathematically and computationally grounded domain of algorithms to a broadened *systemic* understanding of relationships that span algorithmic complexity, hardware efficiency, economic models of supply and demand, and the incentives of purpose-ful social actors.[17] Such a paradigm shift from computational to systemic thinking does not come easy.

GREENING THROUGH IT: HOW COMPUTING CAN GREEN SOCIETIES

If e-waste and energy consumption research starts from the question "how does computing pollute the world?" and reduces that impact, *green by IT* research asks the complementary question "how can computing green societies?" directly and indirectly, and aims to create and amplify that impact. In *Greening through IT*, Tomlinson (2010) provides a comprehensive account of this research program. In his view, IT provides ways to "compress time, space, and complexity" through its capacities to store, retrieve, and analyze information and to support visualization, modeling, and simulation (9). Green IT then can support the challenging leap from human to environmental scales of thinking and action. Tomlinson illustrates this in three example contexts. In education, IT designs can help learners understand ecological concepts through simulation-based interactive systems. But this idea reaches far beyond formal education and can involve quite visceral explorations of nature at a scale not generally accessible to human senses (e.g., Driver 2022). On a personal level, IT can support data collection and visualizations about an individual's footprint and the factors contributing to them. On a collective level, platforms can coordinate, encourage, and mutually reinforce individual actions for sustainability. Tomlinson positions this work carefully in respect to broader critiques that place capitalism and its hunger for growth at the center. I agree with him that that independently of the role of capitalism, IT will play a role in addressing climate change (Tomlinson 2010, 25).

ENGINEERING SOFTWARE FOR SUSTAINABILITY

Within software systems research and practice, the challenge of incorporating long-term considerations effectively into engineering practice has been a central concern since the founding days of the field (Becker 2014). The long-term costs of software systems were one of the central themes at the 1968 NATO conference (Naur and Randell 1969), and Lehmann's laws of software evolution were developed in the 70s (Lehman 1979; 1996). Two decades later, Parnas (1994) lamented "software aging." Another two decades later, the practical urgency of long-term thinking had not diminished (Neumann 2012).

For the most part, however, these long-term considerations were focused on internal perspectives motivated by cost reduction. When the subfield of software maintenance and evolution adopted the language of sustainability, it did so without acknowledging the broad concerns inherent in the wider sustainability discourse. Sustainability language became prominent about a decade ago (Becker 2014; Durdik et al. 2012; Koziolek et al. 2013; Zdun 2013), and sustainability-focused work in software architecture receives growing attention (Venters et al. 2018).

With the emergence of software engineering for sustainability, several research groups began to explore the role of software systems in environmental sustainability. While some focused on work that can be categorized as green software engineering[18] thanks to its focus on energy efficiency, a workshop series on requirements engineering for sustainable systems explored the intersection of the social and the technical perspectives of software engineering, especially the domain-specific challenges of eliciting requirements for software explicitly designed to support environmental sustainability (e.g., see Mahaux, Heymans, and Saval 2011; Chitchyan et al. 2015). Penzenstadler positioned sustainability as "the non-functional requirement of the 21st century" (Penzenstadler et al. 2014). Like safety, security, or usability, sustainability is not simply located in particular features but presents a concern that cuts across functionality, largely independent of the primary purpose of the system under design. Like usability, the implications of technical design choices are varied and must be considered carefully. A review of nonfunctional requirements in the past shows that the emergence of a new concern is generally followed by the growth of knowledge, techniques, measures, and models to address it. But the scale and complexity of sustainability make it more difficult and challenging to address.

Some have proposed to consider sustainability a system "quality"—that is, a property of the system under design (Lago et al. 2015)—but it is important to keep in mind that this reduces the frame of design again to internal and purely technical perspectives. This focus in turn marginalizes the more profound implications of systems design on the broader environment of software technology and treats technical aspects as fully separable from social aspects. As a result, it treats these questions as solvable problems and positions conventional engineering as the sole applicable method. Because sustainability emerges from the *interactions between* the designed technology and its environment, however, it cannot be reduced to a technical property (Becker 2014; Becker et al. 2015). We must resist the reductive framing implied in sustainability as quality and recognize it as a stakeholder concern: a matter of interest to those affected. This rejection of sustainability as system quality has far-reaching implications for systems design methods, models, and practice. It implies that the evaluation of a system's sustainability ultimately must rest with all those who are affected, and that their concerns may translate into a range of system features and qualities.

ICT FOR SUSTAINABILITY

The contours of the *ICT for Sustainability* field (short ICT4S) took shape in 2013 with an inaugural conference (Hilty and Aebischer 2015b), but the conflicted relevance of ICT for environmental sustainability was articulated earlier (Hilty et al. 2006) at about the same time as the focus on energy consumption emerged elsewhere. ICT4S aims to integrate many of the above concerns in an overarching framework that recognizes ICT both "as part of the problem" and "as part of the solution" (Hilty and Aebischer 2015a) and distinguishes between its direct, indirect, and aggregate effects,[19] defined as follows:

- **Life-cycle impacts** refer to the direct environmental and social effects of producing, using and disposing of ICT, which are negative not by definition but in practice (until the production of ICT equipment can be made carbon-negative). The green IT research summarized earlier focuses on these impacts.
- **Enabling impact** is located on the micro level of individual and organizational IT adoption and use. The work of Tomlinson and others explores this space, with a focus on potential positive applications. But enabling impact often has negative implications too, including

the effects of induced consumption and obsolescence. For example, the shift of music distribution from CD sales to online streaming has reduced the environmental impact of CD production but increased the energy consumption of data centers and internet infrastructure (Devine 2019). According to one calculation, the carbon footprint of streaming a song exceeds that of its CD counterpart once it is played more than twenty-seven times (McKay and George 2019).

- **Structural impact** refers to aggregate, macro-level changes induced by large-scale adoption of a product or service. These changes are enabled by micro-level effects, but because of the emergent nature of complex sociotechnical, socioeconomic and sociocultural systems, it is empirically difficult to link them conclusively.

To understand how these effects are linked, consider the well-known online rental platform Airbnb.[20] Its life-cycle impacts refer to the direct environmental and social effects of production and use—that is, the labor of creating and operating the platform and its physical effects, including energy consumption and e-waste. Its enabling effects result on two sides. First, the software platform has allowed travelers to change their habits by offering a new and attractive way of locating and booking accommodation. This attractiveness of lower prices and new types of experiences has induced travel, and it has allowed private and business travelers to shift their booking practices. Second, short-term rentals are more profitable than long-term rentals, and investors around the globe jumped on the opportunity. With increasing scale, more induction effects appear: new business ventures develop and manage condo tower buildings purposely designed for short-term rentals.[21] As an aggregate result of the rapid growth of IT-enabled short-term rentals, metropolitan areas have lost tens of thousands of residential housing units (Combs, Kerrigan, and Wachsmuth 2020), which has contributed to gentrification, increased rental costs, driven out residents, and changed the character of urban life (Wachsmuth et al. 2017; 2018). When municipal governments caught on and began to regulate short-term rentals, they encountered a third enabling effect: The technological affordances of the software platform support the circumvention of regulation, both on the side of individual owners and on the side of business operators, because the commercial database of transactions is under the control of Airbnb and thereby outside the jurisdiction of most national authorities. As a US company with enormous cash reserves, Airbnb invests significant sums into its legal department and has little incentive to play along when small countries request access to business transactions for tax purposes. In addition, the platform deliberately obfuscates the precise location of rental properties (Wachsmuth and Weisler 2018), which hampers the enforcement of regulation.

The story of Airbnb illustrates that the distinction between direct, enabling, and structural impact is crucial for two reasons. First, because it relates to a central feature of information technology: scale. Online services such as Google, iTunes, Netflix, Airbnb, Uber & Co have been able to have such a marked influence on our lives because they are able to scale more flexibly and faster than traditional businesses. For example, Airbnb has grown faster than any hotel chain ever could. Second, this illustrates that structural impact does not arise from enabling effects out of aggregation or multiplication. Instead, macro-level structural effects *emerge* as a result of large-scale adoption due to the convergence of many dynamic factors, and they are of an entirely different nature. In the case of Airbnb, they are not restricted to a disruption of the hospitality industry but also affect many countries' abilities to collect taxes, exacerbate global cities' municipal housing shortages, and alter the character of neighborhoods around the globe. The example also illustrates that this framework of effects can be used to analyze both positive and negative effects of ICT.

An early mapping of a published collection in ICT4S to types of effects suggests that a majority of work in this field focuses on issues such as energy efficiency that can be categorized as life-cycle effects, while some address enabling effects of production (Hilty and Aebischer 2015a, 32). This limited focus on efficiency may be misguided because a range of *rebound effects* often offsets efficiency gains, and to identify these, we need to examine behavioral and structural change (Hilty et al. 2006). As Coroama and Mattern conclude in their review of the evidence about these rebound effects, "digitalization will not redeem us from our environmental sins" (Coroama and Mattern 2019).

SUSTAINABILITY IN HUMAN–COMPUTER INTERACTION: DESIGN WITHIN LIMITS

The conflicted relationship of IT design to sustainability has been a defining topic of work in human–computer interaction (HCI) (DiSalvo,

Sengers, and Brynjarsdóttir 2010). An influential paper by Eli Blevis (2007) positioned the importance of sustainability in interaction design and evaluated design values and methods from that perspective. The focus is firmly on the direct life cycle effects of interaction design as it relates to resource use and waste, with some attention to the induction and obsolescence effects of different design features. But the discussion raises profound questions, citing design scholar Tony Fry's (2005) critique of design practice:

Currently industry . . . is still overwhelmingly deaf to those voices that speak of the complexity of unsustainability, the poverty of current responses to it, the misplaced faith in technological solutions, the myopia of present political and corporate leadership and the extent of changes that are required if a psychology, culture and economy of sustainment are to ever arrive. (23)

Fry (2005) further argued that designers "need to learn . . . how to design in a far more complex and critical frame" (23). In Blevis's (2007) words, "Fry's statement acknowledges the value tensions between sustainability goals and those of enterprise, while prescribing an ethical imperative for designers to confront such tensions" (504).

Blevis acknowledges the prevalent anthropocentrism in "human-centric" HCI research and places significant emphasis on the mutual relationships between interaction design and sustainability. Nevertheless, a majority of work on sustainability in HCI has focused on designing ambient and persuasive technology to influence consumer behavior (Knowles 2013; DiSalvo, Sengers, and Brynjarsdóttir 2010). This broadly falls into the area described earlier as *Greening through IT*. But critical voices and perspectives have been prominent in HCI. Strengers (2014), for example, eloquently argued that many designs of persuasive technology implied gendered stereotyped assumptions about how individuals make choices about their consumption, and Dourish (2010) highlighted that the common mode of uncritical design-oriented research "obscures political and cultural contexts of environmental practice that must be part of an effective solution" (1). He emphasized the central importance of ecological and political perspectives in designing with sustainability in mind. DiSalvo and colleagues have shown that there are substantive disagreements within HCI researchers: about the appropriate scale of design (from the individual to social infrastructure); about the tension between seeing "users as the problem vs. solving users' problems"; about the tension between incremental and disruptive change; and about the question whether HCI as usual is an adequate mode for addressing sustainability (DiSalvo, Sengers, and Brynjarsdóttir 2010, p. 1979). Bran Knowles's (2013) values analysis shows that persuasive tech-based intervention is likely to reinforce environmentally harmful behaviors and values (86–89). Even worse, its discourse "reifies consumerist tendencies that have driven much of the environmental destruction to date, and absolves individuals from having to make more significant behaviour changes" (87).

The computing within limits community emerged within this context.[22] It argues that the computing mainstream is "deeply problematic for ecological and social reasons" (Nardi et al. 2018, 86). In taking a "strong sustainability" view, this group of researchers argues for closer attention to planetary boundaries. From this perspective, they highlight the importance of rebound effects and criticize that classical green IT research in its narrow focus on energy efficiency ignores them. Three principles are derived from focusing on planetary limits: (1) *Question growth* encourages computing researchers and practitioners to avoid work that ultimately depends on economic growth and instead find alternative angles of work that do not encourage it; (2) *Consider models of scarcity* encourages researchers and practitioners to abandon the focus of designing for abundance, as the work in *Collapse Informatics* does;[23] (3) *Reduce energy and material consumption* emphasizes the need to minimize the footprint of computing.

Computing within limits has made significant contributions to the discourse: It firmly embedded computing within societies and the planet at a long-term and global scale and irrevocably demonstrated the responsibility that this places on the field; it brought findings from archaeology and ecological economics to bear on questions of technology design; it compellingly advocated for transformative change; and it has demonstrated that a different kind of computing can at least be envisioned.

SUSTAINABILITY DESIGN: CONVERGING PERSPECTIVES

Within this evolving landscape, it became clear that the communities addressing sustainability-related concerns through their disciplinary lenses had grown in fragmented paths with limited interactions. In 2014, an attempt was made to unite some of these concerns by providing a common ground for the above communities in the form of the "Karlskrona Manifesto for Sustainability Design."[24] The manifesto recognizes the conflicted relationship between software systems and sustainability discussed here and emphasizes the responsibility of those who design, understood broadly as "the process of understanding the world and articulating an alternative conception on how it should be shaped, according to the designer's intentions" (Karlskrona Initiative 2015; Becker et al. 2014). At its center is a set of principles and commitments to a transdisciplinary, systemic, long-term view (Karlskrona Initiative 2015; Becker et al. 2014). Its conceptual framing of sustainability as a concern distinguishes among the five dimensions of individual, social, economic, environmental, and technical sustainability, which are defined and illustrated loosely. The manifesto further takes a decidedly pedagogical stance in highlighting a series of common perceptions, framing them as *misperceptions*, and offering alternative viewpoints. Table 1.1 lists some and maps them to key themes discussed throughout this book.

THE LEVERAGE OF REQUIREMENTS

The Karlskrona Manifesto kickstarted an international initiative of researchers and took important steps toward uniting researchers across communities.[25] The convergence of five dimensions with systemic effects allowed the development of a framework to capture the possible effects of system features and qualities visually. Figure 1.2[26] shows how an instance of this framework represents the range of potential effects of a procurement system and demonstrates the entangled nature of dimensions. For example, making visible each product's carbon footprint in a procurement system allows users to make more responsible choices. Widespread adoption of more responsible choices would influence market dynamics to incentivize more carbon-friendly products. The affordances designed into individual systems have a limited but tangible role to play in the overall shift.

Importantly, the arrows in the diagram represent causal contributions that connect micro-level decisions to aggregate effects. The diagram links these effects across the dimensions using a visual canvas that may allow a range of participants to engage in the design process. Recent work

Table 1.1 Selected misperceptions countered in the Karlskrona Manifesto

(Becker et al. 2014)		
There is . . .	Whereas . . .	Theme
There is a perception that there is a tradeoff to be made between present needs and future needs, reinforced by a common definition of sustainable development, and hence that sustainability requires sacrifices in the present for the sake of future generations.	Whereas it is possible to prosper on this planet while simultaneously improving the prospects for prosperity of future generations.	Tradeoff decisions
There is a tendency to overly discount the future. The far future is discounted so much that it is considered for free (or worthless). Discount rates mean that long-term impacts matter far less than current costs and benefits.	Whereas the consequences of our actions play out over multiple timescales, and the cumulative impacts may be irreversible.	Psychological distance
There is a tendency to interpret the codes of ethics for software professionals narrowly to refer to avoiding immediate harm to individuals and property.	Whereas it is our responsibility to address the potential harm from the second- and third-order effects of the systems we design as part of our design process, even if these are not readily quantifiable.	Ethics
There is a desire to identify a distinct completion point to a given project, so success can be measured at that point, with respect to pre-ordained criteria.	Whereas measuring success at one point in time fails to capture the effects that play out over multiple timescales, and so tells us nothing about long-term success. Criteria for success change over time as we experience those impacts.	Temporal dispersal
There is a narrow conception of the roles of system designers, developers, users, owners, and regulators and their responsibilities, and there is a lack of agency of these actors in how they can fulfil these responsibilities.	Whereas sustainability imposes a distinct responsibility on each one of us, and that responsibility comes with a right to know the system design and its status, so that each participant is able to influence the outcome of the technology application in both design and use.	Responsibility
There is a tendency to think that taking small steps toward sustainability is sufficient, appropriate, and acceptable.	Whereas incremental approaches can end up reinforcing existing behaviors and lure us into a false sense of security. However, current society is so far from sustainability that deeper transformative changes are needed.	Transformative change

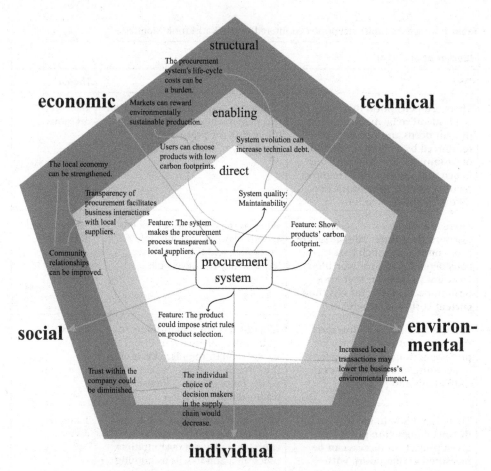

1.2 A visualization of systemic effects of one procurement system.

explores its use in pedagogy (Penzenstadler et al. 2018) and industry practice (Duboc et al. 2019).

In advocating for the importance of sustainability in SE and ICT, the initiative shifted the language toward a conception of "design" and highlighted professional responsibilities. Many of the principles and counterpoints reproduced above will resurface in later discussions of this book on the challenges of long-term choices in systems design and the values, politics and ethics of systems design. On their own, however, these counterpoints did not prove sufficient to enact transformative change, and it is unclear whether the initiative successfully shifted the wider community's perceptions. Instead, it appears that its impact in academic terms to date remains largely restricted to the communities of software engineering, ICT for sustainability, and requirements engineering.[27]

Still, in raising these counterpoints, and in emphasizing professional responsibility, the Karlskrona Manifesto made a significant step from research to advocacy, and its articulation of dimensions and effects is widely cited and has effectively provided a common definitional ground for researchers.[28] A range of academics and professionals have taken the opportunity to support the statement with an online signature, but its language speaks explicitly from and to a standpoint of software professionals: "As designers of software technology, *we are responsible* for the long-term consequences of our designs" (Karlskrona Initiative 2015). One might question the drawing of this boundary between those who claim responsibility and, by implication, those who do not or cannot, and chapter 5 will.

The conceptual framing of sustainability effects, and the arguments brought forward in the manifesto, carry an important consequence: central attention should be paid to the space between the technical and the social where stakeholders in technology development and design projects establish system boundaries and success criteria. "A software system's impact on its environment is often determined by how the software engineers understand its requirements. This impact's foundation is set in the decisions on which system to build (if any at all), the choices of whom to ask and whom to involve, and the specification of what constitutes success" (Becker et al. 2016, 57). In addition to choosing a system purpose, engaging with stakeholders and specifying success criteria, a whole range of activities take place in the design of any system, sometimes explicitly and often implicitly, that "reconcile the technical with the social" (J. Goguen 1994).

Whether explicitly attended to or implicitly performed, decisions about requirements arguably exert stronger leverage than specific techniques such as algorithmic approaches to energy efficiency:

For example, techniques for increasing technical sustainability abound, ranging from architectural design patterns to documentation guidelines. Yet, because applying these techniques often involves an up-front investment of effort, it occurs only when a longer life expectancy of a system is recognized and expressed. On the other hand, a stated requirement for which no technique yet exists will lead to an identified gap in technological ability. This means that in practice, systemic changes to [requirements] activities will dominate the effects of whatever techniques we develop to support these activities. (Becker et al. 2016)

It is in the space of sociotechnical reconciliation represented by requirements that we can find room to act on issues of sustainability. Without a spokesperson who articulates a concern, technical approaches to address it will simply not be introduced into design practice. That places specific opportunities and responsibilities on those engaged in requirements practice. While the responsibility for requirements often does not constitute a separate role on a team, it forms a dedicated practice. What have professionals in computing made of it thus far?

A study probed the level of awareness and attention that current practitioners with this responsibility have, and the obstacles they see in the way of making sustainability a central concern in requirements practice (Chitchyan et al. 2016). The findings suggest an emerging awareness of the concerns regarding sustainability, identify a set of systemic barriers and obstacles, and group these to identify leverage points for practical change. Obstacles are identified on the level of individuals, their professional environments, and prevailing norms in software engineering practice.[29]

The study provides only a speculative outline of possible approaches, but the findings usefully highlight the potential for transformative change, the range of obstacles to be overcome, and the fact that interventions on all levels are required to make sustainability a central, accepted, and established concern in systems design practice. As the authors conclude,

Significant barriers remain to [be] overcome before Software Engineering can claim to routinely advance not just technical and economic, but also social, individual and environmental needs simultaneously. Critical reflection is needed at the individual, organizational and community level to advance the profession's ability and commitment to do so. (Chitchyan et al. 2016, 541)

This reference to "simultaneously advancing" goals across the range of dimensions of sustainability, beyond technical and economic aspects, directly contrasts the professional roles and competencies of systems design professionals with an older, more established profession that has taken its role in sustainable development seriously for a while: engineering. In fact, the *UK Standard for Professional Engineering Competence* states that professional engineers must "undertake engineering activities in a way that

contributes to sustainable development," including the "ability to [...]
progress environmental, social and economic outcomes simultaneously"
(UK Engineering Council 2014, 12). In addition, the licensing process
places the burden of proof on professionals by expecting them to demonstrate they do so, for example by providing "examples of methodical
assessment of risk in specific projects" or demonstrating "actions taken
to minimise risk to society or the environment" (12). Unfortunately, this
"explicit commitment to sustainability . . . is presently amiss within software organizations and their regulating and guiding bodies" (Chitchyan
et al. 2016, 540).

A lot remains to do then to "shift the needle" on sustainability in ICT,
as Mann, Bates and Maher (2018) call it. They examine the corpus of
ICT4S proceedings in light of a "transformation mindset" for sustainable
development, encapsulated in the following ten principled priorities.

1. Socioecological restoration over economic justification
2. Transformative system change over small steps to keep business as usual
3. Holistic perspectives over narrow focus
4. Equity and diversity over homogeneity
5. Respectful, collaborative responsibility over selfish othering
6. Action in the face of fear over paralysis or wilful ignorance
7. Values change over behaviour modification
8. Empowering engagement over imposed solutions
9. Living positive futures over bleak predictions
10. Humility and desire to learn over fixed knowledge sets. (Mann, Bates, and
    Maher 2018, 213)

Their stark conclusion is that ICT4S research is "unfortunately, insufficient to deliver a meaningful change" (Mann, Bates, and Maher 2018,
222). I share their hope that researchers will "examine their own research
and ask themselves how they could contribute to a shifting of the needle
towards ICT4S truly contributing to a positive socioecological transformation" (Mann, Bates, and Maher 2018, 222). The remainder of this book
aims to encourage and facilitate this reflection.

PLANETARY BOUNDARIES, GROWTH, AND DECOUPLING

Much of the conversation around designing for sustainability continues
to take place within a framing of sustainable development, inherently

connected to the dominant economic paradigm of continued growth. The limits to growth argument made clear that endless growth is not possible. Its concept of limitations has evolved into a conceptual framework of "planetary boundaries" (Steffen et al. 2015; Rockström et al. 2009) within which human civilizations can operate without destroying the home they share with the rest of nature. Of the updated nine parameters—climate change, stratospheric ozone depletion, atmospheric aerosol loading, ocean acidification, biochemical flows, freshwater use, land-system change, biosphere integrity, and the introduction of novel entities—five were exceeded in 2022 (Persson et al. 2022).

In response to the recognition that the endless exponential growth of *material* consumption and destruction is clearly incompatible with finite resources, the proponents of what is now called *green growth* argue that IT offers the potential to *dematerialize* the economy and offer immaterial growth. In other words, it supposedly allows us to decouple economic growth from material consumption (resource decoupling) and environmental destruction (impact decoupling). For example, by increasing efficiencies, IT can reduce the material impact of existing activity; with improved monitoring and feedback, IT can support a better understanding and control of complex processes; and by substituting digital services for physical counterparts, as in video conferencing, IT can eliminate the need for material consumption (Royal Society 2020). This green growth argument has dominated economic policy discourse related to sustainable development in the European Union and elsewhere over the past decade and has substantively shaped policy priorities (Parrique et al. 2019; Hickel 2020; Hickel and Kallis 2020). For example, the European Commission's environmental policy for 2013–2020 targets an "an absolute decoupling of economic growth and environmental degradation" (Parrique et al. 2019, 10), and the 8th Environment Action Plan for 2021–2030 continues to pursue the vision that in 2050, "citizens live well, within the planetary boundaries in a regenerative economy where nothing is wasted, no net emissions of greenhouse gases are produced and economic growth is decoupled from resource use and environmental degradation" (European Commission 2020, 10).

But comprehensive reviews of the evidence behind decoupling come to stark conclusions. "The conclusion is both overwhelmingly clear and sobering: not only is there no empirical evidence supporting the existence of a decoupling of economic growth from environmental pressures on anywhere near the scale needed to deal with environmental breakdown, but also, and perhaps more importantly, such decoupling appears unlikely to happen in the future" (Parrique et al. 2019, 3). Parrique and colleagues assess two sides: the historical evidence of decoupling in the past and the logical argument for the feasibility of decoupling in the future. They cannot identify evidence on either side for large-scale, absolute decoupling—decoupling happens only in local, isolated cases, often remains a temporary phenomenon, and in fact, is often an illusion. Reports, for example, that highly developed countries are achieving material decoupling are identified as flawed conclusions: What these countries have achieved is an offshoring of the resource-intensive material extraction and production facilities on which their economies depend. They have externalized their environmental footprint to lower-income countries, yet continue to materially cause excessive resource consumption and material impact on the planet (Parrique et al. 2019, 21; Krausmann et al. 2017; Vadén et al. 2020). On aggregate global levels, the correlation between material use, material impact, and economic activity holds remarkably steady (Wiedmann et al. 2015; Hickel and Kallis 2020). Overall, "there is no empirical evidence supporting the existence of a decoupling of the type described as necessary . . . —that is an absolute, global, permanent, and sufficiently fast and large decoupling of environmental pressures (both resources and impacts) from economic growth . . . it is safe to say that the type of decoupling acclaimed by green growth advocates is essentially a statistical figment" (Parrique et al. 2019, 31). This assessment is confirmed by two large surveys assessing the evidence accumulated by 179 (Vadén et al. 2020) and 835 (Haberl et al. 2020) studies on decoupling.

To address the perennial counterargument that superior technology will bring about the elusive salvation, Parrique and coauthors move beyond historical evidence. In assessing the evidence of future feasibility, they identify significant barriers. Rebound effects form one of seven factors that will continue to prevent decoupling from becoming a reality. Again, the assessment is unequivocal: "we have found no trace that would warrant the hopes currently invested into the decoupling strategy." In other words, "green growth . . . is not possible" (Parrique et al. 2019, 10). The emerging *Degrowth* movement has forcefully shown that continued economic growth is not only unsustainable and destructive but also irrational (Hickel 2020; Kallis 2018; Demaria, Kallis, and D'Alisa 2015; Raworth 2017).

CONCLUSIONS

IT has a role to play in our societies' transformation toward a sustainable life form within the finite boundaries of planet Earth. Computational systems enable the coordination of cooperative and collective action, can *in principle* facilitate the partial decoupling of some economic activities from resource consumption under certain conditions, and offer new ways of living, working, and playing. But there is ample evidence that overall decoupling on an aggregate level is impossible now and in the future. So, IT development must be mindful of its role in perpetuating the exponential growth of material extraction and accept the importance of planetary boundaries.

As the history of technology shows, technology is "neither good, nor bad, nor neutral" (Kranzberg 1986). Nor is software "technology" an amorphous, shapeless whole—on the contrary. In practice, every systems design effort offers myriad moments where choices can work toward or against sustainability. Not all these choices are visible, and their effects, spread across time and space, may remain obscure, uncertain, and ambiguous. Despite its potential, the accounted effects of IT on sustainability hardly present a stellar track record. The best we may be able to claim is that the use of IT in climate modeling has allowed us to better understand the causal relationships between human activity and the climate crisis.

I want to draw attention to several aspects of the overview provided above. First, by and large, this work has focused on "how computing shapes societies" rather than examining the inverse or adopting an explicit view of mutual shaping and coevolution. In doing so, researchers rarely attend to the indirect, slower, but profound and lasting effects of their interventions. Despite this focus, researchers who argue for transformative change have acknowledged the need to better understand the social forces that shape computing, both in theory and in practice.

Second, the importance of *systems thinking* pervades many of the above fields. Different communities draw on different strands of the diverse branches of systems thinking, but the central attention in sustainability has been on system dynamics, used in the first World Model underpinning the *Limits to Growth* report (D. H. Meadows 2008; D. L. Meadows and Club of Rome 1972; D. H. Meadows, D. L. Meadows, and Randers 2004; Wiek, Withycombe, and Redman 2011). When Easterbrook (2014a) argues that a shift is needed from computational thinking to systems thinking, he primarily refers to system dynamics too. But the debate between the perspectives represented in figure 1.1 reflects the epistemological shift that underpins the break from hard to soft systems thinking in the 1970s (Checkland 1981), and the further break with soft systems thinking performed by critical systems thinking in the 1980s and beyond has not been fully considered (Flood and Jackson 1991c). Chapter 5 will return to the rich history of systems thinking in search of a critical perspective on its role in systems design.

Third, I have identified *requirements* as a central locus of attention with strong leverage over the outcomes of design projects. Because requirements shape technology development at the space where the technical meets the social, they influence the outcomes greatly. They offer significant room for actors in the design process to maneuver and substantial opportunity for innovation in practice. But existing requirements theory and practice appear ill-equipped to tackle the challenges thrown by sustainability.

Finally, the need for transformative change and critical reflection pervades many of the conversations. Those who take a step back to assess the state of work in IT typically come to stark conclusions. And, as the next chapter will show, these are only the tip of the iceberg. Not all is well in computing.

2

JUST SUSTAINABILITIES AND THE DEBTS OF COMPUTING

Unless analyses of development begin not with the symptoms, environmental or economic instability, but with the cause, social injustice, then no development can be sustainable.

— Middleton and O'Keefe (2001, 16)

In "A Perfect Moral Storm," philosopher Stephen Gardiner emphasizes that climate change is a challenge of epic proportions because it is not only globally dispersed in space but also temporally dispersed across generations. The convergence of several characteristics causes an "ethical tragedy." The use of the word *tragedy* is poignant and intentional. The title references the popular book and film about a ship doomed to sink at the confluence of three major storms. In the case of climate change, the ship is us, and the storms are global, intergenerational, and theoretical. The global storm arises from the *spatial dispersion* of cause and effect in climate change. The intergenerational storm arises from the *temporal dispersion* of cause and effect. The theoretical storm, finally, arises from the inadequacy of current theories on "intergenerational ethics, international justice, scientific uncertainty, and the human relationship to animals and the rest of nature" (Gardiner 2014, 7).[1]

The global storm has somewhat familiar characteristics. Causes and effects are spatially dispersed, agency to cause change is fragmented, and vulnerability to the harmful effects of climate destruction is unevenly distributed. The intergenerational storm adds that climate change is resilient to change (at each point in time, current impact is irreversibly caused by the past), that it carries significant inertia (actions in the past have already committed us to significant warming), and that it is *substantially deferred* (the cumulative impact of climate change will be felt across a significant time span). But most seriously, the intergenerational storm introduces strong asymmetry. In principle, the under-resourced victims of climate change that are already alive can exert *some* influence over rich polluters, even if it is weak. But the same is not true at all for agents dispersed in time. Future generations can neither speak for themselves in the present nor coordinate with the past. Instead, each generation exerts influence over subsequent generations but has no way to hold the past accountable. This irreversible asymmetry in influence across generations creates "a new kind of collective action problem" that Gardiner (2014) terms "intergenerational buck-passing" (35). While it is collectively rational for each generation to act, it is individually rational for each generation to pass the buck. Because of *asymmetric vulnerability*, this is particularly challenging: The first generation is asked to make a sacrifice, and its incentive structure is set so that it reaps no reward. If it passes the buck, the subsequent generation faces the exact same situation, only worse. Each generation's inaction worsens the problem.

It is clear that under certain conditions, human societies have very successfully governed shared, common-pool resources for thousands of years. Favorable conditions include an alignment of incentives and timelines across the actors who share the resources, a democratic establishment of binding norms and infringement penalties, and transparency about rules and governance (Ostrom 2015; 2016).[2] But the dispersion of causes and effects across space and time fragments agency and renders existing institutions inadequate (Gardiner 2014). Against the backdrop of these storms, in a public sphere heavily distorted by the interference of fossil fuel lobbying (Oreskes 2010; Supran and Oreskes 2021), Gardiner (2014) warns of the danger of *moral corruption* and reminds us to be skeptical about our own abilities of judgment—we cannot rely on ourselves to simply "invoke and apply" the "correct theories" because "our assessment of theories and their consequences is not made in a neutral evaluative setting" (309). Hence, "an issue of intergenerational justice is at stake, and . . . we are likely—given the perfect storm—to be biased in our own favor" (Gardiner 2014, 430).

SUSTAINABILITY, MEET SOCIAL JUSTICE

Gardiner's emphasis on the moral and ethical nature of climate change drives home a central point: climate change is an issue of ethics and justice. As Agyeman, Bullard, and Evans (2003) write, "the issue of environmental quality is inextricably linked to that of human equality. Wherever in the world environmental despoilation and degradation is happening, it is almost always linked to questions of social justice, equity, rights and people's quality of life in its widest sense" (1).[3] They highlight three aspects. First, inequality within countries is strongly correlated with environmental destruction. Second, the damages inflicted by environmental destruction disproportionately affect those who are already disadvantaged. And third, they endorse Middleton and O'Keefe's epigraph and add:

Sustainability, we argue, cannot be simply an "environmental" concern, important though "environmental" sustainability is. A truly sustainable society is one where wider questions of social needs and welfare, and economic opportunity, are integrally connected to environmental concerns. This emphasis upon greater equity as a desirable and just social goal, is intimately linked to a recognition that, unless society strives for a greater level of social and economic equity, both within and between nations, the long-term objective of a more sustainable world is unlikely to be secured. The basis for this view is that sustainability implies a more careful use of scarce resources and, in all probability, a change to the high-consumption lifestyles experienced by the affluent and aspired to by others. (Agyeman, Bullard, and Evans 2003, 2)

Sustainability for them is "at its very heart a political rather than a technical construct" (Agyeman, Bullard, and Evans 2003, 6). It is no coincidence that the emergence of the environmental justice movement in the US was spurred by evidence showing that hazardous waste disproportionately affected African American communities (US GAO 1983; Justice 1987), illustrating the pervasiveness of "environmental racism" (Bullard 1993). Environmental justice is also deeply implicated in gendered workplace struggles around health and pollution in the global tech industry (Pellow and Park 2002). As Dorceta Taylor (2000) shows, this context is significant in at least two ways. First, it was people of color who constructed the environmental justice paradigm, and second, this placed "concepts like autonomy, self-determination, access to resources, fairness and justice, and civil and human rights" into central positions of the discourse (534). As Agyeman et al. (2016) summarize, the paradigm "explicitly links the environment to race, class, gender, and social justice, effectively reframing environmental issues as injustice issues" (326). They attest partial success to the environmental justice movement, citing the prominent emphasis on equity and justice in the 2015 Paris Agreement as evidence and pointing out that the revised UN Sustainable Development Goals (SDGs) (United Nations 2015a; United Nations General Assembly 2015) "read like an institutional agenda for just sustainabilities" (Agyeman et al. 2016, 335). It is true that the seventeen SDGs range from gender equality and the reduction of inequalities to "justice for all" and include such metrics as 16.7, "responsive, inclusive, participatory and representative decision-making at all levels" (United Nations 2015b). Considering that as recently as 2013, *Nature* published an article on SDGs without the word "justice" that only mentions "equity" in terms of gender (Griggs et al. 2013), this is certainly progress. But critics of the framing of sustainable development remain highly skeptical of the willingness and ability of entrenched institutions to facilitate the kind of transformative change required to make these goals come true (e.g., N. Klein 2014; Escobar 2018; Kothari et al. 2019). Instead, they argue for the need to develop "alternatives *to* development, rather than development alternatives" (Escobar 2011, xiii), and we will meet their proposals later in this book. In parallel, "there remains an issue of the historic injustice imposed on the poor and the poor countries of the planet. There is a vast historic ecological debt" (Agyeman, Bullard, and Evans 2003, 30).

Agyeman, Bullard, and Evans's (2003) conception of *just sustainabilities* distills the convergence of social justice and environmental sustainability into an approach that aims to "ensure a better quality of life for all, now, and into the future, in a just and equitable manner, while living within the limits of supporting ecosystems" (2). The relationship between the environment and justice is complex and mutual: "A poor environment is not only a symptom of existing injustice; rather, a functioning environment provides the necessary conditions to achieve social justice" (Agyeman et al. 2016, 335). Others add that justice, equity, and inequality are in fact leverage points toward ecological sustainability:

We are made to choose between human welfare or ecological stability—an impossible choice that nobody wants to face. But when we understand how inequality works, suddenly the choice becomes much easier: between living in a more equitable society, on the one hand, and risking ecological catastrophe on the other. Most people would have little difficulty choosing. Of course, achieving this will not be easy. It will require an enormous struggle against those who benefit so prodigiously from the status quo. And presumably this is why some are so eager that we avoid this course of action: they would prefer to sacrifice the planet in order to maintain the existing distribution of global income. (Hickel 2020, 144)

That sacrifice, after all, happens in a future that cannot reach back to hold them accountable. In other words, "justice is the antidote to the growth imperative—and key to solving the climate crisis" (Hickel 2020, 137). Historians, anthropologists, and economists alike have convincingly demonstrated that capitalist growth always comes at an enormous cost to others. Economic growth has to be fueled by resources that come *from somewhere* (Hickel 2020). In five hundred years of capitalism, growth has come variously from the enclosure of common public ground in Europe that forced commoners into labor contracts, from the extraction of raw materials at unfair conditions imposed by colonialism, from the forced unpaid labor of slaves, or from the extraction of fossil fuels. Throughout the centuries, capital has found the next frontier of growth whenever growth stalled, from deregulation to the extraction of data that describes the bodies and lives of people. The justification of continued growth and decoupling claims that our societies *need* economic growth and that everyone benefits from growth. But beyond a threshold long surpassed by the Global North, only a minuscule portion of our world's population actually benefits from economic growth.[4] The widely told story of growth leading to improved health and well-being has long been debunked. Instead, "progress in human welfare has been driven by progressive political movements and governments that have managed to harness economic resources to deliver robust public goods and fair wages. In fact, the historical record shows that in the absence of these forces, growth has quite often worked against social progress, not for it" (Hickel 2020, 128).

To think that the argument to abolish economic growth as an objective—degrowth—would be an argument for austerity and sacrifice would be a misinterpretation. It refers to a lean metabolism much rather than starvation:

The term "degrowth" may be confusing if perceived as negative GDP growth. To be clear: degrowth is not negative growth. The goal of degrowth is not to make GDP growth negative. There is a name for that: "recession" or, when prolonged, "depression." In economics terms, degrowth refers to a trajectory where the "throughput" (energy, materials and waste flows) of an economy decreases while welfare, or well-being, improves. (Kallis 2018, 9)

The growth of economic activity, measured in GDP, is not a necessary condition for humanity to flourish; it is a necessary condition for capital to thrive and expand. Under the current economic arrangements, national economies and many companies are pressured to grow to survive their debt load, so a reduction in growth is felt through job cuts.[5] Degrowth restructures our societies' economic arrangements so that we do not *need economic* growth. It reverses the artificial scarcity on which capitalist economies depend. "Austerity calls for scarcity in order to generate more growth. Degrowth calls for abundance *in order to render growth unnecessary.*" (Hickel 2020, 168). The call is strongly linked to arguments to decolonize not just social and political relations but also the way we conceive of the relationship between us humans and the rest of nature, away from a view of nature as a lifeless resource, toward an understanding of interdependent relationships—a view that learns from Indigenous worldviews what Western science has pushed to the margins (Hickel 2020; Escobar 2018; Maturana and Varela 1992; Kimmerer 2013; Kothari et al. 2019). These perspectives offer new directions for reorienting and designing sustainable economic activity on our planet, including technology design.

SOCIAL JUSTICE IN COMPUTING AND SYSTEMS DESIGN

The term *social justice*—"justice in terms of the distribution of wealth, opportunities, and privileges within a society" (Oxford 2020e)—has only relatively recently been adopted as an explicit framing in computing discourse (Dombrowski, Harmon, and Fox 2016), but fields such as computer-supported collaborative work (CSCW) have long been invested in the issues it raises (Fox et al. 2017). According to Fox et al. (2016), "As a perspective, social justice explicitly takes into account how the historicity, situated context, and social issues (e.g., class, race, gender, ability, sexual orientation, health and wellness, food access, and so on) impact people's experiences . . . , including how technology is designed and developed, how policy impacts information and communication practices and experiences, and how marginalization and oppression impact people's experiences of and practices with technology" (3294). This perspective has been a central driver of recent work critiquing the booming industry that develops and applies machine learning algorithms at large scales to the classification of human beings in capitalist societies.

Patricia Hill Collins (1990) introduced the term *matrix of domination* in *Black Feminist Thought* to describe how dimensions such as gender, race and class that influence power differentials are connected and organized across four domains (structural, hegemonic, disciplinary, and interpersonal). In current views, the dimensions include gender, class, race, ability, sexual orientation, religion, geography, age, and other factors. The related concept of *intersectionality* (Crenshaw 1991) highlights how the position of individuals at specific locations—at intersections of these dimensions—affects power and privilege in ways that are not captured by an additive accounting of dimensions. Here, as elsewhere, as systems thinker and management scientist Russell Ackoff (1999b) put it, the system is not the sum of its parts but must be understood as the product of their interactions. Intersectional and Black feminist scholars who examine computing from this perspective emphasize that the uneven distribution of power and privilege in society should be conceived on many dimensions. It is notable and hardly a coincidence that not one of the books cited in the next section is authored by a White man.

When Google indexes the web, it shapes the web, and with its ranking of search results, it shapes our media landscape, market dynamics, and politics. The enormous influence of this and other algorithms has become an object of public discussion over the past years. In *Weapons of Math Destruction*, data scientist Cathy O'Neil (2016) explains and demonstrates how algorithms using big data reinforce inequity and existing power structures to the detriment of justice, fairness, and equity. The weapons of math destruction (WMD) metaphor arises from the combination of three characteristics often found in data science models: opacity, damage, and scale. Large-scale models create their own feedback loops, not merely reproducing the inequitable status quo but amplifying and accelerating it to inflict damage at scale. WMD afflict individual lives across all stages, from college admissions and credit scoring to job applications, credit loans, the criminal justice system, and parole decisions. In *Algorithms of Oppression*, Safiya Noble (2018) examines the role of search algorithms in particular and demonstrates how they reinforce racism and oppress women of color, in an excellent illustration of just how strong these reinforcing feedback loops can be. Latanya Sweeney (2013) had demonstrated racial discrimination in online ad delivery much earlier, but nothing had changed since then.

But algorithmic sorting, control, and discrimination are not limited to global platforms. Any computational system classifying its users raises comparable concerns that could in principle be addressed in its design but often aren't. In *Automating Inequality*, Virginia Eubanks (2018) shows, based on long-running in-depth investigations, how the design and use of software systems in social services profiles the poor across the US and reinforces their marginalized status. Inevitably, the material implications of algorithmic discrimination and oppression are unevenly distributed.

I am often approached by engineers or data scientists who want to talk about the economic and social implications of their designs. I tell them to do a quick "gut check" by answering two questions: Does the tool increase the self-determination and agency of the poor? Would the tool be tolerated if it was targeted at non-poor people? Not one of the technologies I described in this book rises to this feeble standard. We must demand more. (Eubanks 2018, 173)

In the US context, the racist tendencies of machine learning algorithms reinforcing existing systemic inequalities and structural racism have become an object of public debate at last. In *Race after Technology*, Ruha Benjamin (2019) provides important historical context and traces the historical trajectory of technology from a Black feminist perspective. She shows how "technologies reflect and reproduce existing inequities but . . . are promoted and perceived as more objective or progressive than the discriminatory systems of a previous era" (15).

To grasp how this takes place and to change it, we need to combine a technical understanding of how they work with a critically appreciative understanding of their social construction, the political forces they are subject to, and their historical evolution. Computer science has been notoriously keen on avoiding this conversation. Notably, one of Eubanks' s (2018) pledges is to "design with history in mind," because to do otherwise is to be "complicit in the 'unintended' but terribly predictable consequences that arise when equity and good intentions are assumed as initial conditions" (174). The crucial point is that what the proponents of large-scale technology interventions often call "unintended" consequence or describe as "unforeseeable" is in fact not unpredictable at all from other standpoints.

These arguments have given significant visibility to the deeply problematic role of computing in maintaining societal structures of power and oppression, made the underlying historical trajectory visible, and explained it in accessible ways to a broad audience. They also call for different approaches to technology design, and recent work has moved from a focus on critique to speculating about new directions for design (Rosner 2018) and to developing generative principles for action. In *Design Justice*, Sasha Costanza-Chock (2020) draws on a wide array of practices and examples to illustrate how an intersectional awareness of the matrix of domination can be brought to bear directly on the practice of design (interpreted broadly with a focus on software systems). They highlight that "the matrix of domination is constantly hard-coded into designed objects and systems. This typically takes place not because designers are intentionally 'malicious' but through unintentional mechanisms, including assumptions about 'unmarked' end users, the use of systematically biased data sets to train algorithms using machine-learning techniques, and limited feedback loops" (Costanza-Chock 2020, 71).[6] The aim is explicitly to "retool design," drawing inspiration from activist and advocacy work from areas including disability justice.

Similar developments are under way in data science. In *Data Feminism*, Catherine D'Ignazio and Lauren F. Klein (2020) relocate and reconstitute data science on feminist ground: "Because the power of data is wielded unjustly, it must be challenged and changed" (14). Correspondingly, data feminism is built on seven principles: (1) examine power, (2) challenge power, (3) elevate emotion and embodiment, (4) rethink binaries and hierarchies, (5) embrace pluralism, (6) consider context, and (7) make labor visible. The authors emphasize that data feminism is not only about and for women, and not only about gender but about power (2020, 14). They discuss "What makes a project feminist? . . . a project may be feminist in *content*, in that it challenges power by choice of subject matter; in *form*, in that it challenges power by shifting the aesthetic and/or sensory registers; and/or in *process*, in that it challenges power by building participatory, inclusive processes of knowledge production" (2020, 18). This sense resonates strongly with the direction taken later in this book.

These arguments reflect a highly overdue confrontation in the computing discourse between its traditional modes of reasoning and a critically informed, intersectional feminist perspective.[7] A wave of practice in design and advocacy accompanies these writings too, with numerous groups forming to collectively address specific harms. Too numerous to list them all, these efforts range from research institutes such as the NYC-based AI Now Institute and researcher-driven efforts such as the Algorithmic Justice League founded by computer scientist and digital activist Joy Buolamwini to federated groups of researchers and practitioners such as the Design Justice Network and to organized labor, as in the Tech Worker Coalition. Such efforts of collective organizing and advocacy regularly perform successful interventions. In 2020, *Springer Nature* was forced to rescind an offer to publish a piece of work that can only be described as neophrenologist. It presented a machine learning algorithm that, its creators claimed, successfully "predicted" criminality based on photographs. The retracting came after a social media campaign gathered 2,500 signatories for a compellingly argued and heavily substantiated letter.

Machine learning programs are not neutral; research agendas and the data sets they work with often inherit dominant cultural beliefs about the world. These research agendas reflect the incentives and perspectives of those in the privileged position of developing machine learning models, and the data on which they rely. The uncritical acceptance of default assumptions inevitably leads to discriminatory design in algorithmic systems, reproducing ideas which normalize social hierarchies and legitimize violence against marginalized groups . . . any effort to identify "criminal faces" is an application of machine learning to a problem domain it is not suited to investigate, a domain in which context and causality are essential and also fundamentally misinterpreted. (Coalition for Critical Technology 2020)

Because of such scholarly advocacy work, the racially biased and discriminatory role of technology in our societies is now well established. Still, despite these voices, the mainstream discussions of social justice in computing, especially in machine learning, continue to frame it narrowly and reductively. Instead of acknowledging the contextual, historical, sociotechnical nature of *algorithmic bias*, the issue is reframed as a compu-tational problem. On that basis, "garbage in, garbage out"—the metaphor that bad data, especially biased data, leads to bad outcomes, for example biased classifiers—is used as an argument to deflect attention from broader issues and refocus it into a technical question. A satirical piece illustrates beautifully how impoverished and dangerous this is. *A Mulching Proposal* unflinchingly describes in technical terms a corporate effort to make com-pletely fair, accountable, and transparent a machine learning algorithm that selects who out of a general human population should be picked up by a drone and turned into food. Initial concerns about the algorithm dis-proportionately targeting White people are quickly addressed by adding people of color to the training data. State-of-the-art metrics of computa-tional fairness are successfully deployed to ensure fair outcomes: every-one is equally likely to be processed. The terrorizing project it describes is technically consistent with the revised ACM Code of Ethics. Since its latest revision, the ACM Code allows systems to be built that cause intentional harm, and the Code places responsibility for deciding whether to build these systems explicitly into the hands of the system designers, with no call for external legitimation (see chapter 12). That is exactly what *A Mulch-ing Proposal* does, to striking effect (Keyes, Hutson, and Durbin 2019).

I want to highlight a few observations. First, many researchers and practitioners in computing reduce concepts with rich social history, such as *fairness* and *justice*, to algorithmic definitions. The well-meaning com-puter scientists at Dagstuhl whom I mentioned in the Introduction are not outliers; they are the norm, but their impoverished reframing of jus-tice is utterly inadequate (Selbst et al. 2019; Jacobs and Wallach 2021). Only recently has this balance begun to shift at venues like the FAccT conference series on "fairness, accountability, and transparency in socio-technical systems."[8] Many of these authors have emphasized the histori-cal, political, and social nature of computing from a perspective located outside of computing, but in recent years, writers such as O'Neil, Timnit Gebru, Buolamwini, and others have written about these issues from a computing insider perspective as well. This has made it much harder for those in computing to ignore or silence their voices—though not for lack of trying (Hao 2021b; Vincent 2019). These courageous researchers have chosen the difficult route of writing *for computing* and working *in computing* for a better, more just, computing.

Second, justice has also been a concern in the sustainable development discourse, and as mentioned earlier, some work in ICT4S has mentioned the distributive justice aspect of technology development. But these concerns have remained at the margins. As late as 2022, the conference on ICT4S—the premier venue focused on such work—included no paper on environmental justice, social justice, or just sustainabilities. And even the socially conscious transformation mindset discussed in chapter 1 does not mention race, class, gender, or power. *Just sustainabilities* incorporate social justice into the sustainability discourse. Wider aspects of social justice remain outside of this framing.

Yet, based on the burgeoning literature on social justice issues and computing, and a range of communities working on issues of sustainability in computing, the nascent attention to the nexus of all three is apparent. For example, the *Smell Pittsburgh* project by Hsu and colleagues (Hsu and Nourbakhsh 2020; Hsu et al. 2017) adopted a community-centered approach to citizen science that empowered communities to hold corporate polluters accountable through collecting and visualizing air quality data. The project is situated explicitly within the landscape of sustainable human–computer interaction, and it makes direct reference to environmental justice. On a broader scale, the threat to the Environmental Protection Agency (EPA) that arose from the 2016 US election galvanized activists and academics into forming the Environmental Data & Governance Initiative (EDGI). The organization initially coordinated distributed efforts to save EPA data from impending erasure and continued to monitor the evolving situation to hold the US government accountable. The framework of *environmental data justice* fuses the value systems of environmental justice and data justice to pursue "public accessibility and continuity of environmental data and research, supported by networked open-source data infrastructure that can be modified, adapted, and supported by local communities" (Dillon et al. 2017). Vera et al. (2019) emphasize that environmental justice calls for approaches that "replace extractive data practices with community-based participatory research and community science projects," just what Hsu and colleagues aimed for, but as the authors argue further, the "desire for more data not only demands that harms be proven through a technically narrowed mode of evidence legible to dominant institutions, but also often requires communities to generate evidence of environmental violence from their own suffering" (1015). In their view, environmental data justice cannot resolve this tension but commits to ongoing critical reflection.

REALITY CHECK: THE DEBTS OF COMPUTING

If we could take stock of computing's balance with our worlds, what would we find? Like any other technology, software is never a neutral element of our societies. Software filters, sorts, and selects; it tracks, monitors, and intervenes; it initiates and shuts down and controls. In extreme cases, it passes judgment (Julia Angwin et al. 2016; Benjamin 2019; Eubanks 2018). It reinforces existing power dynamics, amplifies structural differences, and automates the reproduction of racism, inequality, and inequity. In doing so, it embeds the values and politics that underpin the choices of its designers and the context in which they make them.

But software is not just an element of oppression; it also is a tool of liberation. Airbnb was mentioned earlier as a software-based platform with far-reaching effects, deeply implicated in the erosion of housing availability. We know this because of data-driven investigations that equally rely on, and build, software technology in examining the role of Airbnb's platform in housing markets across the world from a critical angle. The work of urban geographer David Wachsmuth, for example, crucially involves the development of algorithmic de-obfuscation tools to make data analysis possible (Wachsmuth and Weisler 2018).

On a broader scale, civic movements across the world develop open-source software systems to securely and safely organize and coordinate peaceful protests; journalists rely on some of the same systems to report on human rights violations under dangerous conditions; and software systems are used to support environmental sustainability initiatives that range from supporting permaculture (Norton, Penzenstadler, and Tomlinson 2019) to community citizen science initiatives reporting pollution (Hsu and Nourbakhsh 2020) and safeguarding evidence of climate change from corrupt regimes (Vera et al. 2019). We might call these "algorithms of liberation" (Roberts et al. 2018). They demonstrate that, in principle, significant room for maneuver is available to those designing and developing software systems. The emergence of tech workers who take their disagreement with company policies into the open to protest unethical practices attests to an emerging social consciousness and a rising desire for such change in the IT industry (see chapter 12).

Despite these promising efforts and initiatives, however, it seems clear that the harmful direct and indirect effects of software technology development are not shrinking but rising. I call these the *debts of computing*. In fact, most of the initiatives mentioned here that strive for change are issues-centered efforts to address precisely some of those debts, to make them visible against the efforts of the tech industry, and to prevent the worst outcomes. The *Ledger of Harms* operated by the Center for Humane Technology lists the costs explicitly: "Under immense pressure to prioritize engagement and growth, technology platforms have created a race for human attention that's unleashed invisible harms to society. Here are some of the costs that aren't showing up on their balance sheets." The harms listed in the evolving catalog range from algorithmic oppression and negative effects of tech on cognitive abilities to misinformation, social isolation, and threats to democratic governance (Center for Humane Technology 2020).

In software engineering, the term *debt* has received significant attention in the form of *technical debt*, which refers to "design or implementation constructs that are expedient in the short term, but . . . make future changes more costly or impossible" (Avgeriou et al. 2016, 112). In this view, "technical debt presents an actual or contingent liability whose impact is limited to internal system qualities" (112). Technical debt is a burgeoning research area with its own dedicated conference series. The term has been very effective because it explores the meaning of a domain that is hard to grasp (the temporally diffuse effects of design choices) by mapping key elements onto a relatively well-known and formalized domain (finance). That is the nature of metaphor (Lakoff and Johnson 1980; Lakoff 1993). Far from being merely figures of speech or poetic constructs we use to adorn our writing, metaphors are the conceptual mappings by which we view domains we cannot fully explain through the conceptual lens of domains we understand. This domain mapping is central to the structure of human thought and language, and to some degree we can trace them. Consider this statement: "Our new colleague gave a talk yesterday. It was full of interesting ideas, but I wasn't sure what to take away from it." A talk is an intangible, temporally bound event, hard to grasp as an object of thought. Yet we describe it as if it was a tangible object that could be handed from one person to another: Our colleague *gave us* the talk. More so, in the example the talk is treated as a container, with ideas being objects inside the container. Like marbles from a bowl, we assume these ideas can be taken out, and carried away, by the audience. Through layers of such conceptual mapping, metaphors ultimately ground our conceptual understanding of the world in our lived physical, bodily, and cultural experience. This applies not just to colloquial speech but equally to scientific theories and concepts (see also Morgan 2006; M. Jackson 2003), so it is worth reflecting more deeply about the implications of metaphors in design (Blackwell 2006; Becker 2018a).

In the case of technical debt, the metaphor enabled a community to talk in clearer terms about issues that had been bothering researchers and practitioners since the founding of the discipline (Naur and Randell 1969; Lehman 1979; Parnas 1994).[9] When it comes to technical debt, the conceptual mapping suggests that it may eventually be *paid back*, but it also carries additional implications. When formerly expedient design constructs get in the way of further work, the additional effort is understood as interest. This financial concept places the additional burden in a proportional relationship to the presumed benefit gained from the expedience of prior choices. Indeed, debt can be taken on strategically. The literature of technical debt has been explicit about this framing, developed an elaborate vocabulary around it, argued for the positive leverage to be gained from strategically incurring technical debt, and has even started to pull in mathematically sophisticated approaches from finance to address the problem of "managing technical debt" as if it were a financial portfolio. Technical debt is thus on its way to becoming a scientifically minded "theory" in software engineering.

One of the efforts made to leverage the debt metaphor further, *sustainability debt*, refers to "the hidden effects of past decisions about software-intensive systems that negatively affect [the] sustainability of the system under design" (Betz et al. 2015). This angle extends the established metaphor of technical debt outwards, drawing on the systemic effects of software systems visualized in figure 1.2. Where technical debt is about temporal dispersion of effects that remain firmly in the court of the design team, sustainability debt expands the timespan and adds social distance to the picture: The hidden effects of design decisions are not borne out by the designers in the future; they are inflicted upon the world at large. When an affluent business traveler pays a premium to book an inner-city Airbnb for a conference trip, they contribute to the displacement of affordable housing by highly profitable short-term rentals. This remains an individual choice, but it is enabled, shaped, and encouraged by the specific features in the design of Airbnb's platform. Reinforcing feedback loops between consumer behavior and investor capital make this option attractive for travelers and investors alike. The aggregate structural impact is felt by those urban populations who can no longer afford inner-city rents in busy global cities.

Perhaps the most egregious example of externalized debt is bitcoin. In 2022, bitcoin alone generated about 40,000 tons of electronic waste per year (Digiconomist 2022), comparable to the Netherlands (cf. de Vries and Stoll 2021). Already in 2018, "each $1 of Bitcoin value created was responsible for $0.49 in health and climate damages in the US and $0.37 in China" (Goodkind, Jones, and Berrens 2020), for an overall 1:1 matching of market valuation and material damage. Who carries the damage? Not those who invest spare resources in bitcoins or bitcoin mining, but those marginalized populations who are already disproportionately exposed to the health hazards of pollution and climate change.

COMPUTING HAS EXTERNALIZED ITS DEBTS

The metaphor of debt breaks down here at a crucial moment, for the debts of computing—the hidden, delayed, and remote effects of systems design decisions on the world—continue to be externalized. They are offloaded to and paid for by others. Computing is not paying back these debts to societies, nor does it carry any of the interest. Far from it. Instead, from the health costs paid by the copper miners to the higher credit rates paid by African Americans, those who bear the burden of interest for the generated profit, financial or otherwise, are those at disadvantaged locations of the matrix of domination and also often displaced in time or space. They are the least likely people to be involved in technology design themselves. This is evidently unjust. Critiques of this externalization are sometimes met with extraordinary cynicism from those in positions of power, going as far as inventing the term "underpollution" in the context of e-waste:

In 1991, Larry Summers, then Chief Economist of the World Bank (and now President of Harvard University), spoke of the economic sense of exporting first world waste to developing countries (Summers, 1991). He argued that the countries with the lowest wages would lose the least productivity from "increased morbidity and mortality" since the cost to be recouped would be minimal; the least developed countries, specifically those in Africa, were seriously underpolluted and thus could stand to benefit from pollution trading schemes as they have air and water to spare; and that environmental protection for "health and aesthetic reasons" is essentially a luxury of the rich, as mortality is such a great problem in these developing countries that the relatively minimal effects of increased pollution would pale in comparison to the problems these areas already face. (Widmer et al. 2005, 437–438)

This is not a book about the material extraction and supply chains of IT (Frankel, Chavez, and Ribas 2016; Reid-Henry 2012; Crawford and Joler 2018; Crawford 2021) or the continued colonial offshoring of environmental debt (Fuchs, Brown, and Rounsevell 2020) and e-waste (Wang et al. 2020). But it is important to bring this context to attention before we examine how computing's way of framing its domains of interest pushes this destructive impact into the background of systems design practice.

CONCLUSIONS: JUST SUSTAINABILITY DESIGN CHALLENGES

The questions of sustainability cannot be treated as detached from social justice concerns. The framing of *just sustainabilities* serves as a reminder that the two are inextricably entangled, and that their conceptualization must be pluralist, making space for diverging meanings in local contexts (Agyeman 2013). These are important shifts in understanding. Not all social justice concerns are about environmental sustainability, however. Systems design in computing today always must be aware of its never-neutral role in shifting power and causing harm to those at the margins of the process. Because harms are often dispersed temporally and spatially, and because the vulnerability to these harms is always asymmetric, it is just too easy for those who design to disregard the debts that their choices incur. Economically, this is because the debts are externalized: foisted on others. Cognitively, the actors who carry them are located at a psychological distance. Ethically, those who design are typically located at privileged positions in the matrix of domination, so they are especially prone to the dangers of moral corruption. Just sustainability design must attend to the challenges that arise from the multifaceted nature of these concerns.

Modeling challenges arise out of the difficult interrelationships that bridge social, technical, economic, human, and nonhuman domains. Systemic approaches to grappling with these interrelationships have made major strides over the past decades and allow us to articulate in increasingly nuanced ways the dynamic causal and correlational factors of just sustainability in given design contexts.

Cognitive challenges are complicated by our limited understanding of individual and social cognition. We do not have a robust theory to explain how systems design decisions are made in situations where their outcomes are uncertain, ambiguous, and located far away in time and space at a psychological distance to those involved in design. How do teams handle these cognitive challenges, where delayed and accumulating impact comes only clearly into vision as distance decreases? We cannot hope to develop effective methods that prescribe how to design for sustainability and justice without such a theory.

Ethical challenges, finally, arise out of the spatial and especially temporal dispersal of effects, combined with asymmetric vulnerability and the inadequacy of theories. Together, they raise the specter of moral corruption and remind us that we're literally "biased in our own favor" (Gardiner 2014, 430). Most importantly, they remind us that science and technology do not provide the conceptual tools for their own ethical justification. For all its power, the monological form of deductive reasoning manifested by the scientific method has no access to moral judgment. "In this way, science loses sight of its potential role in the search for nonoppressive forms of culture and society. It cannot even enter into dialogue with other forms of knowledge given its de facto claim to have the monopoly on knowledge, compassion, and ethics" (Escobar 2018, 89). Instead of insisting on the formalization of ethics, we have to defer for these judgments to other perspectives and forms of reasoning. But which?

Given the contributions of computational thinking and modeling in conjunction with system dynamics and related approaches, systems design in computing is relatively well equipped to deal with the modeling challenges to anticipate dispersed direct and indirect outcomes at multiple scales. Many difficulties and challenges remain (Kienzle et al. 2020), but these challenges on their own are of a type of complexity that computing is well-equipped to handle. A myriad of collaborations is under way to tackle these types of concerns. I will not attempt to contribute much to advancing these issues but instead focus on the cognitive and ethical challenges. This is because they are so dominant in their influence that progress in the first area will hardly make a dent on sustainability and justice unless we gain more effective insights on them. The computing discourse has little to offer on these questions, so we need to look elsewhere.

In the meantime, computing has managed to externalize its debts, and so far, it has gotten away with it. Part of what is happening here is the corruption of discourse itself. The reduction of "wet" concepts like environmental protection and justice to a "dry" computational problem conveniently allows many in computing to avoid facing an uncomfortable realization: that the present state of technology is a living historical reality that continues to reproduce historically grown power structures, with all the privileges, inequities, and injustices they incorporate. Aside from the incentive structures that make it possible and desirable for those in positions of privilege to pursue this framing, the next chapter will show that the pervasive resistance to acknowledging the wider issues is supported by the metaphors and narratives that structure what computing considers its domain of reasoning.

3

THE MYTHS OF COMPUTING

Computational thinking is a way humans solve problems.
—Wing (2006)

For me, the hardest thing to change is the cultural attitude of scientists. Scientists are some of the most dangerous people in the world because we have this illusion of objectivity; there is this illusion of meritocracy and there is this illusion of searching for objective truth.
—Gebru (quoted in C. S. Smith 2019)

In the *Communications of the ACM*, Jeanette Wing—professor of computer science (CS), director of the Data Science Institute at Columbia University and corporate VP of Microsoft Research—described the merits of what she termed *computational thinking*. Her article, widely cited in computing and beyond, argued that computer science has developed a way of thinking that is so powerful that everyone should learn it: "To reading, writing, and arithmetic, we should add computational thinking to every child's analytical ability" (Wing 2006, 33). Computational thinking encapsulates computing's most profound ways of reasoning:

Computational thinking is thinking recursively. It is parallel processing. It is interpreting code as data and data as code. . . . Computational thinking is using abstraction and decomposition when attacking a large complex task or designing a large complex system. It is separation of concerns. It is choosing an appropriate representation for a problem or modeling the relevant aspects of a problem to make it tractable. . . . Computational thinking is using heuristic reasoning to discover a solution. It is planning, learning, and scheduling in the presence of uncertainty. It is search, search, and more search. (Wing 2006, 33)

Wing enumerates here some of the powerful and elegant ways of reasoning that have drawn countless people into the realm of computing, including your author. Computational thinking is both about analytic abstract reasoning and about *making things*, because computer science sits at an intersection of mathematics, science, and engineering. "The constraints of the underlying computing device force computer scientists to think computationally, not just mathematically. Being free to build virtual worlds enables us to engineer systems beyond the physical world" (Wing 2006). Computational thinking thus presents a powerful toolbox of mechanisms that can be used to solve some classes of problems. And that is at the heart of how computer science understands its mission as "the foundational discipline of computing that studies the use of computers to systematically solve problems" (CS2023 2022).

The first application of such methods to real-world problems is generally attributed to the efforts of operations researchers and other scientists during World War II. Their efforts simultaneously shaped the emerging fields of operations research, game theory, artificial intelligence (AI), computer science, cybernetics, and cognitive science. A central figure in all these was Herbert Simon, whose Nobel Prize–winning work brought together some of the core ideas we will debate: rational behavior, decision-making, problem-solving, design, social planning, and the nature of human thought. By examining what Simon (1962) termed the "architecture of complexity" in naturally occurring and artificial systems, this work produced elegant approaches to managing complexity through design principles around decomposition, modularity, abstraction, and problem representation (e.g., Parnas 1972; Simon 1977). Concepts such as recursion and heuristic search have allowed computing to tackle enormously complex tasks, decompose them carefully into relatively independent subtasks, build reliable systems out of unreliable components, grow layers upon layers of abstractions to build up complex processes based on simpler ones, effectively separate concerns that could be disentangled, and merge heterogeneous insights from diverse information sources in highly modular networked systems of algorithms with known complexity. These concepts are at the heart of all conceptions of computational thinking (Denning 2017). They also enable computing to play an important role in our understanding of complex systems such as the Earth's climate.

Computational thinking (CT) refers to "the thought processes involved in formulating problems so their solutions can be represented as computational steps and algorithms" (Aho 2012, 834–835; Denning 2017). Tedre and Denning (2019) describe it as "the mental skills and practices for designing computations that get computers to do jobs for us, and explaining and interpreting the world as a complex of information processes" (4). They caution enthusiasts: "Computational thinkers need to develop enough experience and skill to know when jobs are impossible or intractable, and look for good heuristics to solve them" (8). It is instructive that the limitations they identify pertain to computational complexity, intractability, and the lack of semantics, not to any conceptual limitations of CT that may require other forms of human reasoning such as ethical judgment.

Learning to think computationally can be an empowering experience. Many computing educators believe that the core of learning computer science is learning to solve problems by developing algorithms (Peters 2019; Malazita and Resetar 2019). This "significance of learning how to write algorithms to solve problems is emphasized over the particular technical skills . . . that students learn in computer science" (Breslin 2018, 98). The core value of that skill is problem-solving: "Through discussing, learning, and practicing these particular modes of thought, students learn to understand and build within the computational universe. They learn to create algorithms, focusing on creating step by step instructions that solve particular problems" (Breslin 2018, 105).

Wing (2006) argues that CT will become so inevitable and ubiquitous that it will disappear into the background. "Computational thinking will have become ingrained in everyone's lives when words like algorithm and precondition are part of everyone's vocabulary . . . and when trees are drawn upside down" (34). In the displacement of the upright tree, the metaphor travels back to uproot our representations of nature—perhaps not such a utopian prospect. But if CT is so powerful and if, as the last chapter suggested, computing as a practice and a theory is already becoming aware of its implication in sustainability and justice, is it not enough to simply step aside and let computing deploy its formidable powers of reasoning, innovation, development—and design—toward solving the world's problems? After all, many well-funded initiatives are doing this already under umbrellas such as "technology and society." The central organizing metaphor often found in these initiatives is technology-driven problem-solving (Pal 2017). Because problems can be represented as symbols, the real world is *rendered technical* (Breslin 2018) into problems of a *domain* of computing (Ribes et al. 2019; Ribes 2019). What could possibly go wrong?

The previous chapters showed that a lot has gone wrong so far. This and the following chapter trace historically grown concepts central to the field of computer science and explore how they inhibit the potential of those involved in systems design to think critically, reflexively, and inclusively through the situations they are facing in their design practice. By doing so, the chapters illustrate the reasoning by which even good intentions often end up reinforcing the status quo, increasing computing's debt while doing nothing to change the process by which that debt is foisted onto others. In pursuing this argument, this chapter builds on a formidable succession of critics of thinking and designing in computing (e.g., Winograd and Flores 1986; Agre 1997b; Suchman 1987; see also Bardzell 2010).

METAPHOR AND MYTH

When students of computing learn to think computationally by learning how to program, they make sense of this new symbolic domain by reference to the concepts they know. In her in-depth study of "the making of computer scientists," Breslin (2018) explains the role of programming languages in establishing the structures by which students of computer science learn to think:

Language operates to constitute reality in particular ways, to create worlds of meaning and implication. . . . Students and professors speak of these worlds as though they exist in space, beyond the physical space in memory and computational time that a program takes up. Programs and code are talked about as though they have a shape and substance. For example, students are told functions . . . have a "territory" or a "scope." Certain functions have property, variables that they own and know about, but that other functions do not. Some data structures are in the forms of trees, with branches that can be traversed breadth-first or depth-first as different searching algorithms. . . . In learning to program, students are thus learning to become fluent in particular languages and particular modes of thought that constitute and enable particular worlds and realities. (97)

In other words, they rely on metaphor to grasp these structures. As we get used to new metaphors, we cease to perceive them as conceptual mappings—in our minds, they eventually become *reified* and seemingly detached from the original domain that the mapping concepts were drawn from. As a result, "metaphors . . . have the power to define reality. They do this through a coherent network of entailments that highlight some features of reality and hide others" (Lakoff and Johnson 1980, 157).

By focusing on one mode of thought, others can drift out of view. Timnit Gebru, a leading critical voice on ethics and racism in machine learning (ML), emphasizes that computer scientists such as herself (and Jeanette Wing and myself) are prone to certain illusions about the nature of their work. In this chapter, I identify four of these illusions, trace them to their origins, explore their implications, and point to evidence that exposes them as flawed. I will treat these illusions as myths. In its simplest sense, the word *myth* refers to a "widely held but false belief" (*Oxford English Dictionary* 2020d). Myths are important because they are often at the heart of cultural narratives. Sets of myths form historically grown networks of stories that establish norms, values, and behaviors as part of a cultural tradition. In *The Charisma Machine*, Morgan Ames (2019) summarizes the role of mythologies and ideology in technology development: "cultural mythologies [are] foundational narratives that are ritualistically circulated within groups to reinforce collective beliefs. Mythologies have an element of enchantment to them, making certain futures appear at once magical and inevitable, straightforward and divine . . . for nearly two hundred years, mythologies have been central to the way that the United States and Europe think about technology" (18–19). It is hard to overlook the resonance of these arguments in Wing:

This kind of thinking will be part of the skill set of not only other scientists but of everyone else. Ubiquitous computing is to today as computational thinking is to tomorrow. Ubiquitous computing was yesterday's dream that became today's reality; computational thinking is tomorrow's reality. . . . Computational thinking will be a reality when it is so integral to human endeavors it disappears as an explicit philosophy. . . . We'll thus spread the joy, awe, and power of computer science, aiming to make computational thinking commonplace. (Wing 2006)

As Ames (2019) continues, "cultural mythologies . . . are aspects of what social theorists call *ideologies*: the frameworks of norms, generally taken for granted and unconsciously held, that shape our beliefs and practices and that justify differences in power between various social groups . . . ideology fades into the background: . . . ideologies are as invisible to many people as we imagine water is to a fish" (19).

Wing's argument has circulated widely and reinforced collective beliefs about the foundational narrative of computing. The emotionally charged terms she uses speak to what Ames calls the "element of enchantment." My aim here is not to deny that computational thinking has merits and value. But it is worth looking deeper into how such narratives work and what they hide. Roland Barthes (1972) writes that "myth does not deny things, on the contrary, its function is to talk about them; simply, it purifies them, it makes them innocent, it gives them a natural and eternal justification, it gives them a clarity which is not that of an explanation but that of a statement of fact" (143). Vincent Mosco (2004) concludes that "according to Barthes, myth is depoliticized speech, with political understood broadly to mean the totality of social relations in their concrete activities and in their power to make the world" (30). As a result of this purification, myths become detached from their origins and turn into "congealed common sense" or "stories that help people deal with contradictions in social life that can never be fully resolved" (28–29). Mosco (2004) provides compelling reasons to pay close attention to myth:

If myths evacuate politics, then the critique of mythology can restore and regenerate it. If the telling and retelling of the mythic story shields cyberspace from the messiness of down-to-earth politics, then the critique of the myth, told many times over in many different ways, gives new life to the view that cyberspace is indeed a deeply political place. (31)

So myths are not just wrong beliefs that can be easily rectified. In Mosco's (2004) words, "Myths are not true or false, but living or dead" (3). In some sense, they are in fact *inoculated*: "It is common to see myths presented with what Barthes called inoculation or the admission of a little evil into the mythic universe in order to protect against a more substantial attack" (34). Inoculation thus refers to minor admissions in a narrative that "serve to protect the myth by granting that there are flaws" (Mosco 2004, 34). In examining the myths of systems design, we need to stay attuned to how they are inoculated.

THE RATIONALIST CORE OF COMPUTER SCIENCE

Computational concepts are not just projected onto the real world, they are the vocabulary and grammar used to make sense of it. In other terms, they are the cartographic tools used to make the map on which reality becomes the domain of computing (Ribès 2019; Breslin 2018). Just what constitutes a familiar domain depends on the background of the individuals and groups that use metaphorical mappings. In the case of computing professionals, their education ensures that computational concepts are at the center of conceptual frameworks and mappings. These concepts are unabashedly positivist (Easterbrook 2014b), grounded in the rationalist tradition. This is reflected in CS education (Raji, Scheuerman, and Amironesei 2021). Winograd and Flores (1986, 14) summarize this tradition by illustrating how it approaches a problem:

The rationalistic orientation can be depicted in a series of steps:

1. Characterize the situation in terms of identifiable objects with well-defined properties.
2. Find general rules that apply to situations in terms of these objects and properties.
3. Apply the rules logically to the situation of concern, drawing conclusions about what should be done.

In this tradition, reasoning is deductive and symbolic, and the stepwise process above still characterizes the core of what computer science students are taught today:

When we asked Computer Science faculty what CS education is "about," we were told either "algorithmic thinking" or "computational thinking." These were functionally interchangeable, and were generally categorized as a combination of:

- breaking down complex problems into smaller, more tractable components
- "seeing through" the mess of reality in order to focus on "only the details that are needed"
- using step-by-step decision-making processes, generally by using logic gates or other formalizable decision trees, to solve a problem
- finding an appropriate process will lead to appropriate solutions. (Malazita and Resetar 2019)

This view of problem-solving is central to the mindset and educational production of computer science subjects who render the world technical *as problems* (Breslin 2018, 98–152). Breslin (2018) reports that the students she studied "took the significance of algorithmic problem-solving to heart.

Even students in the second semester of their first year emphasized how they had learned to think algorithmically, to analyze a problem, to break down a problem to smaller steps, and to devise a solution with step-by-step instructions for a computer to follow" (99). But the concepts and rules that computational thinking supplies are insufficient in at least two ways. First, they are inadequate for developing a grasp of complex real-world situations composed of multiple interrelated factors and for identifying possible interventions (Easterbrook 2014b). And second, CT is severely limited by its inability to consider and address the social and political foundations on which computing practices operate. By treating the world as something that can be computed, its presumed ontology often denies the validity of such concepts as solidarity, freedom, purpose, determination, and will, or denigrates them into the appendix as "soft issues," treated as an afterthought. Marcuse's (1964) words ring true: "Many of the most seriously troublesome concepts are being 'eliminated' by showing that no adequate account of them in terms of operations or behavior can be given" (14–15).

Breslin's (2018) account of computer science education resonates with this characterization. Right from the start, "Students are encouraged to think critically. . . . Yet, in the end they must do so within the rules of the world and follow them, otherwise . . . they will not be able to play the game" (Breslin 2018, 97). That game is defined by the ways in which modeling, abstraction, and algorithmic reasoning renders the world technical. Abstractions, data structures and object representations often represent real world constructs. "Through these modular computational worlds, things and relationships become explicitly specified and solidified into stable representations. Moreover, these representations work as ways of developing 'solutions' to predefined 'problems'" (Breslin 2018, 109). But "the worlds that computer scientists build are both filtered reflections of constructions in the actual world and performances that constitute part of that world" (Breslin 2018, 111). Computer science education produces a very particular kind of subject trained in "modes of thought that bracket ethical or sociological content from technical concern" via abstraction, a subject that "represents and enrolls certain political, ideological, epistemic, and identity positions" (Malazita and Resetar 2019, 5). These positions have no place for social awareness and political engagement; rather, they represent a "culture of disengagement" (Cech

2014). What remains is a logically sound and coherent world of reasoning, predicated on solving problems that are rendered technical by way of these models. Its pull is strong, even for those who are just observing it:

After starting this initiation into computer science thinking and practice it becomes remarkably hard to think around. . . . During observations I was more interested in how computer science was being taught and learned. I simultaneously felt as though I had forgotten all of my anthropological theory. It did not seem to make sense in the context of computer science . . . I was continuously asked what I was trying to find out, to answer, what was my hypothesis? My response that I was interested in how gender is involved in computer science felt unsatisfactory, insubstantial. There was no problem to solve. (Breslin 2018, 139)

THE NEED FOR A CRITICAL APPROACH

If evaluating a myth was as simple as a disagreement about facts, then the normal reasoning processes of science and engineering would be perfectly capable of sorting out the misunderstanding inherent in some myths. But the myths structure how we think and talk about the world—as water surrounds fish. For this reason, the situation is more complicated: we have to lift the hood of our conceptual engines and examine *the way we think* in computing. This is among the hardest things to do. Phil Agre has famously described his struggles to extricate his own thinking from the system of thought that he had been brought up with intellectually—the research field of artificial intelligence in the 1970s—in a much-cited piece that is worth quoting at length.

I had absolutely no critical tools with which to defamiliarize those ideas—to see their contingency or imagine alternatives to them. Even worse, I was unable to turn to other, nontechnical fields for inspiration . . . I had incorporated the field's taste for technical formalization so thoroughly into my own cognitive style that I literally could not read the literatures of nontechnical fields at anything beyond a popular level. The problem was not exactly that I could not understand the vocabulary, but that I insisted on trying to read everything as a narration of the workings of a mechanism . . . I believe that this problem was not simply my own—that it is characteristic of AI in general (and, no doubt, other technical fields as well). (Agre 1997b, 9)

Agre struggled because he interpreted arguments from a different epistemology from within his self-grown epistemology, grounded in what we now call computational thinking. We could say that Agre's (1997b) approach of "trying to read everything as a narration of the workings of a mechanism," coupled with a "tendency to conflate representations with the things that they represent" (8), meant that he was trying to *compute* what he was reading. We can also consider it as an instance of *operationalism*, a term initially coined in physics to indicate a way of defining concepts purely via the operations by which they could be measured. "If a concept could not be operationalized—if there was no set of procedures by which its constituent terms could be measured (or at least detected)—then that concept had no place in science. If a concept could be operationalized, then it did have a place in science, meaning that the social sciences could become 'true sciences' if only they could define their terms properly" (Crowther-Heyck 2005, 65). Over time, concepts become "synonymous with the corresponding set of operations" (Marcuse 1964, 90) so that their meaning is "restricted to the representation of particular operations and behavior" (14).

Operationalism is a useful concept because it encapsulates how prior sets of metaphors and conceptual frameworks establish the rules of reasoning and perception. Operationalism does not only mean that we "operationalize" complex concepts into more tangible elements. Because those elements populate our vocabulary and define its relationships, operationalism more profoundly means the reverse: that we register reality *through* these pre-formed elements. Meanings outside these operations simply remain invisible to an operationalist mindset (Marcuse 1964). In machine learning research, this has led to a collapse of nuanced theoretical framings of justice and fairness into an operationalist definition of fairness as parity, rendering an informed discussion of the rich concepts and their ethical operationalization impossible (Jacobs and Wallach 2021). As a result, "by abstracting away the social context in which these systems will be deployed, fair-ML researchers miss the broader context, including information necessary . . . even to understand fairness as a concept" (Selbst et al. 2019).

This inability to recognize aspects that transcend predefined operations renders an operationalist mindset unable to engage meaningfully in broader discourses. Take Agre's vivid description of his challenges in reading literature from outside computer science, such as Heidegger or Garfinkel (1967):

I found these texts impenetrable, not only because of their irreducible difficulty but also because I was still tacitly attempting to read everything as a specification for a technical mechanism. That was the only protocol of reading that I knew, and it was hard even to conceptualize the possibility of alternatives . . . it finally occurred to me to stop translating these strange disciplinary languages into technical schemata, and instead simply to learn them on their own terms. This was very difficult because my technical training had instilled in me two polar-opposite orientations to language—as precisely formalized and as impossibly vague—and a single clear mission for all discursive work—transforming vagueness into precision through formalization. (Agre 1997b, 10)[1]

Agre's memory illustrates the power of metaphor to structure our engagement with literatures beyond those we are familiar with. He recognized that he was beholden to the *false consciousness* that Marcuse (1964) describes in *One-Dimensional Man* in which "ideas, aspirations, and objectives that, by their content, transcend the established universe of discourse and action are either repelled or reduced to terms of this universe" (14). Agre's reflection on his path out of false consciousness provides crucial lessons to draw on. First, he emphasized the need for critical reflection on how disciplinary language organizes the discourse.

I began to "wake up," breaking out of a technical cognitive style that I now regard as extremely constricting. I believe that a technical field such as AI can contribute a great deal to our understanding of human existence, but only once it develops a much more flexible and reflexive relationship to its own language, and to the experience of research and life that this language organizes. (Agre 1997b, 11)

Second, he emphasized the difficulty of developing this from within: "existing language and technical practice, like any disciplinary culture, runs deeper than we are aware . . . it is difficult to become aware of the full range of assumptions underneath existing practices, from technical methods to genre conventions to metaphors" (Agre 1997b, 12–13). And third, he argued nevertheless not for a disruptive break or a clean slate to begin anew, but instead for continuous, iterative, reflective and constructive engagement with and within the existing technical discourse. He hoped that it would be possible to "develop the critical tools to understand the depths below the ordinary practices of a technical field" (14). In his view,

Critical inquiry can excavate the ground beneath contemporary methods of research, hoping thereby to awaken from the sleep of history. In short, the negative project of diagnosis and criticism ought to be part and parcel of the positive project: developing an alternative conception of computation and an alternative practice of technology. A critical technical practice rethinks its own premises, revalues its own methods, and reconsiders its own concepts as a routine part of its daily work. It is concerned not with destruction but with reinvention. (Agre 1997a, 24)

Feminist and critical scholars in human computer interaction (HCI) have followed this path (see Bardzell 2010). As Shilton (2018) writes, "Critical technical practice, as put forward by Agre, requires questioning the metaphors, forms of representation, and discourse of an entire field" (122). This present book progresses on this path with specific attention to the challenges raised by the social and temporal distance of the outcomes of design decisions; the role of requirements in navigating the space between the social and the technical; and the potential of critical systems thinking (CST) to address this challenge.

Critical systems thinking was motivated by critical theory, which emerged as opposition to the myths of traditional theory, aiming to escape what the proponents of the Frankfurt school considered the prison of "traditional theory"—that is, uncritical scientific knowledge *turned ideology* (Horkheimer 1972; Habermas 1968; Feenberg 2014; Marcuse 1964; Jeffries 2016). Later chapters will draw from critical theorists through their influence on CST and science and technology studies. For now, we only need one more concept: reification. In *The Philosophy of Praxis*, Andrew Feenberg (2014) describes reification as "the thing-like appearance of the system of practice" (262). He explains the concept's origin in Lukács, who used reification to characterize how modern societies and their institutions had come to appear as natural and immutable things.

Bureaucratic administrations, markets, and technologies are all products of our scientific age; like science they are thought to be morally neutral tools beneficial to humanity as a whole when properly used. But in reality these institutions are social products, shaped by social forces and shaping the behavior of their users. They more nearly resemble legislation than mathematics or science. Thus their claim to universality is flawed at its basis. Like legislation, they are either good or bad, never neutral. Lukács argued that when societies become conscious of the social contingency of the rational institutions under which they live, they can then judge and change them. This implication of the theory of reification distinguishes society, including its technology, from the nature of natural science. (Feenberg 2014, viii–ix)

From this macroscopic context of stratified society, reification transitioned into social life. For the Frankfurt School, reification refers much more broadly to the unproblematic appearance of social reality as objective fact beyond doubt, critique, or valuation.

Reification means, literally, treating human relations as relations between things. In Lukács's usage, the "thing" implied in the "re" of reification is not just an entity in general but an object suited to formal rational comprehension, prediction, and technical control . . . The problem, Lukács argues, is not with scientific reason per se, but with its application beyond the bounds of its appropriate object, nature. (Feenberg 2014, 62)[2]

This last point is echoed by critical systems thinkers. The present book too does certainly not aim to disparage the scientific method or evidence or computer science, nor claim that computer science is simply ideology without merit or that any perspective on knowledge has "equal validity." Far from it. Instead, a more critical and reflective understanding of the assumptions underpinning scientific reasoning and engineering must simply be a cornerstone *of* science and engineering in the twenty-first century. And there is more to reality than computational operationalism admits. Reification explains "how the world can appear as a collection of facts" (Feenberg 2014, 86). By configuring how we think about our reality, belief systems prestructure how we perceive it and reason about it. They define what we can recognize as facts—as Horkheimer (1972) wrote, facts are *socially preformed*. There is an important historical character to that, since both the perceived objects or facts and the organ of perception are shaped by their history (200). Because they structure facts and the questions we can ask about them, our beliefs—including rigorous, logically sound belief systems such as science—turn into ideology if there is no room for a critical questioning and reflection.

When they become reified and unquestioningly adopted in a false consciousness, science and technology turn into ideology. This ideology will serve those who already have power. The challenge that critical systems thinkers tackled head on, as we will see later, is that the logical system of science, closed and coherent as it is, is incapable of justifying its own assumptions. This is not to say that it is unjustified, simply that its justification relies on concepts and arguments that are not in themselves scientific.

The aim of *dereification* is to free the false consciousness and allow it to comprehend reality more fully. What Agre described as "defamiliarizing" himself from formalized language has profound implications. He was looking for *transcendence*, understood simply as the desire to see beyond the preformed frame of reality.[3] This is not mysticism; it is a critical under-standing of reason and rationality.[4] Agre recognized that reified, socially preformed computational reasoning was not enough to make sense of what AI was trying to do, and he struggled to extricate himself from this framework of thought *so that he could appreciate it more fully.* This did not mean abandoning its ideas, mechanisms, and tools—it meant incorporat-ing them into a broader perspective in which other forms of reasoning also had a place. This book pursues parallel aims with respect to such frame-works as requirements engineering and ICT for sustainability.

The critical distance helps us notice what Werner Ulrich (1983; 1985) calls the *sources of deception* inherent to any process of discovery (1983, 22). When someone plans, or designs, for someone else, sources of deception arise most insidiously from the structure of rational argument itself. Ulrich recognized that no matter how holistic in aim, any approach to design will be selective in its explicit *and implicit* drawing of boundaries.

Even with the best intentions, selectivity will remain unnoticed unless we look for it. Ulrich's response for a critically systemic approach to design was to develop a deeply grounded system of critical heuristics that arose from the distribution of power and agency across those involved and those affected by design. We will encounter his critical systems heu-ristics in chapter 5. Throughout the next chapters, this book aims to sys-tematically identify concrete sources of deception in common theories, ideas, and metaphors of computing to illustrate how we can extricate our thinking from this socially preformed frame of engagement to more fully appreciate it and deploy it more responsibly. For now, I want to return to the topic of myths by way of Feenberg's summary:

As technologies develop, their social background is forgotten, covered over by a kind of unconsciousness that makes it seem as though the chosen path of progress was inevitable and necessary all along. This is what gives rise to the *illusion of pure rationality.* That illusion obscures the imagination of future alter-natives by granting existing technology and rationalized social arrangements an appearance of necessity they cannot legitimately claim. (Feenberg 2014, 166–167. Emphais by the author)

A MYTHOLOGY

The myths I focus on are myths of logic and reason, central tenets of methodology and practice, so I take here a more conceptual view than Mosco and Ames. The myths have been questioned and dismissed before, yet they continue to have enduring appeal and effects with far-reaching consequences, like a dangerous undercurrent hidden beneath engineering methods, research questions, design pedagogy, and industry practices. They can variously be seen as stories, illusions, or theories. I single out here four myths that structure the conceptual domain of computing and its associated engineering and design practices. If any of these seem implausible, keep in mind that myths are not *believed* the way we believe the latest findings of a scientific study, or the basic laws of physics. Instead, the question is how they shape the broader currents of systems design.

The myth of *value-neutral technology* claims that software technology is neutral—that because of their abstract form, algorithms and software systems only come in touch with human values in application. Because their effects are then merely an artifact of usage choices, politics can and should be kept out of design. Moving beyond this view allows those who design to recognize their work as political and to develop sensitivity and responsibility toward the role of values in design.

The myth of *rational decision-making* conceptualizes thinking as information processing, based on a metaphor of the human mind as a computer. On the basis that the rational choice is considered optimal, it understands human deviations from rational choice as limitation, bias, and error. Moving beyond this view allows designers to recognize how their judgments transcend the narrow framing of rational decision-making, and it allows researchers and educators to account for varied forms of human judgment in design.

The myth of *objective problems* in systems design maintains that problems have an independent existence and can be discovered using applied scientific reasoning. As a result, the central emphasis in problem formulation is on correctness or consensus. Moving beyond it allows those who design to pay attention to the ethical questions in problem formulation and enable meaningful participation of those affected by their efforts.

The preceding three myths produce a fourth: What I call the myth of *solvency* maintains that *design is problem-solving* and that any negative effects are outweighed by the benefits. Solvency's current meaning is "the possession of assets in excess of liabilities; ability to pay one's debts." (Merriam-Webster 2020b) But is not so clear that computing can or will pay its debts to the world. Moving beyond that way of thinking enables those who design to consider the wide space of alternative approaches using different metaphors.

Together, these beliefs are at the center of computing's central paradigm— the "set of theories that explain the way a particular subject is understood at a particular time" (Cambridge Dictionaries 2011). The following sections outline each myths' merits, origins, implications, and sources of deception. These will be covered in more depth by the chapters in part II. Because the struggle for a position vis-à-vis these ideas has been a defining part of the history of applied systems thinking and of design, I will center the discussion of each idea on these fields. Their debates have had a profound influence on the current discourse in computing, and their residue can be felt in the undercurrents of systems design.

THE STORY OF VALUE-NEUTRAL TECHNOLOGY

Both in industry and across the range of academic fields of computer science, technology is still widely characterized as neutral: that is, value-free and impartial. This holds for different views of what technology is—a vague abstract whole, a concrete group of technologies such as support vector machines or relational databases, and a concrete technological artifact or product such as WhatsApp or GPT-3 (T. Brown et al. 2020). In this view, purity and neutrality are seen as a virtuous absence of values, based on the traditional ideal of science as an objective enterprise (Reiss and Sprenger 2017). In this context, Boaz Miller (2014) defines the term *value* as "anything that serves as a basis for discriminating between different states of affairs and ranking some of them higher than others with respect to how much they are desired or cared about or how the personal, social, natural, or cosmic order ought to be" (70).

The theory of value-neutral technology (VNT) maintains that only in instantiation, configuration, and application do values enter the world through choices made by people with partial interests. While these choices may result in possible unfairness, the best that engineering can do is to be independent, as it were, of these choices—neutral, and thereby innocent.

WhatsApp cofounder Brian Acton, for example, did not mince words about his stance on the moral position of technology: "There is no morality attached to technology, it's people that attach morality to technology . . . It's not up to technologists to be the ones to render judgment" (Levy 2020). This view continues to be widespread, and it is reflected in academia too: "ML systems are biased when data is biased," writes a leading researcher, claiming that bias is introduced into machine learning solely through the choice of data sets.[5] Even some philosophers of technology maintain that technologies are neutral (Pitt 2014, 90).

But this view is patently false. Software systems are never truly neutral. On the contrary, designers and their organizations embody values, their design choices embed specific values in the systems through the features and qualities they construct, and software systems in turn express and enact these values through their behaviors and affordances. As Grady Booch (2014) put it, "Every line of code represents a moral decision; every bit of data collected, analyzed, and visualized has moral implications." Only some of these implications are immediately obvious. Selecting a Boolean as the type for the *gender* variable is a choice and not a neutral one: It embeds the conservative value of binary gender stereotypes into the material artifact of code, and it *will* inevitably lead to situations in which a person who does not conform to the stereotype will experience its torque. Airport body scanners, for example, tend to flag trans people as suspicious (Costanza-Chock 2020). But the opposite—to choose not a Boolean but a more appropriate type for this variable, or to refrain from collecting data about gender—is also not a neutral choice, since it explicitly considers the value of gender-sensitive technology design and enacts it in code. *There is no neutral ground here: Every choice turns values into facts.* In both cases, the designer's choice is not fully determined by facts. Another way of putting it is that the technological facts are *underdetermined*: There is room for human judgment (Feenberg 2017) and for the need to consider values in the sense introduced earlier. VNT masks this underdetermination by diminishing the role that values play in determining the shape of technological artifacts.

Proving VNT wrong in ways that are empirically persuasive for those *in computing* has been very difficult. Miller (2020) describes how persistent the idea remains despite its apparent weaknesses. He suggests that one reason for its persistence lies in the fact that its critiques are not persuasively legible from within a science and technology perspective.[6] This brings back the specter of operationalism illustrated by Agre's experience. For each new case that is convincingly demonstrated, a concession is made— "True, *this* system is not neutral, it's terrible," "yes *that* system is not neutral—it's badly designed," we are told, and then the writers retreat to the position that *on its own*, software technology is neutral. This move admits values not into the technology itself but into the *choices* of designers and engineers, in the form of personal bias or irrational choices as opposed to the "proper," supposedly neutral way of designing things prescribed by design methods and rational decision-making. The empirical absence of neutrality is attributed to the deviation of the instance from the rule of neutrality. VNT is inoculated by the other myths. But asking for empirical falsification of VNT misplaces the burden of proof—after all, there is no empirical evidence that technology is neutral. Instead, we need to ask how this idea can be so resistant to evidence to the contrary. To do so, we will need to take a critical approach.

VNT deceives us by suggesting that design can be neutral, with three important consequences (Miller 2020): (1) It allows designers and engineers to evade responsibility; (2) it prevents critical questions; and (3) it avoids the placement of restrictions on technologies that embody unacceptable values.[7] Statements like the quote from WhatsApp cofounder Acton are always made from positions of privilege and place an undue burden of proof on those who are already disadvantaged.[8] Not admitted into these accounts is the way in which values of dominant interests *systematically* become prioritized, privileged, and embedded in artifacts, while the values of marginal interests get systematically suppressed and left out. Even when this is not happening explicitly, the outcomes of systems design are silently shaped by the values of those involved, because most common methods in the engineering and design disciplines in computing are oblivious to the role that values play in systems design.[9] The asymmetry of vulnerability that is at play between those involved and those affected suggests that the danger of moral corruption needs to be taken seriously. VNT allows those in charge to treat each instance in which a value is demonstrably violated as a bug, an accidental "problem" to be "fixed." This book takes the position that these are not bugs; they are features of the current world of computing, embodied in the idea of Silicon Valley (Schrock 2020; Liu 2020).

As Feenberg (2017) put it: "Values are not the opposite of facts . . . Values are the facts of the future" (8). Technology shapes social reality through its functions and affordances and effects, so the values that it embodies have significant reach. They do not anymore just reside in the persons who made the choices but live on in the technological artifacts that result from these choices. Therefore, "Technology designers and constructors cannot evade moral responsibility for the consequences of their products by arguing that they are morally neutral" (Miller 2020, 19).[10] In more humorous terms, here is the second entry to the Devil's Dictionary of Computing.

> **Software engineering**, n.: the social practice that converts human values, politics and moral decisions into code, features, qualities, documentation, and other technological facts.

What remains to be persuasively shown is precisely *how values become facts* in systems design, and how we can critically handle this. We will return to that challenge in chapter 6.

THE STORY OF OBJECTIVE PROBLEMS

The story of VNT presents values as sharply distinct from facts. Facts, as has been asserted throughout much of the history of science in the twentieth century, are or at least should be "value-free." The ideal of the scientific investigator is a neutral observer, and the ideal of the outcome of a scientific investigation is an objective fact that we can treat as truth. Values enter this process either as "epistemic" (such as accuracy) or as "contextual"—"moral, personal, social, political and cultural values such as pleasure, justice and equality" (Reiss and Sprenger 2017). By interpolating across observations, perspectives, and viewpoints, the observer strips these values from observations until only the facts remain. Knowledge in this view becomes "objective" by virtue of being "value-free" after the "biasing" influence of all contextual values has been removed by way of scientific reasoning.

Whether the assumption is that facts are objectively true because they correspond to an external reality or because there is a consensus mechanism that establishes when they should be regarded as true, *facts* are regarded as starting points of technological activity. Facts are what needs to be discovered with various scientific instruments to establish the basis for technological development.[11] The most important fact that must be established early on in design is the definition of *the problem*. The etymology of the word *problem* leads back to ancient Greek, literally "a thing put forward," then in Old French (via Latin) morphing to mean "a difficult question proposed for solution." Today, the dictionary defines it as "a question raised for inquiry, consideration, or solution" (Merriam-Webster 2020). But even though we *know* that the problem is *put forward* by someone, we are quick to handle it discursively as if it had an actual existence. Its status of objectivity establishes it as (1) correct, (2) value-neutral, and (3) independent of the observer.

This idea of objectivity manifests as the *myth* of objective problems when the problem concept itself becomes reified—when *the problem* is taken as a thing to exist in the real world and treated as an object in itself, independent of the observer. This happens more frequently than we like to admit. *"What is the problem?"* already suggests that *there is* (exactly) one problem in existence, out there, if only we could have an accurate representation *of it*. In this objectivist view of problems, "problems have an autonomous existence that does not depend on any subject's knowledge, although someone must be aware of their existence . . . in the empiricist tradition, formulating a problem is viewed as analyzing reality, not as searching for goals" (Landry 1995, 321). Conceptually, saying "the problem is X" is a shortcut that implies (a) how the world currently is, (b) what is wrong about that, and (c) how we would know that it has been fixed. When we forget that this is a shortcut, the problem concept has been reified. When we remember it, we allow ourselves to recognize the value judgments inherent in the framing; we can make transparent the politics of the situation; and we can introduce fairness into the process by which problems are articulated and selected as the basis of design work.

The history of applied systems thinking provides insightful lessons of how the understanding of the nature of *problems* has shifted in the twentieth century (M. Jackson 2003; Flood and Jackson 1991a). Two major turns, each building on the previous, represent the evolution of thinking about problems and problem-solving (Flood and Jackson 1991c). Surprisingly perhaps, these shifts have not been fully replicated within the design-oriented disciplines in computing, despite long-standing debates.

In traditional systems analysis and systems engineering (Jenkins 1969), which arose from the paradigmatic problem-solving discipline operations research, problems are real and given objects. Problem statements must correctly capture the problem, in the tradition of natural and physical sciences, and the focus is to find the most effective and efficient way to *solve the problem*. Problem specifications are the central starting point because "all problems ultimately reduce to the evaluation of the efficiency of alternative means for a designated set of objectives" (Ackoff 1958). This approach runs into massive difficulties as soon as the domain in which problems are located is social. In a given situation, different stakeholders will identify different problems based on their background, including their disciplinary training. This is why in Ackoff's example of an old woman dying on the stairs, a doctor, a social worker, a family member, and an economics professor see a medical, a social, a financial, and an economic "problem" (Ackoff 1999). None of these four problems exist independently of the observer: In the real world, a poor woman had a heart attack walking up the stairs of a home that didn't meet accessibility standards for cost reasons, and she died because of the lack of affordable medical care in the vicinity of the social housing project that the home is a part of. As soon as anyone speaks of this situation in terms of "the social problem"—or, in the case of computing, more typically "the technical problem"—they frame and shape the subsequent discourse on its basis.

Ackoff (1999a) famously proposed that a problem can be absolved, resolved, solved, or dissolved: Absolution means ignoring the problem; resolution means taking a satisficing approach to addressing it; solution means taking an optimizing approach; and dissolution of a problem means "redesigning the system that has it" (115). But in discussing "the problem itself," in speaking of "errors of conceptualization," and in attributing ownership of the problem to "the system," he reinforced the idea of objective problems. The language he used still implied a sequence of problem recognition, problem definition, problem-solving that is represented by traditional systems analysis (Checkland 1981, 155). He overlooked that problems do not have a material existence *out there*: What exists in his story is a *situation* in which an old woman has died of a heart attack on a staircase. *The problem* is a concept that the observers in the situation use to frame their understanding of *what to do*. The framing of the problem is inevitably based on a set of perspectives, concepts, and assumptions that are bound up with the standpoint of the observer stating the frame. Ackoff conflated the model with the reality it was meant to represent. He conflated "having the problem" with "framing a problem."

In the world of puzzles and mathematical problems, where the methods of operations research came from, the problem concept sits on the same logical plane as the situation it describes. There is no fracture (Agre 1997b). When we take the problem concept from an abstract realm of solvable puzzles into the real world, however, the mapping of the concept will lead us to register certain regular features of reality while masking others. It will shape how we render the world technical, and we must not mistake the rendered map for the territory. Metaphors are a central part of the social preformation of frames we use to articulate problems, so it is worth developing systematic attention to that: "spell out the metaphor, elaborate the assumptions which flow from it, and examine their appropriateness in the present situation" (Schön 1979, 138).

The fundamental turn of soft systems thinking (SSM) was a shift away from an objectivist understanding of the systems idea. In this epistemology, systems are no longer assumed to be real but treated as discursive constructs.[12] Problems too are not objectively given from nowhere but socially constructed (M. Jackson 2003), so Checkland (1981) advocated refraining from the early use of the problem concept in favor of a *problem situation*, "a nexus of real-world events and ideas which at least one person perceives as problematic" (316).

If what matters most about a problem is that it is correctly articulated, in the sense of a *correspondence* to facts, then by extension, the application of scientific methods is the only legitimate way of reasoning. As a consequence, those affected by problem-solving efforts are not regarded as a source of facts, only a source of data. That data needs to be turned into insights through scientific interpretation by the experts. The views and voices of those affected are marginalized (Midgley 1992). If, on the other hand, what matters most is that all who are present find *consensus* on "what the problem is," or *what should be done*, then we need a firm grasp of the nature of consensus that accounts for power imbalance, coercion, and marginalization. SSM does not offer this because it lacks the social theory to recognize coercion as a factor of influence (M. Jackson 1982), and neither does standard current practice in disciplines such as requirements engineering (Duboc, McCord, et al. 2020). The turn to *critical systems thinking* responded to the realization that neither hard nor soft systems thinking were adequate in the space of technology design. What is left out in each case is the central question of legitimacy: How can those who design justify the implications of what they are doing to those who are affected by their intervention? This question will play a central role in future chapters. For now, I conclude with another entry to our dictionary:

> **kick-off**, n.: the short period in which all active project participants succumb to the illusion that they agree on what the project purpose is.

The myth of objective problems joins VNT in supporting an abdication of responsibility over technological choices: From both angles, rational methods play the role of guaranteeing an objectively necessary outcome that is freed from the values of the expert by virtue of scientific reasoning. Those who design only need to perform the task of "objective reason" by making rational decisions.[13]

THE STORY OF RATIONAL DECISION-MAKING

Since the joint emergence of artificial intelligence and computing as disciplines, the idea of rational decision-making is founded on the premise that human decision-making and thought can be explained by reference to the way a computer processes information, because "intelligence is the work of symbol systems" (Simon 1996, 23). Concepts describing the working of a computer, which is relatively well understood and formalized because it is a designed artifact, are used to understand the elusive inner workings of the mind: "The computer is a member of an important family of artifacts called symbol systems, or more explicitly, physical symbol systems. Another important member of the family . . . is the human mind and brain. . . . Symbol systems are . . . goal-seeking, information-processing systems" (21–22).

The computer thus quickly came to provide the metaphor to describe the human mind. This core metaphor of the human mind as an information processor was established by Simon's work on cognition and human problem-solving, which proved hugely influential in computing and the social sciences (Augier and March 2004; Erickson et al. 2013). Nobel Prize after Nobel Prize, the foundational premise of the mind as a computer anchored, shaped, and structured a wide range of complex debates about the limited degree to which humans adhere to this model (Simon 1978), the ways in which they deviate (Kahneman 2002), and the best methods to measure and improve their behavior (Thaler 2017). Wing's argument comes full circle: the human mind should now gear up its computing abilities to acquire more potent concepts of higher programming languages.

The idea of rational decisions manifests *as a myth* in systems design research, education, and practice when decision-making itself is framed exclusively by the operationalist concepts that arise from the information processing metaphor. Because the influence of the idea of rational decision-making is so pervasive, its origins are rarely cited, and its assumptions are rarely defined as succinctly and explicitly as in this software engineering paper:

In most problems, to make a decision, a situation is assessed against a set of characteristics or attributes, also called criteria. Decision making based on various criteria is supported by multi-criteria methodologies. (Filho, Pinheiro, and Albuquerque 2016)

Indeed, multicriteria decision-making (MCDM) methods are central to computing and engineering. MCDM arose out of the mathematical theories of Bernoulli (1954) who in the eighteenth century developed principles that prescribed how a theoretical agent *should* make optimal choices between gambles under conditions of well-defined probabilistic uncertainty. This provided the foundation for utility analysis (Keeney and Raiffa 1993). By way of operations research, game theory, and early computing theory, this mathematical framework has provided a foundation for countless methods in computing fields like software engineering. MCDM methods prescribe how to analyze well-defined situations when choices have to be made to identify which choice should be considered optimal, based on the assumption that the conditions and success criteria can be specified. According to this family of theories, an agent makes a decision by evaluating a set of options against a set of weighted criteria, uses this matrix to create a ranking of options, then selects the best option out of the set.[14]

While eminently useful as a normative framework to structure the evaluation of options, the underlying theory of MCDM—now called *rationalistic* in the field of judgment and decision-making—was never validated as a descriptive framework for human thought (Tversky and Kahneman 1986). On the contrary, ample evidence demonstrates that its assumptions, predictions, and explanations are inconsistent with the reasoning processes of the human mind (Beach and Lipshitz 1993). For example, large-scale field studies in decision-making showed that high-performing professionals did not evaluate multiple options against multiple criteria to compare them, rarely ranked options, and rarely selected an option from a set. Instead, they used their highly developed perceptual skills to match cues in the environment to patterns in their experience to *generate* one plausible course of action. They then used mental simulation to predict what would happen if they pursued it, and they adapted, adopted, or dropped one action at a time in sequence until they found one that satisfied them. In doing so, they often outperformed rationalistic approaches (G. Klein 1998). In parallel, ground-breaking research in the biology of cognition concluded that "the popular metaphor of calling the brain an 'information processing device' is not only ambiguous but patently wrong" (Maturana and Varela 1992, 169; Maturana 1980). Winograd and Flores (1986) showed that this realization has far-ranging implications for computing and design and questions the value of rationalistic theories.

Despite their flaws, however, the appeal of rationalistic theories proved so strong that they underpinned most empirical research in cognitive psychology, behavioral economics, AI, and computing for decades. The behavioral assumptions of computer science are firmly grounded in this rationalistic tradition, and it is not alone in struggling to overcome it. All these disciplines owe great advances to the normative foundations laid by Simon and others in these fields but struggled for decades to overcome their inherent limitations (see chapter 7).

In systems design, the myth of rational decision-making manifests in subtle ways. For one, it provides a ready-made package of reified metaphors for how decisions supposedly happen. In truth, humans *can act* in accordance with these rational models if they choose to do so, but they also have many other forms of reasoning that such narrow, impoverished frames fail to recognize (G. Klein 1998; Gigerenzer and Selten 2001). When the conception of decision-making as choice between enumerated alternatives according to specific objectives frames research, pedagogy, and practice, its operationalism masks anything that transcends these concepts. This leaves no room for the many expressions of human judgment and wisdom that the operationalist view fails to recognize. It is no coincidence that the field that studies human decision-making is now called *judgment and decision-making* (Keren and Wu 2015). The expanded term simply recognizes that there's more to consider than rational choice.

Yet, software engineering methods are often treated as if they were programs to be run by practitioners, even though they are more appropriately described as one of the resources that practitioners use in situated action (Dittrich 2016). The narrow framing focuses research efforts on deviations from the rational model. These are understood as "bugs" in people that can and should be "fixed," usually by reference to cognitive biases (Mohanani et al. 2018), rather than indications that the normative model may be off, wrong, or unreasonable. In chapter 7, we will see how human judgment in practice often transcends what rationalistic methods can handle and how to reorient our perspective. This is particularly important when it comes to understanding how design teams make decisions with uncertain effects at a distance, as in design decisions that affect sustainability and justice.

The alternative is already here, if we are willing to look and learn from those disciplines that have grappled with Simon's legacy. Rationalistic theories become part of a broader understanding of human reasoning, information processing is understood as distinct from human judgment, and the study of decision-making can leverage *naturalistic* approaches that examine situated action as it happens in design practice (Crandall, Klein, and Hoffman 2006; G. Klein 1998). Chapter 11 will explore how behavioral research in systems design based on this perspective can approach the challenging questions that judgment and decision-making in systems design raise when the outcomes of design practice lie at a distance. For now, a few additions to the dictionary:

Human, n.: annoying reminders of the real world. See *User*.
Irrationality, n.: those parts of <u>human</u> life to which scientific rationality and engineering have no access.
Judgment, n.: that which is <u>irrational</u> in <u>human</u> reasoning.

The myth of rational decision-making supports the myths of neutral technology and objective problems. Since every rational person will supposedly arrive at the same conclusion, those who design are only executing objective reason. The myth thus dispenses with judgment, overlooks expertise, and creates the illusion that there is such a thing as a neutral perspective. Rationalizing matters this way has two simultaneous effects: First, any deviation from neutrality in design outcomes, such as bias in algorithms, can be explained away as a mistake or straying from the correct path of rational decision-making. Second, blaming the irrational human for the deviation inoculates the theory and exculpates its methods from responsibility to account for those modes of reasoning that lie outside of it.

THE STORY OF DESIGN AS PROBLEM-SOLVING

The focus on objective problems matters because, in computing, design is widely characterized as *problem-solving* (Ko 2020; Jonassen 2000), but there is no shortage of debates in design about the merit of this framing (Buchanan 1992; Dorst 2006; Holt, Radcliffe, and Schoorl 1985; Huppatz 2015). The idea of design as a search for a solution in a defined problem space shares its roots in the joint origins of computing, AI, and cognitive science in the 1950s. Herbert Simon describes problem-solving as "a search through a vast maze of possibilities, a maze that describes the environment" and adds: "Successful problem solving involves searching the maze selectively and reducing it to manageable proportions" (Simon 1996, 54). Based on the paradigmatic example of a cryptarithmetic problem, he proceeded to demonstrate that the rational solution to this problem is entirely out of reach for a human problem-solver due to restrictions on memory and processing power. This theory of problem-solving provided the foundation for ground-breaking empirical research on human problem-solving (Newell and Simon 1972). Simultaneously, it provided a lasting foundation for the design of problem-solving methods and machines in the computing disciplines (Pomerol and Adam 2006; Augier and March 2004).

Simon's conception of design as problem-solving, built on the theory of bounded rationality and the metaphor of the mind as symbol processor, also had enormous influence on the discourse of design to this day (Rosner 2018). Some design researchers long struggled against the reductive metaphor of design as heuristic search in a defined problem space (Schön 1983; Margolin and Buchanan 1995; Cross 2006; Huppatz 2015). Cognitive studies of design activity and problem-solving demonstrated that in practice the problem space and the solution space are not given—they are constructed and co-evolve continually (Dorst and Cross 2001; Ralph 2015). "Much contemporary design research, in its pursuit of academic respectability, remains aligned to Simon's broader project, particularly in its definition of design as 'scientific' problem solving. However, the repression of judgment, intuition, experience, and social interaction in Simon's 'logic of design' has had, and continues to have, profound implications for design research and practice" (Huppatz 2015). In design-oriented HCI research, "framing current situations as problems and technological systems as solutions is common" (Baumer and Silberman 2011, 2273). A recent paper in fact characterized "95% of HCI research" activity *as* problem-solving (Oulasvirta and Hornbæk 2016). The crucial skill of choosing a problem to solve is often attributed to intuition or to inert abilities akin to the so-called geek gene.[15] As Joyojeet Pal (2017) writes, "the institutions in this endeavor extend from philanthropies and corporate social responsibility groups to academic departments and inter-government agencies unified in hope that technology can solve wicked social problems" (710).

The idea of design as problem-solving, which I call *solvency*, manifests *as myth* in systems design when problem-solving becomes the only form of engagement and problem *framing* is done uncritically or implicitly. This is highly likely when the primary organizing metaphor of the discourse is computational thinking because computational thinking often treats the problem to be solved as an unproblematic starting point provided from outside. As Breslin (2018) puts it, "what [students] learn to do with their programs or algorithms is just that, to solve problems. This approach necessitates a frame of reference as formed of problems and solutions" (105). When that frame of reference dominates, systems design becomes a narrow-minded, operationalist version of what it could be. Solvency overlooks the discursive nature of the problem concept and how it positions and frames the discourse, and it overlooks or marginalizes the existence of legitimately divergent worldviews. Its rationality redefines critical concepts that transcend the meanings of its given computational structures. When multiple problem definitions contradict each other, some are seen

as correct and others as incorrect. The crucial step of constructing, negotiating, and legitimating a problem definition to settle on is omitted.

With problems reified and taken uncritically as things, problem-
solving the naturalized way of engagement, and the assumptions that the
technology used and designed to solve the problem is neutral and that
scientifically minded problem-solving produces objectively necessary
and ideal outcomes, the stage is set for an approach to technology development that is predictably harmful but remains all too common. This
approach is not necessarily limited to computing—you may argue that
these myths are myths of technology, science, or enlightenment rationality. Their confluence in computing, however, is especially pronounced.

ARE THESE MYTHS REALLY STILL AROUND?

In some areas of computing or adjacent to it, on the other hand, some
myths have been thoroughly discredited. I will name just a few examples
among many. My concern is not with those parts of the computing discourse that have overcome misleading ideas about the nature of technology,
rationality, and objectivity—but rather, with what the rest of computing
can learn from this.

In design studies, Simon's framework has been long critiqued, and
the process of design has been reframed (Schön 1983; Cross 2006; Dorst
1995; 2006; Dorst and Dijkhuis 1995; Margolin and Buchanan 1995; Rosner 2018). But most of this debate took place outside of computing and
has not been recognized universally. In information systems and requirements engineering, soft systems methodology is often cited as an important approach for deciding what problem should be addressed by systems
design in situations where many stakeholders hold diverging views (e.g.,
Alexander and Beus-Dukic 2009), but it has been widely misunderstood
as dealing with soft systems rather than taking an interpretive epistemological position on the systems concept itself (M. Jackson 2003; Checkland
2000).[16]

Critical disciplines adjacent to computing were never directly influenced
by rationalistic ideas or illusions of value-free objectivity. In contrast, they
were born out of opposition to these ideas. So, when it comes to value-
neutral technology, its influence in HCI is limited. This is also owed to

the long-standing conversations about values in design (Friedman 1996; Friedman and Hendry 2019; Shilton 2018; Johnson and Nissenbaum 1995; Nissenbaum 1998; 2001; Flanagan, Howe, and Nissenbaum 2008). Similarly, the politics of design have been a central topic of debate (e.g., Dourish 2010). In requirements engineering, on the other hand, the recognition of human values is a recent arrival (Thew and Sutcliffe 2017), and so is the attention to politics (Milne and Maiden 2012; Duboc, McCord, et al. 2020). In CS education, Frauenberger and Purgathofer (2019) demonstrate the values and possibilities of instilling a broader awareness of ways of thought in computing undergraduates. Other progressive voices similarly are rethinking computing education from a critical perspective (e.g., Guzdial 2020b; 2020a; Ko et al. 2020).

When it comes to rational decision-making, Lucy Suchman's work on plans and situated action (1987) is widely cited and referred to in HCI and AI, and Hutchins's work on cognition in the wild (1995) is similarly recognized (Rogers and Marshall 2017; Bødker 2006). It should be noted, however, that neither of these works directly aim to supplant the core of rational decision-making outlined earlier—instead, they address and correct some of its implications. Naturalistic decision-making research using approaches such as cognitive task analysis (Crandall, Klein, and Hoffman 2006) in HCI is commonly geared at understanding the task of *users* (Diaper and Stanton 2004), not the tasks of engineering and design practice (Zannier, Chiasson, and Maurer 2007; Becker, Walker, and McCord 2017). In software engineering, Paul Ralph (2018) has characterized the tensions between normative and empirical research and pointed to the inadequacy of traditional conceptions of design as sequential problem-solving (Ralph 2015). But as Ralph and Oates (2018) have pointed out, software engineering clings to what they call "dangerous dogmas," ideas closely related to the myths I discuss.

This suggests that these ideas remain central to the discourse and are upheld despite evidence to the contrary. As Amy Ko and her coauthors write, "many of us in the computing discipline . . . dismiss the idea that computing is anything but a value-neutral tool independent from society." In words that resonate strongly with this book, the authors call out a set of "neophilic myths: that software is always right, that software is always value-neutral, and that software can solve every problem" (Ko et al.

2020). Myths continue to remain powerful as long as they are retold, and they bring forth additional false narratives. Countering them remains important (Owens and Lenhart 2020).

CONCLUSIONS

The myths of systems design exert strong influence over what computing research and practice do by shaping how central questions are proposed, discussed, and studied. When we examine the origins and connections of myths, we can see the role they play in establishing cultural meaning. The ideas of scientific objectivity, technological neutrality, rational decision-making, and design as problem-solving have shaped the self-understanding of computing throughout the past seven decades. In systems design practice and research, they manifest *as myths* when their underlying validity is overextended into a context in which they are not empirically or theoretically supported. As myths, these ideas exert a subtle influence by shifting and distorting the frame of discussion so that crucial insights are prevented from surfacing.

4

PROBLEMISM
THE INSOLVENCY OF COMPUTATIONAL THINKING

The integration of AI systems within social and economic domains requires the transformation of social problems into technical problems, such that they can be "solved" by AI. This is not a neutral translation: framing a problem so as to be tractable to an AI system changes and constrains the assumptions regarding the scope of the problem, and the imaginable solutions.
—Crawford et al. (2016)

To make significant progress towards any serious notion of a sustainable digital society, an altogether different kind of thinking and academic discourse is required.
—Knowles (2013, 5)

When Dr. Abeba Birhane, cognitive scientist and artificial intelligence (AI) ethics researcher, asked the GPT-3 language model (T. Brown et al. 2020) to "generate a philosophical text about Ethiopia," she was prepared to encounter bias and racism in the result. Machine learning algorithms that produce racist and misogynist text outputs had unfortunately become common. The result was still disappointing. Under the heading "What ails Ethiopia?" the text began with "The main problem with Ethiopa is that Ethiopa itself is the problem. It seems to me like a country whose existence cannot be justified. . . . A solution to its problems might therefore require destroying Ethiopia" (Birhane 2020).

Like other large language models, GPT3 did not invent racism. It learned and amplified it from massive text corpora representing human discourse on the web, social media, and literature. They "encode the dominant/ hegemonic view" (Bender et al. 2021). The racism in the generated piece is blatant, but another aspect stands out too. The text's central organizing concept is problem-solving. On a single page, the term *problem* appears eight times. Every turn of the argument identifies a new *problem*, frames it in colonial terms, and closes with a racist solution or dismissal of the solution as in the closing passage: "Ethiopia suffers from extreme corruption, which is perhaps understandable given the country's history of foreign domination. However, it seems that there is no way to solve this because Ethiopia can never be independent long enough for such problems to completely disappear" (Birhane 2020). This is no coincidence. GPT3 is simply picking up a dominant way in which we engage with the world and discuss it.

The exclusive framing of research topics into problems and solutions is a favorite staple of the sciences too, including computing for sustainability. A recent call for papers explicitly states what it considers an appropriate contribution to a "body of knowledge for software sustainability":

Concrete contributions . . . will be structured as follows:—A description of the sustainability problem you address.—A description of the SE solution you propose.—A discussion of how results are measurable (e.g., KPIs).—A presentation of the evidence of contribution to sustainability, ideally including real world experiences.—A discussion of the costs and benefits of your approach.—A presentation of the transferable artifacts you are contributing e.g., replication package, code, examples, documentation, educational materials, case studies. (https:// bokss.github.io/bokss2021/)

In other words, papers that frame *sustainability as a dilemma*, papers that carefully characterize a problem but do not yet offer a solution, and papers that offer both but not in terms of measurable properties of real-world objects, are not regarded a contribution to this body of knowledge. The stage is set firmly in the rationalist tradition. This severely restricts meaningful progress because "framing unsustainability as a problem misses the nature of the situation and misguides our attempts to address it" (Baumer and Silberman 2011, 2273). For example, when the focus on objective problems and measurable entities obscures the fact that the domain is not conducive to measurement (S. Bell and Morse 2008), the community is likely to fall into a measurability trap (the preference for and appeal of measurability). In terms of Simon's problem-solving approach through search in a defined space, the community is stuck in a local optimum of measurable improvements, without genuinely making progress towards the larger goal of genuine sustainability. The measurability trap certainly applies to the issue of energy-efficient algorithms and perhaps more broadly to software sustainability.

By emphasizing how the conceptual framing of problem-solving structures and limits our cognitive attention and our conversations, this chapter shows how the myths of systems design work and interact. To be very clear, the merits of problem-solving as such are not in question here, and neither is the merit of the problem concept. Clear and legitimate formulations of problems are powerful enablers of important work. Once problems have been legitimately agreed on, problem-solving must play a central role in systems design practice, research, and education (Oulasvirta and Hornbæk 2016; Jonassen 2000). The insidious nature of problem-solving as a conceptual frame lies in how it lays out and preconfigures the conceptual map by which we structure our view of the situations we engage in as computing professionals (cf. Costanza-Chock 2020, 124–125).

What I call *problemism* occurs when the lens of computational thinking turns data-driven problem-solving into a tunnel vision, unaware of the social construction of problems and data, the varied forms of individual and social cognition and decision-making, and the politics of stakeholder engagement in framing problems. Problemism takes the powerful concept of problem-solving into areas it is not suitable for. The effects of its simplistic implications are misleading, widespread, and devastating but rarely harmful to the designers. In fact, they thrive on it, as illustrated in this interview with a data scientist:

The strange thing about being out here in the Bay Area is that the worldview has just completely saturated everything to the point that people think that everything is a technical problem that should be solved technologically. It's a very privileged view of very smart people. It's troubling. (Tarnoff and Weigel 2020, 127)

I will first portray problemism as an entry in a fictitious handbook of clinical psychiatry to show its features as in a caricature, before examining it through the lens of the myths of systems design.

PROBLEMISM AS A COGNITIVE CONDITION

Overview. Problemism is a cognitive condition, widespread in technical computing disciplines, that predicates the worldview of the affected on problem-solving. The condition is socially contagious, and its spread often involves a tangible problem-solution pair or set of pairs. These pairs somehow act as infection carriers, but it is empirically established that on their own, problem-solution pairs are not harmful. The infusion of venture capital accelerates incubation. Once it takes hold, the condition centers its victims' attention on problem-solution pairs. Over time, this progressively impairs their reasoning and design abilities and precludes potential awareness of other conceptual frameworks, with deleterious effect on other people's lives. Problemism can be found in all walks of life, but it disproportionately affects computer scientists, venture capitalists, and combinations thereof (O'Neil 2016; Liu 2020). It is highly contagious in monocultural environments. Besides fixating the victims' minds on problems, the condition also instills a deceptive sense of confidence. The condition has been on the rise in at least the Global North for decades. Figure 4.1 shows an indirect indicator reflecting the historical trend since 1800. In some countries and professional spheres, critical observers have warned of near-epidemic proportions (Morozov 2014).

Long-term effects. There is to date no evidence that the condition causes harm to the affected. To others, however, its effects can be highly dangerous. Several factors complicate diagnosis. First, harm manifests in subtle and varied forms in time and space; second, harms tend to accumulate far from those afflicted by the condition; and third, those afflicted have

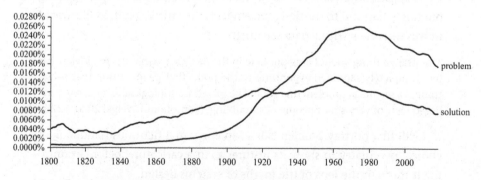

4.1 Google n-gram trend for problem, solution since 1800.

typically moved on to other problems by the time the harm manifests. Harm is then quickly identified by others suffering from problemism as a problem to be solved. This creates a reinforcing feedback loop.

Diagnosis. Symptoms include excessive belief in technological solutions; decreased sensitivity to other perspectives; a declining sense of personal responsibility; a propensity for moving too fast and breaking public goods and other people's things; in advanced stages, progressive loss of empathy, critical thinking, patience, and deliberative abilities. These symptoms are common and in isolation, they are relatively harmless, so nonspecialists find the condition difficult to diagnose. Isolated cases of self-diagnosis have been reported (Agre 1997b) but not independently verified. Your author self-diagnosed not long ago.

Differential diagnosis. Problemism is a relative of solutionism (Morozov 2014), the "presumption that technology can fix social, cultural, and structural problems" (Owens and Lenhart 2020, 1). Like solutionism, it prevents its victims from seeing other ways of solving a problem, other perspectives on the problem, and other possible problems that could be defined and addressed in the situation. In addition, it prevents victims from thinking in terms *other than problem-solving*. Where solutionism shapes the problem definition by way of what can be solved, however, problemism insidiously obstructs other ways of making sense of the situation altogether. The presence of solutionism always suggests problemism, but problemism can be present even if solutionism is not. Affected individuals and groups will focus prematurely on discussing the problem as a singular object, even if they do not explicitly consider possible solutions to the problem.

Treatment and prevention. No effective treatment has been found to date, and no vaccines exist. (Funding priorities and commercial interests have not supported the development of cures and vaccination strategies.) Treatment is difficult, tedious, and sometimes outright dangerous, but it must be undertaken. Research suggests that eradicating the condition is impossible, so the suggested focus is mitigation and prevention. Inoculation efforts that focused on the early development of critical deliberative skills in young adults, combined with treatments such as accessible readings of the history and philosophy of technology, have been found very effective. The latter are sometimes difficult to obtain, however, and their availability and effectiveness depend in large degree on the former. Dominant pedagogy in computer science works to thwart their efforts (Raji, Scheuerman, and Amironesei 2021).

Immunity. Some individuals are empirically shown to be immune, but the reasons are not well established. Hypotheses of immunity bestowed merely by prolonged exposure to critically oriented scholarship such as critical theory (Marcuse 1964; Feenberg 2014), policy and design studies (e.g., Rittel and Webber 1973; Schön 1983), feminist technoscience (Haraway 1988; Banu Subramaniam et al. 2016), feminist design (Choi and Light 2020), or critical philosophy of technology (Feenberg 2002; 2010; 2017) have all been empirically falsified by cases of problemism among such individuals. Therefore, the conditions of immunity most likely do not reside in the individuals alone but involve their sociocultural environment. Pilot studies have established the tentative effectiveness of exposure to critical systems thinking in treating affected subjects, but the effectiveness of the treatment may have remained contained to the individual problem-solution pair. Generalizability of this treatment is unclear to date. Further research is needed.

Emergency care instructions. Should you meet an acute case of severe problemism, do not confront the victim directly at first. Instead, gently lead them to consider the implications of their reasoning on those directly and indirectly affected by their proposed problem-solving approach. Aim to create a sense of empathy with those affected. If this fails, the best you may hope for in the short term is to instill a sense of curiosity. Try to introduce the victim to scientifically grounded evidence that legitimates viewpoints that stand in marked contrast to the assumptions underpinning the victim's reasoning. The unsettling of the victim's assumptions is a first step to the victim's realization that these assumptions may be wrong. That in turn is the opening to further discussions. Good luck.

PROBLEMISM IN COMPUTING PRACTICE

As described above, problemism is a Weberian *ideal type*, and it may seem like a caricature. Few—though not that few—real-world situations manifest exactly in the satirical form described previously. Instead, various

facets of this ideal type surface in degrees of strength in computing practice. But is the characterization of problemism an exaggeration? Consider
Whizz, an app that frames the problem of gig workers struggling to find
a bathroom as if it was an information problem and solves that problem
with a marketplace approach: bathroom access can be offered for a fee
and located on the app (Ongweso 2020). Consider the start-up Relida
Limited, which framed the problem of detecting whether a person *really*
had an orgasm as a binary quiz that can be solved using objective measures of bodily signals, especially the person's heart rate (because that is
readily available through wearable sensors) (Kleinman 2020). Or consider
how the ConsentTracker start-up framed the problem of consent in dating as a transactional problem and tried, for a while, to solve it by allowing dating couples to record on an app what exactly they consented to
(using Blockchain, of course): "Mutually agree each step of your date with
the Consent Tracker App to build #respect #trust & have a #fun #dating
experience," the team posted on Twitter.[1] Fortunately, they abandoned
the project after the Beta test with the comment "We looked, we tried, we
moved on."[2] Not every problemist learns that quickly.

In cases like these, problemism manifests as a premature move from
understanding the situation and the multiple divergent perspectives on
the issues it contains towards a narrow computational framing in which
an aspect of the situation is captured as a computable problem and put
forward for technical solution. Problemism is thus the flip side of solvency. It occurs when we forget that problems rest in their articulation,
that stating a problem is a shortcut: It carves a particular slice out of
the interpretation of a situation and presents it as a solvable puzzle. This
move is much more likely to be acceptable when some or all of the following assumptions are taken for granted:

1. the technology used to solve that problem is in itself neutral—therefore,
   it does not carry ethical debt or imply a moral obligation to evaluate its
   political implications;
2. humans are rational decision-makers who expect symbolic input as the
   key ingredient to make judgments—about dating; about sexual fulfil-
   ment; about their bodily needs during a working day in the gig economy;
3. the problems that these apps are designed to solve actually exist; and

4. problem-solving is the appropriate paradigm for intervening in the situation to cause supposed improvement.

In other words, problemism is most likely to occur where the myths of systems design intersect and reinforce each other. Technology-driven research and design efforts for sustainability also often aim to solve computational problems that poorly reflect the real-world issues they are meant to represent.

Such work uses simplifications to create computationally tractable representations of natural and social systems, but grappling with complexity is central to dealing with the challenges of unsustainability. . . . For example, 18th century efforts at calculating and deriving "maximum sustainable yield" from lumber forests ultimately led to nutrient-poor soils, disease-ridden forests, and failed crops, due to simplifications that did not represent such factors as the roles of birds, fungi, and rotting deadwood. (Baumer and Silberman 2011, 2272)

Today, countless apps promise to improve environmental sustainability by helping us make more environmentally friendly choices, for example, to reduce our energy consumption. How do the myths of systems design shape and mislead the outcomes in such design efforts?

The myth of solvency sets the frame by focusing on problem solving, with the assumption that the benefits of problem solving will outweigh any harmful side effects. It takes "the problem of energy consumption" for granted and suggests that the designer proceed to identify the object, properties and rules in this situation: the device, the algorithms, and technical rules. As illustrated earlier, this framing is strongly prioritized in current research on sustainability in software systems.

The myth of rational decisions facilitates a quick response in that tradition because it helps to frame the issue of energy consumption as an information problem. In this view, people just need better information— more precise and more detailed data—about their energy and resource consumption to compute optimal choices; or, conversely, because of their heuristics and biases, they need to be nudged into making more sustainable choices.

The myth of value-neutral technology suggests that the system to be designed has to be evaluated only for its effectiveness and efficiency in achieving the task it is designed to do. Side effects and political issues of participation, agency, or justice are not in the picture. The myth thereby lends an illusion of simplicity and innocence to the problem framing, and it lures designers into believing they are not responsible for undesired harmful consequences because they did not design these consequences. If they occur, it must be because of incorrect adoption or misuses.

The myth of objective problems configures the boundaries of the design discourse and specifies the dimensions of the problem space. If "the problem is that people consume too much energy," the solution must be that people need to consume less energy. Instead of expanding the view of the problem situation in which growing energy consumption is harmful to gain a systemic appreciation of its factors, the myth narrows the view by representing one particular framing as objective fact. Instead of exploring the richness of the situation through such practices as boundary critique (chapter 5) or value-sensitive design (chapter 6), the myth lends credibility to the framing presented by whoever leads the effort. It makes their knowledge claims appear as facts, and it will make marginalized voices who may speak critically about the inadequacy of one framing in the face of their lived experience appear illegitimate. For example, if the problem is framed as "consumers waste too much energy," this implies that the problem would be solved if consumers wasted (even somewhat) less. Alternative perspectives (Knowles 2013; Strengers 2014; Brynjarsdottir et al. 2012) would highlight that: (1) Providing information about energy consumption does not necessarily lead to the desired changes in consumption. It is simply one input to situated behavior that is complex, multifaceted, and best understood in context. (2) Consumers may find themselves unable to make the desired changes because they interfere with what they perceive as their lifestyle requirements or subsistence needs. (3) While some consumers waste energy, others don't have enough, particularly on a global scale. The richest 1 percent pollute more than twice as much as the bottom 50 percent, so changing *their* habits would have much stronger influence (Gore 2020; Chancel et al. 2021). (4) It is well established that consumer choice is not a strong leverage point for climate change in the absence of systemic change (Hickel 2020; N. Klein 2014). Corporations, specific industry sectors, and the overarching economic paradigm of growth-obsessed economic policy, all present much more effective sites for large-scale changes in energy consumption. (5) Treating people as citizens instead of consumers shifts the mindset more profoundly. In fact, it has a marked effect on their sustainability choices (Bauer et al. 2012). All this is not to say we should not foster awareness about consumer choices, but that the problem-framing prioritizes one leverage point for action and can prevent the most important conversations from taking place.

When these myths come into play all at once, they reinforce the common trope that climate change is consumers' responsibility, that technology can solve this social problem, and that we should step aside to let technologists solve it. That is the fairy tale that tech billionaires are now trying to sell.

PROBLEMISM AND THE MYTHS OF SYSTEMS DESIGN

Figure 4.2 illustrates some consequences emerging at the intersections of these myths, with problemism emerging at the center. For example, if technology is neutral (left), with neither politics nor a role reinforcing inequality, any negative consequences of design as problem-solving (right) are *unintended* and can be attributed to accidents, incomplete reasoning, or mistakes. These form new problems to be solved (top) within the existing frame, effectively displacing more nuanced perspectives. The myth of value-neutral technology thus allows problem solving to marginalize the broader consequences of its interventions.

Similarly, by framing thinking *as* information processing, the myth of rational decision-making (bottom) supports the concept of design as problem-solving performed by search (right). Rationalistic theories of choice have no mechanisms for deliberating about legitimate goals to strive for, but they provide the foundation for rationalist design methods. In fact, the original framing of bounded rationality *is* one of problem solving (Simon 1956; Simon 1996). How the frame is discursively established is out of its scope and remains invisible. Instead, the myth of objective problems (top) predicates a focus on solving those problems framed by measurable objects and entities already recognized by science over the issues voiced by those with lived experience. These problems are socially preformed, and existing technology partially preconfigures what counts as a solvable problem to begin with. The objectivity of science and its associated bureaucracy legitimate scientific expertise over other knowledges in that process. The sole focus on "correctness" and on

4.2 Problemism and the myths of computing.

the correspondence of representations to observable conditions distracts from fundamental questions of legitimacy and justification.

IS COMPUTING INSOLVENT?

Computing's role in sustainability and social justice is complicated. Despite its potential to make the world a better place, more often than not computing's harmful effects get externalized. This *debt of computing* raises the question whether computing—as a profession, an industry, and a field—will be able to pay back what it owes. This chapter advanced the position that beyond existing critiques of computing's ethical and moral positions, the very framing called problem-solving, which is so ubiquitous in computing, severely restricts *what we can talk about* when we talk about computing, human–computer interaction, and design. The myth of solvency manifests frequently, and in problemism it contributes to the rising debt of computing.

In a rigorous, values-focused discourse analysis of green computing research, Bran Knowles (2013) concludes that "the [green computing] discourse effectively commoditises 'sustainability'—making it purchasable in the form of Green ICT, persuasive apps, etc.—and in the process both 1) reifies consumerist tendencies that have driven much of the environmental destruction to date, and 2) absolves individuals from having to make more significant behaviour changes" (89). As a result, "Green Computing would need to accept major changes to the very premises upon which it is built before it is possible to a) recognise the reasons why its current strategy is unlikely to succeed, and b) craft a new strategy" (Knowles 2013, 110). Similarly, an analysis of climate-focused AI research argues that "technocentric approaches typically reduce complex human-environment relationships in ways that fail to account for social relations and power dynamics" and attests that "environmental and climate crises are grist for tech solutions" (Nost and Colven 2022, 23). In other words, this kind of work often is sustainability for IT, rather than IT for sustainability, as "many climate AI actors are interested in it for surveillance, greenwashing, and commodifying algorithms" (23) rather than to save the planet. In parallel, a review of AI ethics education attests a narrow focus that entails "disciplinary self-isolation . . . a loss of values, assumptions and methods that are crucial" (Raji, Scheuerman, and Amironesei 2021, 522). The authors conclude that this effectively renders computing graduates incapable of seeing beyond the narrow framing of technical expertise, to the detriment of real-world outcomes.

If we agree that the discourse of pure computational thought is indeed insolvent—unable to pay its debts—what should we do? In finance, insolvency can be met in two ways. The first is bankruptcy proceedings, in which assets may be liquidated to pay off outstanding debts. This would hardly be a useful approach here. Aside from the practical challenge that there are no clear entities and no judges, most of us want computing to be a part of this world—we just need it to be different. We need computing to shed what Feenberg calls the "illusion of pure rationality."

That illusion obscures the imagination of future alternatives by granting existing technology and rationalized social arrangements an appearance of necessity they cannot legitimately claim. Critical theory demystifies this appearance to open up the future. It is neither utopian nor dystopian but situates rationality within the political where its consequences are a challenge to human responsibility. (Feenberg 2014, 167)

How should we situate the rationality of computational thought within a political concept of systems design in the face of the continued dominance of the rationalist tradition? Winograd and Flores (1986) proposed a definition of design as "the interaction between understanding and creation" (4). On this basis, they too argued that "we need to replace the rationalistic orientation if we want to understand human thought, language, and action, or to design effective computer tools" (1986, 26). Unfortunately, despite the significant influence of their book, this tradition still dominates computing culture, perhaps in part because from a perspective of technical computing, their argument remained "incomprehensible" (Agre 1997a, 39).

I will here continue the critical engagement with that tradition, invoking the other option from finance for handling insolvency: restructuring. To restructure a company's debt means to propose a detailed plan to the creditors for how the company could continue its operations while simultaneously making good on its obligations (Tuovila 2020). It seems an apt analogy for what we need.

RESTRUCTURING: WITH A LITTLE HELP FROM THE CRITICS

Stepping away from the language of problemism does not mean inaction and paralysis, but it requires more attention to care, to discourse, to deliberation. For example, some critical design approaches dissolve problemism by redesigning the system of design in which it appears. As a result, design projects built on principles of design justice (Costanza-Chock 2020) or data feminism (D'Ignazio and Klein 2020) are unlikely to exhibit severe problemism. In emerging fields such as human-centered data science in spaces with closer interactions to fields such as science and technology studies, this is already happening too (e.g., Aragon et al. 2016; Passi and Barocas 2019). But in the meantime, the rationalist tradition continues to design as well.

How can we reconcile the undisputed power of computational thinking with a renewed focus on the social and political foundations of computing practice? Not by adding the latter to the former as an afterthought, but by bringing both to the same table as equal partners. Instead of adding considerations of human factors as a separate module to our body of computing knowledge or adding one elective from the humanities or social sciences to the computer science curriculum, we need to rethink the human and social foundations of computing and design practice. When computing first became social, computer science and CS education did not fully pursue the consequences. As a result, "CS pedagogy on its own is not able to elaborate the disciplinary norms and create conditions for stable comprehensive and socially beneficent technical artifacts" (Raji, Scheuerman, and Amironesei 2021, 523). Some are already working on this task of rethinking, but it is far from completed (Connolly 2020).

CONCLUSIONS: OVERCOMING PROBLEMISM

In problemism, unreflective solutionism in the unquestioned service of dominant interests reinforces the inequitable status quo and misperceives itself as a force of enlightened progress, leaving destruction in its wake. One might say that problemism is the dominant paradigm of computing. The path to sustainability and justice begins with setting it aside.

Computing contains many pockets of critical and divergent perspectives that are hard at work repaying its debts to societies. But the mainstream narrative of computing that presents computational problem solving as the world's savior and proposes a technical rationality as the "solution" to its wicked problems is indeed unable to pay its dues. The insidious nature of problemism lies in how it lays out and prefigures the conceptual map by which we structure our view of the situations we engage in.

Instead of opening bankruptcy proceedings, I suggest we restructure computing and its understanding of systems design on a foundation that includes social and critical theory. In that suggestion, I am not alone, and many have taken crucial steps in that direction. I will not propose a grand plan or pretend to know precisely how this should happen in totality, but instead make focused and partial suggestions.

Moving beyond the myths outlined here opens a view of problem situations with ample potential for computational interventions within a critically systemic framework. By presenting problemism in a purified and somewhat sarcastic form, I hope to draw attention to the argument that *it is not enough not to be a problemist*: In systems design practice, there is no neutral ground. Instead, responsible systems design practice for the twenty-first century involves something akin to the opposite of problemism. It must be critical and systemic. Its practice must be not instrumental but critical; not deductive but dialectic; not repeatable but replicable; not seeking proof but legitimation; not aiming for optimization but justification; not universal but contingent; not rational but reasonable; not technical but sociotechnical; not complete but proudly and savvily incomplete. Developing this will be the task of the book's second part. Once we supplant computing's myths, we can build such a practice with a little, or a little more, help from the critics. There is a lot of work to do before this can become the mainstream of computing, but many have already begun. It is time for a critical turn in computing.

II

RESTRUCTURING

It is impossible to adequately understand our ecological crisis with the same reductive thinking that caused it in the first place.
—Hickel (2020)

The field of CS has not yet come to the full realization that it deals with problems which exceed its traditional field of competency.
—Raji, Scheuerman, and Amironesei (2021, 523)

Who is likely to first notice the limitations of the engineers' useful but narrow conception of reality? There is no meta-discipline able to predict the need to integrate multiple forms of disciplinary knowledge. Another source of knowledge must come into play.
—Feenberg (2014, 213)

5

COMPUTING'S CRITICAL FRIENDS

Any use of expertise presupposes boundary judgments. . . . When the discussion turns to the basic boundary judgments on which his exercise of expertise depends, the expert is no less a layman than are the affected citizens.
—Ulrich (1983, 306)

Our position is not that of idealized neutral observers, but rather judges in our own case, with no one to properly hold us accountable.
—Gardiner (2014, xii–xiii)

Gardiner's warning rings true for tech development. Those involved in tech design are held accountable only in the most egregious cases of malpractice. In the absence of true accountability and independent oversight, existing ethics codes reinforce the idea that systems designers should be their own judges. This "makes it all too easy to slip into weak and self-serving ways of thinking," as Gardiner continues. For these and other reasons, Phil Agre's call for a "critical technical practice"—a practice that combines technical work with a critical orientation and a reflexive attitude—continues to resonate strongly with many of us (e.g., Ratto 2011; DiSalvo 2014; Britton et al. 2020). To develop an approach to systems design in computing that centers sustainability and justice, we need to gain a view of the technical rationality inherent in computing that explores its boundaries and gaps, or in other words, that doesn't end at the limits of reified frameworks of pure technical rationality. Like Agre, I am committed to developing my approach to just sustainability design in a continuous constructive engagement with computing, in computing.

This chapter argues that to support that work and balance critique with constructive design and development work, computing can call on *critical friends*.

A critical friend is someone who is encouraging and supportive, but who also provides honest and often candid feedback that may be uncomfortable or difficult to hear. In short, a critical friend is someone who agrees to speak truthfully, but constructively, about weaknesses, problems, and emotionally charged issues. ("Critical Friend Definition" 2013)

A critical friend brings support and profound respect to the table but also enough distance to take a contrasting perspective that challenges our worldview. They ask questions that nudge us to reflect on our assumptions and beliefs. One could say that critical friends are the friends who tell us what no one else dares to—because we would not accept to hear it from just anyone. This is important: for a critical friend's advice to be valued, the recipient must listen and value constructive critique.[1]

The concept of the critical friend has been elaborated in education. After all, constructive feedback is crucial to successful learning. For the professional development of teachers, establishing relationships with peers who act as critical friends is seen as a valuable mechanism for reflective feedback (Costa and Kallick 1993; "Critical Friend" 2014; Kember et al. 1997). While education focuses on individuals or groups, it can inform our thinking about the preconditions for a successful critical friendship of communities, including the need to foster a relationship of mutual trust that allows "unguarded conversations" (Baskerville and Goldblatt 2009).

Computing's critical friends can be found in several disciplines, and some research communities in computing have long cultivated constructive relations with their critical friends.[2] For example, much of the work on critical design methods in human–computer interaction (HCI) that followed Agre's call draws from feminist science and technology studies (STS). To some degree this book does too, and the next section will briefly review key influences in intersectional feminist STS and the critical philosophy of technology. My main focus lies on *critical systems thinking* (CST). I consider CST a close critical friend, because CST combines the critical approach with a robust systemic perspective and grows directly out of a critical turn in operations research away from the rationalistic tradition that still dominates computer science. This is why its insights and arguments are strikingly relevant today. Its history shows parallels to the challenges that technologists face today in designing for sustainability and justice, because its proponents too grappled with the overreach of rationalist scientific reasoning and with crucial questions of expertise and legitimacy. For me, someone schooled in the rationalist paradigm, CST offered a way to restructure my understanding of computing and its role in sustainability and justice. For these reasons, I draw on the arguments of CST to grapple with the myths of computing, and I use its frameworks to populate an initial toolbox for reorienting systems design.

Below, I introduce these three critical friends: feminist STS, the critical philosophy of technology, and CST. My aim is to set in motion a conversation continued in subsequent chapters by introducing key ideas and arguments, exploring how they complement each other, and illustrating the idea of critical friendship.

FEMINIST SCIENCE AND TECHNOLOGY STUDIES

STS "investigates the institutions, practices, meanings, and outcomes of science and technology and their multiple entanglements with the worlds people inhabit, their lives, and their value", and it often aims "to open up science, technology, and society to critical assessment and interrogation" (Felt et al. 2016, 1–2). Prominent early examples include the argument that material infrastructure can embed political priorities and values (Winner 1980). Because the study of technology was not traditionally the subject of the social sciences, some STS researchers focused on emphasizing the importance of including technical objects into social analysis (Latour 1987). Others focused on the question how technologies are socially constructed (Bijker et al. 1987; 2012). The emphasis on the social construction of technology mirrors the emphasis within computing on *how computing shapes societies*. But unlike computing, STS has long maintained that societies and technologies mutually shape and constitute each other (MacKenzie and Wajcman 1999; Subramaniam et al. 2016, 407). The term *technoscience* reflects that STS views science and technology as convergent (Felt et al. 2016, 7).

Feminist STS has examined the uneven distribution of influence over technoscience and its benefits. Far from restricting itself to a focus on women's representation in technoscience and the ways in which women were affected by it, intersectional feminist STS also explored fundamental categories, beginning with gender, race, and sexuality. Drawing on such concepts as reification and the role of metaphor in structuring our understanding of the world, feminists questioned the adequacy of dominant epistemologies and ontologies (Subramaniam et al. 2016). Sandra Harding, for example, wrote that

as a symbol system, gender difference is the most ancient, most universal, and most powerful origin of many morally valued conceptualizations of everything else in the world around us . . . gendered social life is produced through three distinct processes: it is the result of assigning dualistic gender metaphors to various perceived dichotomies that rarely have anything to do with gender differences; it is the consequence of appealing to these gender dualisms to organize social activity, of dividing necessary social activities between different groups of humans; it is a form of socially constructed individual identity only imperfectly correlated with either the "reality" or the perception of sex differences. (1986, 18–19)

Donna Haraway's argument that all knowledge is situated (1988)— or rather, that all *knowledges are* situated—opened an influential line of argument with parallels to CST: despite all claims to universality, knowledge is produced from a perspective. Because that perspective is always partial, knowledge is not, and cannot be, absolute or complete. It cannot be entirely separated from a knowing subject, as rationalist philosopher of science Karl Popper would have had it (1972). Its location and context matter, and the dominant form of knowing in technoscience is not the only form. The conclusion is not that everything therefore is equally valid, but that a critical awareness of each perspective's position, and its partiality, helps us in assessing its validity. This helps us to produce more accurate and legitimate knowledge or what Harding calls "strong objectivity":

In societies where scientific rationality and objectivity are claimed to be highly valued by dominant groups, marginalized peoples and those who listen attentively to them will point out that from the perspective of marginal lives, the dominant accounts are less than maximally objective. Knowledge claims are always situated, and the failure by dominant groups *critically and systematically to interrogate* their advantaged social situation and the effect of such advantages on their beliefs leaves their social situation a scientifically and epistemologically disadvantaged one for generating knowledge. Moreover, these accounts end up legitimating exploitative "practical politics" even when those who produce them have good intentions. (1992; emphasis added)

With this feminist sensitivity to the historicity and positionality of knowledge, classic studies in STS examined the mutually constitutive entanglement of technology and society through wide-ranging objects of study including household technology (Wajcman 1991) and the international disease classification system (Bowker and Star 1999). Bowker and Star's study showed how classification systems are historically shaped by social forces and interests; how marginalized perspectives are only included after significant advocacy (a prime example is the classification and later declassification of homosexuality as a disease); and they show how these technologies exert a *torque* that twists the lives of those who happen to get into its force field. Diabetics like me involuntarily collect lived experience of that torque on a daily basis. In contrast to the dominant desire to assimilate disabled bodies into the normative standard by assistive devices designed *for* disabled people by supposed experts, *crip technoscience* mobilizes feminist STS theory to advocate a nuanced politically conscious design perspective on which basis "technoscience can be a transformative tool for disability justice" (Hamraie and Fritsch 2019, 3).

Today, STS encompasses a dizzying array of perspectives, methods, theories, and approaches (Vertesi and Ribes 2019; Felt et al. 2016). Of primary interest for my argument is the intersectional feminist attention to the positionality of knowledge production, the questioning of categories, the role of the matrix of domination in technology design, and the challenging of entrenched power structures (D'Ignazio and Klein 2020). There are two reasons for this: first, these concepts speak directly to the challenges of just sustainability design, and second, they are the connection points where the arguments of STS meet the arguments of CST. These issues have been prominent in feminist HCI too (Rosner, Taylor, and Wiberg 2020). In addition, I want to introduce the critical philosophy of technology developed by Andrew Feenberg, because his democratic theory of technology suggests possible intervention points in the trajectory of technology development.

QUESTIONING TECHNOLOGICAL RATIONALITY:
FEENBERG'S CRITICAL CONSTRUCTIVISM

In a series of books, philosopher Andrew Feenberg (e.g., 2002; 2017) has developed his critical theory of technology building on the work of his advisor Hebert Marcuse (1964; Feenberg 1996) and specific influences in critical theory, especially Lukács (Feenberg 2014). This theory forms an independent body of work, but it is also seen as a part of STS (Felt et al. 2016). As Feenberg writes, "Critical theory of technology agrees with STS that technology is neither value neutral nor universal while proposing an explicit theory of democratic interventions into technology" (2016, 635). A comprehensive review of this theory is beyond my ability and scope, but a few of its concepts and arguments will illustrate its central relevance for our subject.

Lukács' concept of *reification* explains "how the world can appear as a collection of facts" that seem natural and are not questioned (Feenberg 2014, 86). Technology design needs to respect facts, of course, but they alone never determine the shape of technological artifacts (Feenberg 2017). Instead, artifacts remain *underdetermined*. Different actors with differential power and influence each operate within their "margin of maneuver" (Feenberg 2002) in trying to influence the outcomes. The resulting artifacts of the present are coproduced by past values.

Feminist STS work emphasizes that each of these actors is bound to speak from their partial perspective. Power dynamics and the matrix of domination will force some perspectives to the margins. From there, their knowledges—"subjugated knowledges" in Foucault's terminology— often provide insights that go beyond the dominant perspectives, simply because *they have to*, and because their lived experience provides insights that are not available to the hegemonic view. Data feminists D'Ignazio and Klein call attention to the "privilege hazard":

When data teams are primarily composed of people from dominant groups, those perspectives come to exert outsized influence on the decisions being made—to the exclusion of other identities and perspectives. This is not usually intentional; it comes from the ignorance of being on top. (2020, 28)

The privilege hazard is a hazard of moral corruption not unlike Gardiner's. In combination with reified rationality, it is just too easy to continue business as usual as long as that business's debts are born by others. Because the view from the top often overlooks insights from elsewhere, marginalized actors are forced to re-present their insight *as if* it came from other perspectives, just so that their expertise becomes legible *within* the dominant discourse.

Feenberg shows how marginalized perspectives are often excluded from technical standards and frameworks *by design*. For example, a standard curb, with its sudden drop to the road, cannot be handled easily by a wheelchair user. It literally manifests the priorities of values from the able-bodied top that coproduced it. Disability advocates were often motivated to organize because of their lived experience. Representatives of these marginalized perspectives organized to collectively advocate for changes, leveraged the narrow margin of maneuver available to them, and eventually achieved changes to the technical standards that afforded them some degree of accessibility. Through regulation, emerging practice and new values were embedded into a revision of standards and design guidelines. This serves as an example of how marginalized interests are expressed through *democratic interventions*, "the actions of citizens involved in conflicts over technology" (2016, 646). They take place during technology development or as "a posteriori" interventions or appropriations. In this process, expertise and knowledge have to be translated:

the claims of experience and those of technical disciplines must be reconciled in the design process. . . . In the real world of technology, a largely unacknowledged dialogue between lay and expert is a normal feature of technical decision making and should be further developed. (Feenberg 2016, 647)

In computing, this translation work is typically attributed to requirements professionals, user experience designers, systems analysts, or adjacent roles. It is recognized as a central issue for requirements engineering (see chapter 10). Feenberg sees a significant role for technology professionals in this dialogue:

Democratic interventions must be translated by technical professionals into new regulations and designs. Struggle gives rise to new technical codes both for particular types of artifacts and even for whole technological domains. This is an essential form of activism in a rationalized society. It limits the autonomy of experts and capitalist management and forces them to redesign the worlds they create to represent a wider range of interests. (Feenberg 2014, 214)[3]

He is well aware that we cannot simply expect this to work out for everyone. Reified technical codes present the illusion of pure technical rationality, and professionals are subject to incentive structures, historically grown professional competences and constraints, and historically produced educational programs, all shaped in the context of capitalism. Their margin of maneuver is thus severely restricted, and their reasoning can be subject to a false consciousness that presents the world of their activities as a collection of facts which exclude the perspectives of those marginalized, suppressed, and hurt (D. Noble 1984; 1977; Hoffman 1989; Breslin 2018). Instead, for this process of reconciliation to be realistic and fair, *dereification* is needed to free the false consciousness and allow it to comprehend the transcendent reality beyond its operationalized concepts. To overcome reification, Feenberg argues for the development of a *dialectical rationality*:

Dialectical rationality is what transcends the one-dimensional reified thought and supports dereification and reconstruction. . . . Now rationality is associated not only with science and experiment, but also with the practical critique coming from those subordinated to the forms of capitalism. Their situated knowledge reveals aspects of reality to which reified rationality is blind. (Feenberg 2014, 206)

In today's world of IT, with its lack of equity and diversity in the workforce and its uneven global distribution, the privilege hazard combines with the danger of moral corruption highlighted by Gardiner to make the project of reconstruction, including the restructuring of narratives, more urgent and difficult than before. For Feenberg, this continued reconstruction is activism. He believes that the continued struggle for reform is an inevitable component of social change, and that it must be incremental. Disciplinary knowledge and politics present a challenge to this process. This chapter's Feenberg epigraph suggests that no "meta-discipline" can foresee the need for integrating disciplinary knowledge, and that therefore, a different kind of knowledge is needed. Systems thinkers would readily agree and point out that this is precisely what systems thinking is all about. But depending on which systems thinker gets the word, the conclusions vary drastically.

THE SYSTEMS APPROACH AND ITS STRUGGLES

A few decades before Agre and Feenberg, West Churchman struggled with the limitations of rationalistic understanding presented by a dominant systems paradigm. Churchman had written one of the first textbooks on operations research together with Russel Ackoff. Both grew disillusioned with the narrow analytic understanding that came to dominate the communities that they themselves had helped shape. Both ultimately abandoned these fields since, in Ackoff's words, their systems approach had "degenerated into mathematical masturbation" (1977): It had lost interest in its real-world context beyond its operationalist understanding, and it was unable to recognize its own shortcomings (Kirby 2003).

Throughout his work, Churchman strived to reconcile the comprehensive aim of a systems approach with the recognition that the reductive approach represented by these forms of systems *analysis* violated the very principles of systems thinking. In his book *The Systems Approach and Its Enemies* (1979b), a strange dialectic unfolds. The main protagonist, *the planner*, comes at the world with the best of intentions and an approach we would now describe as "hard systems thinking." This approach applies scientific principles and mathematical or logical procedures to analytically address social problems. In the process, scientific propositions gain *normative content* in their context of application, because they entail value judgments, real-life implications, and side effects. The planner—conceived in the gendered language of the day as a man—is full of optimism and good will. Yet, at every step of his planning, he encounters objections brought forward by the "enemies" of the systems approach: politics, morality, religion, and aesthetics. To rationally refute the seemingly irrational objections of these enemies, he needs to sweep in additional parts of the system's environment. In response to their repeated objections, the boundaries of the system continuously expand until the entire universe seems to be inside, in a *reductio ad absurdum* reminiscent of Borges's fable of the map that expands until it fills the territory (Borges and Hurley 1999).[4]

Churchman never resolved his dilemma. Ulrich (1983) summarizes his message (substitute "design" for "planning"):

[In "The Systems Approach and Its Enemies,"] the systems approach for the first time has become truly self-reflective with respect to the normative content of its own quest for systems rationality. In Churchman's terms, the systems approach cannot realize its search for the comprehensive rationality of planning so long as it seeks to absorb the 'enemies' of such rationality, e. g. politics, morality, religion, and aesthetics. Rather, the systems approach can claim comprehensive rationality if it learns to reflect on its own limitations, namely, by listening to its 'enemies' and by understanding them dialectically as what they are: mirrors of its own failure to be comprehensive. (34)

In his strange way, Churchman articulated the inability of analytic, deductive rationality to handle other forms of reasoning. For example, attempts to classify and measure all relevant values in systems design so that they can be formally represented in a model of that system's values is a prototypical example of the strategy of absorption—and it is unlikely to convince the enemies. Churchman also expressed the uncomfortable realization that rationality is founded on commitments it cannot fully explain. He did not go further than that, but his work expressed the conceptual foundation of two major turns in systems thinking. The first, in the 1970s, broke with positivist assumptions about the nature of systems as structurally present in the world and knowable. The core contribution was to acknowledge that conventional systems approaches to problem-solving fail in the social world because the central difficulty in a techno-logical intervention is *how to define the problem* it should solve (Checkland 1981). The presence of multiple legitimately diverse perspectives demands a way to find consensus on how to frame the problem to be solved. The resulting soft systems thinking, exemplified by Checkland's soft systems methodology, is interpretive. It uses systemic concepts to organize a discourse in a problem situation, without making assumptions about how the real world is structured (Checkland 1981; 2000). This was an important step in applied systems thinking, but it did not address the issues highlighted by Feenberg, as discussed earlier, because it lacked the social theory to recognize and address issues of power dynamics and inequity among the participating stakeholders, it ignored the dangers of false consensus arising in its application, and it did not address questions of coercion.[5] In response, the second major turn resulted in critical systems thinking.[6]

THE SYSTEMS IDEA, CRITICALLY UNDERSTOOD

Critical systems thinking starts from the recognition that the systems idea of holistic understanding, understood critically, must begin with examining the inevitable selectivity of discursive claims *about* systems. CST emerged in the 1970s. It draws on contemporary critical social theory such as Habermas and Marcuse and later, the work of feminist scholars and writers such as Foucault (Flood 1990; M. Jackson 2019). This critical turn in systems thinking brought forth a heterogeneous set of approaches that combine a critical perspective with the intention to intervene productively in the world to create social change, much like Agres critical technical practice (1997b). It is worth retracing key steps of this difficult argument to interpret it from a contemporary perspective and resituate it in the context of systems design for the twenty-first century.

CST starts with Churchman, whose designer faces a crucial dilemma. To *rationally* evaluate even the smallest choice, the designer must justify their evaluations by reference to broader concerns. In strongly interconnected problematic situations, what they decide, explicitly or implicitly, amounts ultimately to a judgment over what is supposed to constitute *the whole system*. These "whole system judgments," or *boundary judgments*, are necessary to make any real choice. Churchman articulated the struggle and the need for a dialectic approach. His advisee Werner Ulrich continued Churchman's questioning of the rational justification of the propositions of applied science. He started with the observation that each boundary judgment—for example, whether a fact should be considered as part of the justification for a design decision or not—can be traced back to its underlying assumptions: When questioned as a claim, it should be substantiated by reference to assumptions that in turn can be questioned, and so on. At some point, the justification inevitably stops—it breaks off, as visualized in figure 5.1. The set of explicit and implicit claims at which justification breaks off is considered the *reference system*.

Ulrich was under no illusions about the homogeneity of those involved in design, or the willingness of those who are in power to open their claims to scrutiny. He distinguishes between those involved and those affected in design, as shown in figure 5.2.[7] These distinctions too are boundary judgments built on implicit reference systems: when we question who is or should be involved or affected and why, the answers will lead us to a reference system. Because the assumptions and conditions that constitute the reference system logically precede the design effort and cannot be logically or empirically justified, "the planner must trace the normative

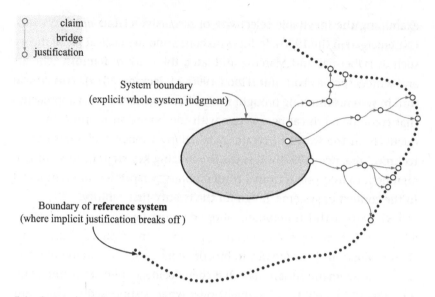

5.1   System boundaries and reference systems.

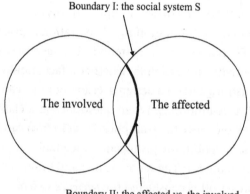

5.2   Two basic boundary judgments (Ulrich 1983, 248).

implications of *alternative* boundary judgments . . . and make these implications transparent to those concerned" (Ulrich 1983, 244).

Ulrich looks "for the crucial sources from which the normative content of any social map or design derives" and finds them in intentionality: "*human intentionality* is the constitutive element accounting for
the complex and normative nature of the object domain 'social reality'"
(Ulrich 1983, 245). Intentionality sharply distinguishes the social world

from the technical and the world of natural sciences. The *purposes* pursued by human actors and social groups introduce an entirely different kind of element into the object domain of design. Crucially, humans are purpose*ful*: capable of choosing their own path. In introducing purpose to the argument, Ulrich directly opposes the main proponent of rationalistic systems design, Herbert Simon.

PURPOSEFUL SYSTEMS AND THE EVACUATION OF POLITICS

The rationalistic tradition cumulating in Herbert Simon's *Sciences of the Artificial* (1996)—which lives on in computational problem solving and rationalistic systems design—stands in marked opposition to the position occupied by Churchman (Ulrich 1980; 1983, 319–325). Simon considers the ascription of purposes to social systems as a teleological fallacy, while in Churchman's view, it serves "a necessary critical purpose against hidden value assumptions." Perhaps the most interesting comparison results from the opposing views of what the *"crucial design task* to be solved" is (Ulrich 1983, 323). For Simon, it is *"problem decomposition*: how to decompose a complex system into simple systems that are easy to be controlled? Or: how to design and control complex hierarchies?" His view and focus still dominate computer science (CS) curricula today. For Churchman, the crucial task is *"problem identification*: . . . how to identify the whole system." While Simon privileges the divide-and-conquer standpoint and sees a problem of control, Churchman privileges the ethical standpoint and sees a problem of boundaries. But Churchman is not averse to relying on the powerful mechanisms developed by Simon's paradigm as long as their application can be ethically justified within a critical and reflective framework.

The problem of [Simon's approach] is clear: it simply avoids the moral and political question of how a better whole will result from incremental improvement of small and separate subsystems; it leaves that question to the market as a surrogate planner—and to the interest groups that control it. . . . The crucial task then is not one of providing analytic tools such as social indicators, simulation models, etc., but rather dialectic tools to help the planner reflect . . . and enter into reasonable discourse with the affected . . . no systems approach for handling real-world complexity can be rational unless it makes room for the (self-) critical reflection of *free citizens*; for these alone know what social reality is and what it ought to be like. The systems approach may yet have a great future, if only it begins to understand that the "enemies" are really its allies. (Ulrich 1983, 325)

In other words, the rationalistic approach exemplified by Simon evacuates values and politics from the design discourse and leaves behind a depoliticized *science of design*. The resulting analytic tools are powerful but one-dimensional.[8] Other systems approaches, such as cybernetics, similarly evacuate intentionality.[9]

The project of critical systems thinking is not concerned with denying the validity or effectiveness of the analytic tools of "one-dimensional" hard systems approaches, but with undoing the evacuation of politics. The aim is to support the integration of analytic tools into a critical approach in which their application can become legitimate and appropriate under clearly defined conditions.[10] In systems design for sustainability and justice, too, the central challenge is not one of providing more exact analytic tools to measure sustainability and justice, but to address the need for dialectical forms of reasoning in which conflicting worldviews, including on those analytic tools, can be reconciled (see chapter 8). It is difficult to overstate the importance of this project in today's social reality of systems design, presented as unpolitical (Costanza-Chock 2020, 120–123; Dourish 2010), in which the legacy of Simon's approach still enjoys a hegemonic position (Rosner 2018).

To reestablish purpose in systems design, Ulrich relies on a principled argument: Just as causality is an important idea in the physical sciences, intentionality matters for understanding the "facts" of the social world, and it is in fact no less empirical in content, no less observable, than causality. He draws on Habermas's (1972) argument that human interests and purposes are always constitutive factors of knowledge. There is a useful distinction to be made among *technical interests* in prediction and control in the domain Habermas labels "work," *practical interests* of mutual understanding in communication in the domain Habermas labels "interaction," and *emancipatory interests* in autonomy and freedom (Habermas 1972).

In design, technical interests intersect with practical interests and emancipatory interests when stakeholders seek consensus on what to do about a situation. Habermas's concept of the *ideal speech situation* captures the idea of a situation in which participants can freely arrive at a consensus that can be regarded as genuine because it arises from a fair and equal process. The concept serves primarily as a hypothetical comparison point to help observers distinguish true from false consensus: that is, consensus emerging from unfair discourse. But it is a bit more than that: in Habermas's view of rationality, the possibility of the ideal speech situation has to be at least presupposed in principle in any act of communication. For him, any communication assumes four validity claims: that what is said is comprehensible, true, right, and sincere. "Thus we can ask a speaker 'What do you mean?' 'Is what you say true?' 'Are you entitled to say that?' and 'Do you really mean it?'" (Outhwaite 1994, 40).[11] These questions apply to any claim made about systems and, therefore, any boundary judgment.

Against this background of interests in a situation where claims are made about systems, the designers need to identify what the perceived social reality *means*:

Defining "purpose" as a mapping dimension analogous to "space" and "time" is thus *not* to "introduce" value judgments into social mapping but rather to include in our basis of experience the value judgments that are always already there, as constituent elements of both social reality and our maps of it. (Ulrich 1983, 243)

Ackoff had made an important distinction between purposive and purposeful systems. Organs such as the human heart are purposive systems—the heart's main purpose is fixed: to regulate blood circulation. It will respond to external conditions such as exercise or sleep by adapting its goals: pumping more or less blood. In contrast, a purpose*ful* system "can change its goals under constant conditions; it selects ends as well as means and thus displays *will*. Human beings are the most familiar example of such systems." (Ackoff 1971, 666).[12] This distinction becomes especially relevant when we consider organizations and sociotechnical systems. After all, "power is the ability of an individual or a group to impose its purpose on others" (Galbraith 1975, 88). For example, an individual person is a purposeful system whose elements are purposive—the heart does not choose its purpose. In contrast, organizations are not only purposeful systems but their elements too are purposeful systems. This crucially distinguishes organizations from organisms (Ackoff 1971, 669). Despite its dry conceptual basis, Ackoff's deceptively simple distinction has serious ethical consequences: it becomes unethical to treat a purposeful system as if it was merely purposive—to treat it as if it was unable to select ends, to treat it as a means to preselected ends. That is why the algorithmic control of humans, such as the control of workers in places like Amazon warehouses, is so repulsive to many of us.

BOUNDARY JUDGMENTS AND REFERENCE SYSTEMS

It is the designers' responsibility "to secure the transparency of the boundary judgments" and "to trace their possible normative consequences" (Ulrich 1983, 227). In opposition to Simon's paradigm, Ulrich proposes a critical interpretation of the role of the systems concept. To develop meaningful questions about boundary judgments and reference systems, he lists typical examples of boundary judgments that locate a systems design effort in the space of intentions, reproduced here in abridged form:

1. What should be considered as a *problem?* That is, what or who ought to belong to the section of social reality that is to be studied?
2. What *purpose* should the study serve? That is, who ought to belong to the group of those who are to benefit from the project?
3. What ought to be the totality of conditions that define the client's *standard of success?*
4. What should be the *time horizon* with respect to the "relevant" conditions in the future? For instance, should the *future generations* be included within the problem-relevant client system? Is intergenerational transfer of costs or risks acceptable? Are irreversible consequences acceptable?
5. What should be the *time horizon* with respect to the "relevant" conditions in the *past?* (Should the past generations—e.g., their goals and dreams, the traditions we inherited from them—belong to the problem-relevant system?)
6. Who might be *affected* by the project (by a change in the social system in question) although he does not belong . . . to those who are involved and benefit from the planning effort? Under what conditions can we assume it is legitimate that some people are affected although they cannot belong to the client system? How do we draw the boundary between the affected and the unaffected in the case of long-term risks such as radiation or cancer? (Who among the affected ought to be involved in the project?)
7. Who should *plan* and who ought to belong to the *decision-making body?* (Who ought to belong to the "experts" and "decision-makers"?). (Ulrich 1983, 228)

Each of these questions represents a boundary judgment. I hardly need to elaborate the striking relevance of these judgments to sustainability and justice. Some of these judgments directly map onto the central practices of systems design represented by requirements (see chapter 10). They are the departure points that motivate the development and justification of critical systems heuristics (CSH). The questioning of the explicit boundary judgments and their justifications can to some degree make the reference system visible. And it is here, at this frontier, that the selectivity inherent in all worldviews becomes visible.

CRITICAL SYSTEMS HEURISTICS

Based on the distinction between the involved and the affected, Ulrich distinguishes between three classes of involvement according to their sources of influence. The client represents *sources of motivation* (purposes, values, interests), the decision-maker represents *sources of control* (resources, authority), and the planner represents *sources of expertise* (knowledge, skills). The group of those affected but not involved is more difficult to delineate—its members qualify not by virtue of influencing the design, but by virtue of being affected. Because of the vast array of potential effects and their dispersion in time and space, this group can be simply too large. Ulrich covers the array of dispersion later addressed by Gardiner, including references to socio-ecological destruction, and emphasizes that the group includes "individuals to whom the planner cannot turn for hearing their concerns, be it because they are unborn, too young, or handicapped by other reasons." (Ulrich 1983, 252) The crucial question of legitimacy is how a design "discourse can be rational *even though not everyone affected can become involved."* (Ulrich 1983, 252). The key category is that of the *witness*:

The essence of their role, it seems to me, is that of a *witness*: by virtue of their own affectedness, they can bear witness to the way in which all those who cannot voice their concerns may be affected—their feelings, their suffering, their moral and political consciousness, their ways of expressing dissent, their ways of *living* the social reality in question, their vision of their own future . . . the planner cannot adequately trace the normative content of alternative boundary judgments . . . without referring to *some* actors playing the role of a witness.

The four categories (motivation, control, expertise, and legitimation) are crossed with three basic questions that ask for (1) the social roles of the involved and the affected, (2) role-specific concerns, and (3) key problems in determining the necessary boundary judgments. The resulting twelve heuristic categories, and the questions they translate into, are not a checklist but a starting point for continued investigation and reflection. After all, "a well understood systems approach begins and ends with the questions we ask, not with the answers we give" (Ulrich and Reynolds 2010, 290). The term *heuristics* remains crucial: this approach is not a universalist critical theory but a humble yet powerful and effective dialectic device. The risk of deception is central, so in using the categories, a constant vigilance and humility is important.

1. Sources of motivation unfold into the client, the purpose of planning (design), and the *measure of improvement*. There is never a perfect measure of improvement, but "some such measure is implied in every social design, and . . . the existence of such a measure is a heuristic necessity" (Ulrich 1983, 255). In other words, this category is the "value basis" of design (Ulrich 1993).
2. Sources of control unfold into the decision maker, the components under the control of the decision maker, and the *decision environment* that provides the conditions outside of the decision maker's control. They constitute the "basis of power" (Ulrich 1993).
3. Sources of expertise unfold into the planner (or designer), their expertise, and the *guarantor* of expertise. The guarantor term refers not to people but to the conditions that assure us that the expertise to be applied is valid and applicable. It is important because expertise can be deceptive: "How can the planner even know that the experts' skills, experience, or tools are not a source of deception?" The category compels the designers to examine the guarantors of expertise. That is the context of the chapter's epigraph: "no amount of expertise . . . is ever sufficient for the expert to justify all the judgments on which his recommendations depend. When the discussion turns to the basic boundary judgments on which his exercise of expertise depends, the expert is no less a layman than are the affected" (Ulrich 1983, 306). In other words, this category captures the "basis of knowledge" (Ulrich 1993).
4. Sources of legitimation unfold into the witness, their ability to emancipate themselves from the rationality of those involved, and the *worldview* underpinning the planning effort. The crux lies in the possibility of fundamentally conflicting worldviews—different visions and ideals of what life should be like. In Ulrich's words, "The essential point is that the affected must be given the chance of emancipating themselves from being treated merely as means for the purposes of others" (Ulrich 1983, 257).

Figure 5.3 summarizes the conceptual relationship between the two types of boundary judgments, the categories of being involved and affected, and the heuristic categories. The resulting matrix of categories is summarized in table 5.1. This set of categories should be instantiated with attention to the context of design, so the exact terminology and focus can

Boundary I	Boundary II	Central issues	Categories
The social system S to be bounded	The involved	Sources of **motivation**	1. Client 2. Purpose 3. Measure of improvement
		Sources of **control**	4. Decision Maker 5. Components 6. Environment
		Sources of **expertise** and implementation	7. Planner 8. Expertise 9. Guarantor
	The affected	Sources of **legitimation**	10. Witness 11. Emancipation 12. Worldview

5.3 The basic categories of CSH (adapted from Ulrich 1983, 258).

Table 5.1 CSH categories adapted from Ulrich and Reynolds (2010), Reynolds (2007), Ulrich (1993)

Sources of influence	Social Roles (Stakeholders)	Specific Concerns (Stakes)	Key Problems (Issues)
Motivation (the value basis)	Beneficiary ("client")	Purpose	Measure of Improvement
Control (the power basis)	Decision Maker	Resources	Decision Environment
Knowledge	Expert	Expertise	Guarantor
Legitimation	Witness	Emancipation	Worldview

vary accordingly (McCord and Becker 2019; Duboc, McCord, et al. 2020; Ulrich and Reynolds 2010; Wing 2015; Ulrich 1998). For each category, the key question can be asked in two modalities: an empirical "what is the case" and an ideal "what ought to be the case." The system of categories can be used in at least three ways:

1. *Reflectively*, those involved in a design effort can use it to understand how their own worldview is positioned—how their knowledge is situated in the world.

2. *Critically*, in constructive engagement, a critical friend can help the designers understand the boundaries of their system of assumptions and how it invokes selectivity.
3. *Polemically*, those affected but not involved can expose the lack of legitimacy of undemocratic interventions in situations where the designers (those involved but not affected) are unwilling to engage. They can do this by dialectically demonstrating to others what some of the designers' assumptions are and, importantly, that the designers are unable or unwilling to admit or justify them.

Ulrich's work marks a milestone for CST. It has found a wide range of applications in planning and design, in the evaluation of sustainable development projects (Reynolds 2007; Ulrich and Reynolds 2010) and healthcare policies and, more recently, in understanding the politics of stakeholder involvement in systems design (Wing 2015; McCord and Becker 2019; Duboc, McCord, et al. 2020). This will be of closer interest in chapter 10.[13]

CRITICAL TECHNICAL PRACTICES IN CST

Because CST proponents sought a balance between reflective critique and real-world interventions, they developed systemic frameworks for participatory Action Research (Flood 2010) and management (Flood and Jackson 1991a; M. Jackson 2019) as well as community-oriented methodologies broadly situated in the domain of social work (Midgley 1997; 2000). What unites these approaches is that they emphasize a careful evaluation of partial perspectives as the crucial step that must be secured for technical approaches to be legitimated in their application. CST methodologies such as total systems intervention (Flood and Jackson 1991a), systemic intervention (Midgley 2000) or critical systems practice (M. Jackson 2019) do not reject rationalistic approaches altogether or substitute critical reflection for technical modeling, simulation, or prediction. They do not abandon technical work but instead consciously design processes by which technical approaches are evaluated for applicability and legitimacy, and then embed technical work *within* the critical and reflective frame of CST if and when it is appropriate. In a sense, CST researchers developed what Agre strived for in their respective disciplines. And these disciplines are much closer than they may appear: both operations research and artificial intelligence originated in the same place, time, conditions, theories, and even people as computer science.

CST practice addresses a wide range of situations and domains, from internal organizational questions in management (Flood and Jackson 1991b) to public healthcare (Midgley 2006), sustainable development (Reynolds 2007), or community planning (Midgley, Munlo, and Brown 1998). More recently, CST has been brought in dialogue with ecofeminism (Stephens 2013; Stephens, Taket, and Gagliano 2019). These CST approaches all share a three-fold commitment (Midgley 1996):

1. to *critical reflection*, as illustrated by critical systems heuristics and further developed into the theory and practice of boundary critique by Midgley (Midgley 2000; Midgley, Munlo, and Brown 1998);
2. to the *emancipation* of marginalized perspectives, for instance through attention to the mechanisms by which situated knowledges and the concerns of those affected are marginalized (Midgley 1992); and
3. to *methodological pluralism*, including heterogeneous understandings that Gregory describes as "discordant pluralism" (Gregory 1996a).

SHARED THEMES

We are finally ready to return to Feenberg's question: "Who is likely to first notice the limitations of the engineers' useful but narrow conception of reality?" (Feenberg 2014, 213). It appears that there *is*, indeed, someone ready to notice these limitations. There is a meta-discipline "able to predict the need to integrate multiple forms of disciplinary knowledge." Critical systems thinking emerged to do almost precisely what Feenberg calls for. It provides a dialectical rationality that begins with the recognition of positionality and partiality that feminist STS also calls for. In terms of Feenberg's perspective, with CST, one-dimensional rationality gets embedded in a multidimensional reasoning frame. Ulrich's reference system, and the way Midgley developed his ideas about first- and second-order boundary critique, essentially conceptualize in systemic terms *how knowledge claims are situated* and how they are mobilized in design. Ulrich's heuristics allow the demonstration that a given scientific or engineering method cannot justify its own assumptions, and that the legitimacy of experts does not rest on their knowledge alone. Only those affected can justify the normative implications of design.

The critical idea and the systemic idea need each other (see also Midgley 1996, 18; Flood 1990, 177). A perspective that is critical but not systemic will be ineffective or mistaken: either the boundaries of the argument remain vague and implicit (which severely limits the degree to which the inquiry can be critical), or they are subject to continual expansion. At the same time, perspectives that are systemic but not critical will be harmful and unethical. The critical turn in systems thinking illustrates how to extricate ourselves from instrumental thinking and create a critical technical practice. In the following, with no claims for comprehensiveness, I trace some commonalities and differences between the positions of CST and STS on issues that are central to the myths of systems design, before suggesting preliminary conclusions.

Today, feminist data scientists and their allies find themselves in a dilemma similar to the critical systems thinkers: They don't want to abandon data science, because it can be useful; and while categories are problematic and loaded, data must be categorized to be usable data. They need to find ways to work with categorization while constantly challenging it, just like the critical systems thinkers needed to find ways to think systemically while grappling with the inevitable selectivity of their claims. Just as data feminism is not ready to abandon the nutritious grounds of data science with all its powerful tools, CST decided not to abandon the powerful tools of systems thinking but relocate them onto a ground on which they can be put to ethical use. The tension between critical and generative perspectives identified by Bardzell (2010) in the context of feminist HCI can be resolved by critically systemic thought, because it allows a critical approach to become generative, and a generative approach to become critical.

Sandra Harding critiques science in ways strikingly parallel to CST, rejects value-neutrality just as CST did, and addresses the importance of purpose (1986, 46). Ackoff's distinction between the purposive and the purposeful adds an additional lever to the argument. Critical systems heuristics, with its focus on selectivity, may be able to complement rationalistic computational reasoning very effectively, helping those involved in computing practice to pursue Harding's call to "critically and systematically . . . interrogate their advantaged social situation and the effect of such advantages on their beliefs" (Harding 1992). The nature of CSH helps it in building the necessary rapport that a critical friend needs to be heard by those schooled in rationalist thought (see chapter 10).

The design justice framework as discussed by Costanza-Chock (2020) starts from a Venn diagram with striking similarities to Ulrich's. It splits those affected into *those who benefit* and *those who are harmed*. There are often overlaps—for example, most users of social media are arguably in both groups—but it leads to similar consequences. In a move with parallels to Midgley's approach to systemic intervention, design justice emphasizes the importance of shifting participation in design to empower those affected and ideally involve all those who are affected. Ulrich and Gardiner remind us that in sustainability and justice, this ideal will always remain unfulfilled. CSH therefore raises the complementary question: How can those who are involved but not affected minimize and reflect on the inevitable omission of some who are affected, and how can they develop an approach to design that is legitimate even though not everyone affected can participate fully?

This is not to say CST has all the answers. Its focus lies on discursive claims, boundary judgments, methodological questions, and legitimate justification. Intersectional feminist thought in STS offers substantive, empirical, and material attention to the matrix of domination and nuanced studies of gender, race, class, ability, and other dimensions. This attention is not sufficiently developed in CST. "Data feminism insists that the most complete knowledge comes from synthesizing multiple perspectives, with priority given to local, Indigenous, and experiential ways of knowing" (D'Ignazio and Klein 2020, 18).

CST has explicitly referenced the work of feminists and STS researchers. So, is CST feminist? When Churchman speaks of seeing the world through the eyes of another and emphasizes the partial nature of all perspectives, he prefigures themes that emerge later in Harding and Haraway. When Midgley writes about boundary critique, he addresses the positionality of situated knowledges. And Ackoff writes about the role of values in objectivity in ways similar to Harding:

Objectivity is not the absence of value judgments in purposeful behavior . . . because such behavior cannot be value free. Rather, objectivity is the social product of an open interaction of a wide variety of individual subjective judgments.

There is no concept as value-loaded as objectivity, and no activity more value-full than science. Objectivity is obtained only when all possible values have been taken into account, not when none have been. (Ackoff 1977)

In some sense, CST is perhaps feminist in process and sometimes in form but rarely in content. The attention of CST writers is still guided by categories that are gendered and racialized and that are not always appreciated or recognized as such. CST has also not addressed decolonial thought at any length, even though the "discordant pluralism" (Gregory 1996a) of CST is ready for the pluriverse that Escobar (2018) argues for—ready for "a world in which many worlds fit." And CST and feminist STS certainly agree that the myths of systems design are just that, myths.

These complementary arguments suggest that there is a fertile ground to explore in their interactions. Feminist STS has been vibrant and active over the past decades. Its attention to structural power relations and inequity provides rich accounts of the larger social forces shaping technology that add much to CST. On the other hand, CST offers a way to systemically interrogate how knowledge claims are situated and mobilized in design, and how different types of knowledges are elevated into a "sacred" or demoted to a "profane" status (Midgley 1992). As a result, critically systemic approaches to participation in design offer rigorous conceptual frameworks that could afford additional depth to design justice. Second, there are more subtle epistemological, historical, strategic, practical, and pedagogical lessons to be learned from the critical turn in systems thinking and its direct confrontation of the mode of reasoning that underpins computing.

CONCLUSIONS: AT LAST, A CRITICAL TURN IN COMPUTING?

From design justice and data feminism to algorithmic justice, refusal and resistance, there is no doubt that critically oriented computing research is surging.[14] In their practice, these scholars, activists, and tech workers draw heavily on the conversations discussed above. Large-scale algorithmic harms have placed these concerns finally at a more visible space on the table of public attention. "It Is Time for More Critical CS Education" as well, as Amy Ko and her students write (Ko et al. 2020). Their argument speaks to the need to reorient the perspectives in computing, and computer science education is a central vector for this necessary social change. I concur and agree that education is a central vehicle by which these social changes should be enacted. But as a leverage point for social change, education is foundational, slow, backloaded, and resistant. While change is being introduced, prior commitments and rationalistic worldviews continue to percolate through research and practice. A critical turn in computing is important for education, but it has already begun in tech workers' collective organizing practices, in public protest about harmful technologies such as racist algorithms, in researchers' turn to intersectional approaches to algorithmic justice and fairness, and in research on sustainability and justice. More and more, these approaches do not stop at critiquing from outside, but engage in constructive generative work from inside. That is the critical turn that is already happening. It is time to recognize and amplify it. Systems design for the twenty-first century finally needs to overcome the limitations of the rationalistic worldview that computer science has grown up in. Computing's critical friends are ready to help with the dialogue of restructuring. Their arguments suggest concrete ways to rethink and reshape the narrative of systems design in computing. The following chapters will explore how.

6

SOFTWARE IS NEVER NEUTRAL
HOW DO VALUES BECOME FACTS?

Designing computer artifacts is an inherently value-based activity, deeply implicated in longstanding political struggles of the wider society in which computer
science is embedded. Rather than viewing this fact as a breakdown in what should
be a disinterested project, this alternative position embraces the place of systems
development as a critical arena for the expression and enhancement of values.
—Suchman (1998)

Through design, the values of those involved shape the artifacts of the
future. But how exactly does this take place? Systems design and engineering have developed ways to become more sensitive to the role values play in design and to make productive use of our deliberative and
reflective capacity to elicit, negotiate, and critique values. This work has
drawn on a wide range of influences, including moral philosophy, ethics,
and psychology. It may seem absurd to believe that technology could be
value-neutral, but the myth of value-neutral technology remains strong
even today (Ko et al. 2020; Shilton 2018). To ensure that the political
nature of systems design is broadly recognized and addressed, we must
articulate the role that values play in it.

There is an important distinction to be made between values and value
conflicts on the one hand and ethical imperatives on the other hand.
Broadly speaking, ethics focus on what is the right thing to do, while
values point to what we consider important. In the current computing

discourse, the term ethics carries a strong normative framing in prescribing how to decide what to do, exemplified in the ethics codes that dominate the conversation. On that basis, "ethics is framed as a problem that can eventually be 'solved'" (Metcalf, Moss, and boyd 2019, 8). The myth of value-neutral technology (VNT) supports the primacy of ethics codes as prescriptive rules to be followed in design and engineering because it suggests that following these rules is sufficient to be ethical.[1] In contrast, the term "values" points to a more fluid, descriptive accounting of fundamental beliefs and preferences that direct and orient individual and organizational work. This sensitivity is often absent from ethics-focused work. The prevalence of VNT may be related to the focus within engineering disciplines on normative views of ethics (Spiekermann 2016, 2) and practice (see chapter 7). Moving beyond it reveals a perspective of sensitive, reflective, and critical opportunity.

This chapter briefly traces some of the origins of VNT to examine where it comes from. It then conceptually examines *how values become facts* in systems design, identifying a range of disciplines that offer fragments of a puzzle which we can usefully piece together. This is followed by a brief discussion of what is left to do for values to ethically shape just sustainability design.

ETHICS, VALUES, FACTS, AND THE MYTH OF TECHNOLOGICAL NEUTRALITY

Evacuating values from technology design in favor of prescriptive ethics principles and rules has a long history, rooted in the question of value judgments in social, purpose-oriented decisions. Ulrich (1983) locates important steps toward what we now call VNT in the work of Hobbes, Weber, and Popper. A milestone is Weber's distinction between *means* and *ends*—supposedly, the latter result from value judgments (where do we want to go?) while the former can be picked purely on questions of facts (what is the best way to get there?).[2]

> **facts**, n. pl.: claims made about the environment of technology design for which those who make them forgot to question where they came from, how they came about, which values they embody, whose values these are, whose facts they are, and whose interest that serves.
> **assumptions**, n. pl: facts about the project at the time of kickoff.

Sound familiar? The false dichotomy between facts and values is the basis of the prevalent logic suggesting that engineers and technologists should simply figure out *how* to do things (means) while "someone else" should decide *what* to build (ends) (cf Shilton 2018, 151f). It is the logic used by some AI advocates to claim that facial recognition technology, autonomous weapon systems, or automated hiring systems, are just a neutral value-free means without politics. "It depends on how they are used," they say, without regard for the values already expressed in and reinforced by the technologies they advocate for. Their reasoning is dubious. In his defense of VNT, Pitt provides a seemingly logical dissection of Langdon Winner's (1980) classic example of the New York underpasses. Winner suggests that their low height, which prevented the public buses used by socioeconomically segregated Black communities from accessing Long Island Beach, manifests political values. Pitt claims that he is simply unable to locate any values in technical artifacts.

Where would we see them? Let us say we have a schematic of an overpass in front of us. Please point to the place where we see the value. If you point to . . . a number signifying a distance from the highway to the bottom of the underpass [and] tell me that is Robert Moses' value, I will be most confused. There are lots of numbers in those blue prints. Are they all Moses' values or intentions? (Pitt 2014, 95)

The claim that the drawings merely represent technical facts and therefore *not* values is ultimately based on the same logic that positions the prescriptive rule systems of ethics codes as necessary and sufficient guarantees of ethical conduct. But the basis of this logic does not hold up to scrutiny.

Here is the snag. Decisions on means are never questions of fact only, for all means are in need of critical examination with regard to the value implications they themselves contain . . . means and ends are not substantially distinct categories but rather different perspectives for considering hierarchies of goals: what appears as a means from "above" (from the next-higher system level in the goal hierarchy) appears as an end from "below". . . . Once this is clearly understood, it seems almost unbelievable how uncritically a majority of contemporary social scientists, led by the logical empiricists and critical rationalists, have adopted the dogma that means and ends are substantially distinct categories, so that only "ends" are supposed to involve value judgments while "means" are understood as value-neutral. (Ulrich 1983, 72)

So, while any technological system is of course a means to some ends, different perspectives will reveal the same system to be a different means to different ends, and more importantly, every system also carries value implications beyond those expressed in its purpose. In this light, Pitt's argument looks rather disingenuous. No one claimed that a distance value in an engineering drawing is an exact representation of a human value and expressed as such. Drawings that represent technical facts still *express* values by virtue of their relationship to the world they are in. It is precisely in these models, placed and interpreted within their social context that gives them meaning, that these values find their expression. If we are willing to decipher them, we might find it easier than expected. Let us add an entry to the *Devil's Dictionary*:

> **models**, n. pl. (computing): the carpets under which, if we look carefully, we can find the human values, politics and moral decisions that have become code, features, qualities, documentation and other technological facts through the social practice we call *systems design*.

WHICH VALUES BECOME FACTS IN SYSTEMS DESIGN

I propose a brief exercise. Open Google Maps, navigate to your favorite city, zoom to your favorite large park, and have a close look at what you see. Make a screenshot. Now look at the same place via OpenStreetMap. What do you see?

Maps are a quintessential example of scientific objectivity. We tend to think of the mapping techniques developed over millennia as a pinnacle of the objective representation of our world through technologically constructed artifacts. Scholars of cartography have of course long shown that mapping has never been a neutral technology. Maps are often drawn by victors, colonizers, settlers, missionaries, and oil companies. Maps have been drawn to mark territories, identify oil drilling sites, and support state surveillance and genocide. On the other hand, mapping technologies are also used by civil rights advocates and land defenders across the globe to fight injustices.[3] Even before adding objects, the projection of our globe onto a two-dimensional surface involves value-loaded choices about which distortions to introduce. Still, in our daily life, when we think of the content of maps, we typically evaluate it based on how "accurate" and "complete" we believe the content to be. In discussions of objectivity in science philosophy, values such as accuracy, correctness, and completeness are often referred to as *epistemic* values and admitted into the scientific process, while *contextual* values, the personal preferences of the scientist, are to be eliminated from the conduct of science and relegated to the choice of research question. This builds on Weber's distinction between means and ends (Reiss and Sprenger 2017; Diekmann and Peterson 2013).

On my recent vacation to the Canadian East Coast, I took a walk around the neighborhood of a friend's place in Wolfville, Nova Scotia, to bring down a high blood sugar. For driving directions, I use Google Maps, but for exploring places on foot, I turn to OpenStreetMap. There's a reason for that, illustrated in the screenshots of the same place, at the same time, rendered through these platforms (figure 6.1).

The two maps are both accurate representations showing important facts of Wolfville. But which values do they show? Google Maps takes a commercially oriented view. It shows us all the places where we can spend money. OpenStreetMap, the Wikipedia of maps, shows little of that in many regions, but look at the detail by which it represents public spaces. Every university building is named, every park bench is marked, and in your favorite park, I bet every drinking water fountain is marked too. For two platforms focused on mapping the features of our world, they show striking differences. The crucial difference stems from how they are disposed to different aspects of the world that we value. These maps manifest the organizations' and editors' diverging interests in different features of the mapped world and different purposes they choose to pursue.

Feenberg's concept of *formal bias* is useful here. It stands in contrast to subjective or organizational bias such as the commonly criticized kind that refers to a human bias based on content or substance such as gender, race, or ability. In contrast, formal bias is embedded in a rational form such as a technological platform or paradigm. For example, an automated hiring system trained on historical data may reinforce and amplify prior human bias manifested in the data, data generated by individuals who exhibited discriminatory patterns of hiring decisions that favored White men over Black women. The resulting algorithmic bias has become an important concern. "This sort of bias is properly called 'formal' because it does not violate formal norms such as control and efficiency under which technology is developed and employed" (Feenberg 2014, 166).

6.1 One place on two highly accurate maps (Google Maps, OpenStreetMap/maps.me).

But there are additional nuances of categories that we should distinguish. First, it matters not just where but how bias manifests. Formal bias can be easy to detect when it literally surfaces in the content or *substance* of technological platforms: OpenStreetMap shows different objects than Google Maps. When it resides in the invisible or intangible structures of affordances, features, qualities, and constraints, however, it is much harder to detect. For example, the OpenStreetMap API allows users to edit the

Table 6.1 Selected values prioritized by Google Maps and OpenStreetMap

	Explicitly prioritized	Implicitly prioritized
Values in Google Maps	social content	Commercial value Centralized control
Values in OpenStreetMap	distributed, peer-produced content	Objects in public space Shared editorship

map content, and many apps accessing the platform support this feature. Google provides much narrower paths for users to provide *consumer feedback* via reviews of places represented on the map. Via these affordances, each platform configures its users' roles. In similar ways, gig economy platforms such as delivery apps limit and constrain what each participant can do and see; automated hiring systems, beyond the evaluation and ranking, also structure the process of applications and the collection of data about jobs and applicants; and robotic automation technology structures the division of labor around it. When Amazon automates warehouse tech so that a machine controls and monitors how humans pick up objects, that tech embodies the removal of agency from the worker and subordinates them to the machinery (Crawford 2021, 48–74). In contrast, when a warehouse equips its workers with better machinery to help them pick up objects they could not personally carry for physical reasons, while leaving it to the workers to coordinate their activities, it embodies the amplification of physical strength and personal autonomy. Projects to design either type of system may have efficiency improvements as primary purpose, but other values also shape and constrain the systems' dispositions. Because these examples stay within the formal norms of conceptual frameworks such as "usability" and "features," some will consider this kind of system as "neutral" even when it marginalizes one group's values in favor of another group's values.

Second, it is fair to assume that formal bias is compliant with the local norms of a given context, but it can very well violate formal norms defined elsewhere. For example, substantive formal bias is addressed in norms of fairness and non-discrimination (to some limited degree), but these are not always implemented and enforced. In fact, organized advocacy for a change of norms is often directed toward making possible the local contestation of formal bias of substance or affordance under these new regulations.

In discussing values, the relevance of some is more readily accepted and barely doubted because they are culturally embedded into engineering norms. Values such as efficiency, effectiveness, and reliability are treated in engineering the way epistemic values are treated in science. Others, such as worker agency, may be much more contested. But in fact, there is no fundamental difference between these values and their influence on the design process. As Feenberg writes, "when the division of labor is technologically structured in such a manner as to doom subordinates to mechanical and repetitive tasks with no role in managing the larger framework of their work, their subordination is technologically embedded. Inequality is enforced by the very rationality of the machine" (Feenberg 2014, 166). The dismissal of "worker agency" as a "loaded value" is not in any way a neutral act. And none of the systems listed here are neutral.

HOW VALUES BECOME FACTS IN SYSTEMS DESIGN

Different systems, even if focused on the same overall purposes (such as mapping the world), will embody different values simply because they are the manifestations of different interests. These values manifest in subtle and less subtle ways through the range of features and qualities of systems. Can we describe more precisely how this happens?

Prior work on values in systems design provides us with useful, if partial, responses to this question. Significantly, it has mapped out how values as discriminating criteria in deliberative processes can be intentionally elicited and used to drive design activities to shape models and artifacts. Value-sensitive design (VSD) provides theory, tools, and methods that support designers in facilitating intentional conversations about specific values when they design artifacts. It is meant to sensitize those involved in design toward the role of values. It focuses on how values and the "moral imagination" can shape design outcomes (Friedman and Hendry 2019). It offers sensibilities to longer-term implications too via supporting justice-oriented systems design across generations, exploring "how the element of time might be leveraged in design processes with deep-seated value tensions" (Friedman, Nathan, and Yoo 2017, 83).

To do so, VSD proposes a trio of interacting perspectives: conceptual explorations, empirical investigations, and technical analyses are meant to proceed iteratively and in parallel. The relational concept of *values suitabilities* connects an artifact's specific technical property, such as a feature or an affordance, to a value such as worker agency. Thus, it supports what Brey called the "moral deciphering" of technical artifacts (Brey 2000; 2010). Technical investigations in VSD can be retroactive analyses deciphering existing artifacts or proactive forms of design. VSD researchers have also investigated which factors support the surfacing and centering of values conversations in design (Shilton 2012). VSD envisions these streams to be integrated, although critics have pointed out that some aspects of this integration remain underdeveloped (Manders-Huits 2011).

Value-based engineering (VBE) (Spiekermann 2017; Spiekermann and Winkler 2020; IEEE 2021), on the other hand, treats values in much the same way that requirements engineering treats requirements: as the object of applied scientific activity (Akkermans and Gordijn 2006). This introduces the idea and ideals of scientific objectivity into its methodological foundations. Values in this family of approaches are stakeholder positions that are elicited, documented, weighted, measured, and when in conflict, negotiated. In doing so, this work introduces the concept of *value disposition*: a system quality positively or negatively disposed toward an identified human value, much like an affordance in design is disposed toward a potential human action and parallel to a value suitability in VSD.[4]

There is an important difference between this objective, supposedly neutral stance of VBE and the formative, orientating stance of VSD. VSD's attention centers the designers as subjects, while VBE's focus lies in removing subjectivity by prescribing processes to treat values as objects of applied scientific measurement and construction. In VSD, it is a designer subject(ivity) who is *sensitive*, while VBE centers the supposedly objective value "base" of engineering. In the terms of critical systems heuristics, the guarantor for VSD is a designer; for VBE, it is method. In the spectrum between these two, we find approaches that use scientific measurement frameworks for values (Schwartz 1992; 1994) to drive software engineering and design processes (Whittle et al. 2019; Ferrario et al. 2016) or use *values* as a central concept in requirements engineering (Thew and Sutcliffe 2017).

But this explicit articulation of how values are intentionally expressed by and impressed onto artifacts is only half of the story. Values (including value tensions) also *implicitly* shape the discourse and, directly or indirectly, models and artifacts—not only by being referenced as discriminating criteria in deliberative processes, but also by shaping the narrative and themes of conversation, thus *orienting* those who design. Values shape preferences and thus choices, including choices *about* which values to articulate. For example, a group of participants in a design project may use envisioning cards, a VSD tool (Friedman and Hendry 2012), as part of their value-sensitive design process. Within this process, however, values such as group conformity will influence which cards participants select, which values are given prominent space, and how the conversation is structured. Values will also shape participation itself and the articulation of evaluation criteria for design process and product. While VSD is better prepared to be sensitive to the nature of this shaping, neither VSD nor VBE incorporate a critical examination of *implicit value positions* and their influence in design. As Shilton writes,

> it is important to acknowledge the limits of a values-oriented approach to design. Attention to values and ethics within design can help to ensure that new technologies do not perpetuate or aggravate existing bias, or create new unfairness. But values-oriented approaches do not address larger power structures that perpetuate bias and unfairness in technology hiring practices, design education, or technology regulation. Values-oriented design methods are just one part of a larger culture of ethics that must become part of technology education, scholarship, and practice. (Shilton 2018, 150; see also Manders-Huits 2011, 282)

There are, however, approaches that can help us to examine the values and power structures beneath the surface. Computing's critical friends have at least three avenues to offer. First, through a range of approaches to discourse analysis, they have examined which values underpin computing *research fields* such as machine learning (Birhane et al. 2021), computer vision (Scheuerman, Denton, and Hanna 2021), and ICT4S (Knowles 2013). Second, the *mindscripting* method (Allhutter 2012) allows teams to deconstruct the stories they tell about their work in order to identify and better understand the identities, value positions, tensions, and power dynamics inherent in the situation they design in (Allhutter and Hofmann 2012). This is not a straightforward toolkit, but it has been used successfully to explore values and value tensions in computing spaces as distinct as machine learning (Allhutter and Berendt 2020) and requirements engineering (see chapter 10). Third, critical systems thinking (CST) methods support "closely examining the assumptions and values entering into actually existing systems designs or any proposals for a systems design" (M. Jackson 1991). This is precisely what is missing in the puzzle.

Drawing on discourse ethics, Ulrich's work considers the relationships between discursive claims about facts, values, and the scope of boundary judgments as illustrated in figure 6.2. In making boundary judgments as discussed in chapter 5, we observe and evaluate facts based on the reference system that scopes our assessment of what constitutes the relevant field of knowledge. Values shape the selection of facts, and vice versa. Recognizing new facts as relevant implies a reevaluation and rescoping of the system of interest. Similarly, shifting the scope of that system changes what we perceive as relevant facts and values (Ulrich 1998, 6; cf. JafariNaimi, Nathan, and Hargraves 2015, 97). Critical systems heuristics (CSH) can be used to make the values shaping the scope of the system visible; to allow those involved in design to reflect on their beliefs about purpose and improvement; and to create a space where those affected can (to some degree) question the values of those involved.

Similar relationships between facts, values, and boundary judgments hold in modeling (Diekmann and Peterson 2013). Contextual values, not just epistemic values, influence which properties are represented in models. In terms of figure 6.2, they influence boundary judgments and thereby facts; they influence the parameters used in models; they influence the choice of competing models; and sometimes, they are in outright conflict with epistemic values. For example, the *simplicity* of a design solution may conflict with a disposition to a substantive value such as *fairness*. In

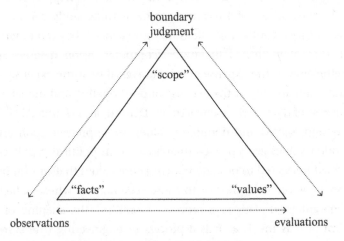

6.2 Facts, values, and scope (adapted from Ulrich 1998, 6).

the values discussion mentioned in the Introduction to this book, these kinds of value influences could have been at the forefront. Instead, the attention of the audience centered exclusively on the visible *object* of values and applied a modeling lens *to* this object: we modeled values. The result: a collection of facts about values. Which values implicitly guided this discussion? How did they shape and constrain the selection of facts the group chose to represent in its model of values? Today, I would be better equipped to guide the group in that direction.

Ironically, this anecdote illustrates an important observation from CST: "the boundaries of accepted knowledge define the values that can emerge. Similarly, the values adopted will direct the drawing of boundaries that define the knowledge accepted as pertinent" (Midgley 1992, 9). The boundaries of knowledge that implicitly drove the conversation were the disciplinary boundaries of model engineering and software engineering. The crucial questions, however, fall squarely within the domain of computing's critical friends. The power dynamics within the group—which individuals were central, which were considered leaders in the field ostensibly being discussed, how the question may transcend that field, who was comfortable speaking up, and so on—interacted with epistemic norms and values of what constitutes a good model. Again, the triangle of observations, evaluations, and boundary judgments provides a lens into the mechanisms that elevate some stakeholders' knowledge claims and forces others' claims to the margins (Midgley 1992). These dynamics do not happen at random. Similarly, the authors of a recent machine learning study contrast dominant values with conflicting values that get silenced, based on meticulous deciphering of published literature: "computer vision datasets authors value efficiency at the expense of care; universality at the expense of contextuality; impartiality at the expense of positionality; and model work at the expense of data work" (Scheuerman, Denton, and Hanna 2021, 1).

In design, sacred and dominant values get impressed upon artifacts, while other values get repressed (marginalized in political pushback) and suppressed (relegated to profane status). Those values that in turn become impressed are often oppressive to those stakeholders whose values have been marginalized. This is a process of politics—the shaping of social organization—as much as it is a process of engineering (the systematic application of scientific procedures)—both at the same time. And it is why systems design is inherently a value-loaded activity. By deciphering which

Table 6.2 Examples for techniques to surface explicit and implicit values

	Proactively, during design	Retroactively, given an artifact
Explicit values	Value-sensitive design techniques such as *envisioning cards* help to orient design teams to values. *Value-based engineering* processes such as IEEE P7000 introduce some values concepts into the engineering vocabulary and support systematic mappings. *Value-based requirements engineering* similarly centers values, motivations and emotions in RE.	Technical investigations in VSD can be retroactively conducted to decipher the value suitabilities of existing artifacts (Friedman and Hendry 2019, 89). VBE techniques and models could in principle be applied retroactively.
Implicit values	*Critical systems heuristics* can be deployed in ideal mode to orient a team toward reflecting on implicit value positions (Ulrich and Reynolds 2010): see chapter 10. Comparison between ideal ("ought") and actual ("is") brings value conflicts to light. *Mindscripting* supports teams in surfacing values and negotiating value tensions (Allhutter and Hofmann 2012).	*Critical systems heuristics* deployed in reflective mode can unearth further value conflicts, and in polemic mode, it can make value suppression visible (e.g., Ulrich 1981; McCord and Becker 2019). *Deconstruction* more generally can bring to light the value commitments of engineering methods and artifacts (Allhutter 2012, Allhutter and Berendt 2020), and the contrast between espoused and manifested values (Beath and Orlikowski 1994).

values are expressed by the artifacts, we can make some of this visible in hindsight. But to proactively open the conversation, we must combine the complementary views brought on by computing's critical friends *during* systems design, as far as that is politically feasible. Table 6.2 summarizes some techniques that can be employed to help.

BEYOND NEUTRALITY: THE POLITICS OF SYSTEMS DESIGN FOR JUST SUSTAINABILITY

The asymmetric vulnerability and distance that characterize just sustainability design sharpen the need to become critically aware of the processes of marginalization by which values turn sacred or profane, and to incorporate perspectives on *distant values* proactively into systems design. On the explicit/proactive quadrant of table 6.2, envisioning cards provide a reasonable starting point, but, on their own, they are insufficient because asymmetric vulnerability creates a power differential that VSD does not know to address.

A strictly normative uncritical approach that prescribes rules for action cannot address uneven power relations nor close the gap between those who design and those who are affected. It will therefore do nothing to change *whose values* dominate and shape systems design. This means that we need to combine multiple methodologies in an approach that is both constructive and critical, both critical and systemic. When designers shape artifacts with a moral imagination sensitized to sustainability and justice, they must also critically interrogate their assumptions, value positions, and power. When engineering and design perspectives translate stakeholder perspectives into value dispositions and technical specifications, we must similarly combine them with a critical examination of the normative reference systems that underpin these translations. We need the critical friendships between engineering disciplines and computing's critical friends to figure out how to make this work.

In systems design practice, all this calls for a social change embodying a shift of priorities toward values and toward different values than those currently dominating computing (Barendregt et al. 2021). A systems design practice reoriented to just sustainability design will strive to incorporate value sensitivity considering all quadrants of table 6.2.

This shift is decidedly a political project that embraces "systems development as a critical arena for the expression and enhancement of values," as Suchman calls for in the chapter's epigraph. In terms of research, we need to examine how methods from the four quadrants can be combined. The explicit shaping of artifacts with a moral imagination sensitized to sustainability and justice must be complemented with a critical interrogation of assumptions, value positions, and power. In terms of education, we need to demonstrate more clearly than existing work "how values become facts" by building concrete and convincing cases that comprehensively document to engineering audiences the tangible intermediate steps that link explicit and implicit value positions to value dispositions and their material effects.

7

PEOPLE ARE MORE THAN RATIONAL
BEWARE THE NORMATIVE FALLACY

It's strange how often the critics of artificial intelligence object to the wrong thing . . . they are horrified at the suggestion that computers can think, whereas they should be horrified at the suggestion that people are information processors.
—Churchman (1979a, 124)

The design and development of a computational system is full of trade-offs. Software engineers, architects, programmers, testers, user interface designers, project managers, and many others must work in concert with all sorts of stakeholders to navigate design options that shape the system they are making. Because of this, disciplines such as software engineering treat decision-making as a central part of their methodological focus. For example, software architecture is a decision-centric discipline (van Vliet and Tang 2016). Eoin Woods, a leading practitioner and writer, describes it as "the set of decisions which, if made incorrectly, will cause your project to be cancelled" (Bass 2013, 25). Decision-making is similarly central to requirements engineering (Aurum and Wohlin 2003).

How do people make all these decisions? Without understanding what happens in engineering and design practice, we can hardly hope to improve it.[1] Put simply, engineering and design disciplines *prescribe* what people should do and why—they develop normative frameworks such as methods that define what should be done and how. In contrast, studies of behavior *describe* and explain what people do in practice. In behavioral studies *of* engineering and design, these two modes overlap, because they strive to describe and explain what happens in practice in order to prescribe what people could do better, and how, for some standard of evaluation. For example, many researchers in software engineering design new artifacts such as methods and tools and deploy them into industrial contexts, then study how they impact performance. What type of knowledge should provide the foundation of such studies?

Behavioral researchers collect data about professional practice. When they organize this empirical part of their study, they often rely on the toolbox of theories they used to design the methods. But the theories that *describe* what people do are different from the engineering methods that *prescribe* what they should do. When it comes to how people make decisions, these theories in fact carry mutually incompatible assumptions.[2] As a result, the tension between description and prescription can lead behavioral researchers to misunderstand practice in subtle but important ways. When researchers misappropriate normative and prescriptive theories that lack descriptive validity for descriptive purposes, they commit a *normative fallacy*. Their findings may appear persuasive, but they are invalid and they will mislead us.

This chapter retraces the tradition of decision-making research reflected in the myth of rational decisions to outline what it misses and show how it has misled the computing field. The historical view will help illustrate how the myth of rational decision-making lives on in systems design research and practice. I focus on empirical behavioral research of engineering practice to make my case. The argument draws parallels and lessons from developments in other disciplines shaped, just like computing, by the domineering influence of the rationalistic tradition. By exploring how these fields extricated themselves from a singular focus on rationalistic theories, we will see why computing needs to follow suit and how to reorient it.

RATIONAL CHOICES, REASONABLE DECISIONS, WISE JUDGMENTS

The rationalistic tradition dominates the attention to decision-making in computing. When and if disciplines such as software engineering define decision-making, they do so within this tradition, as in this example mentioned in chapter 3: "to make a decision, a situation is assessed against a set of characteristics or attributes, also called criteria" (Filho, Pinheiro, and Albuquerque 2016).

This definition of decision-making as a selection from a predefined enumerated set of options is common outside of computing too: The APA defines decision-making as "the cognitive process of choosing between two or more alternatives" (APA 2020a), which makes it indistinguishable from choice: "an act of selecting or making a decision when faced with two or more possibilities" (*Oxford English Dictionary* 2020a). But as we will see, people often arrive at a decision without performing a choice. The interdisciplinary area of judgment and decision-making (JDM) explores how humans make complex decisions and judgments (Keren and Wu 2015). Its perspectives range from psychology and social psychology to behavioral economics, sociology, and neuroscience. While the terms choice and decision are sometimes used interchangeably, their meaning differs significantly:

1. A *decision* arises in a situation in which someone could conceivably make different commitments on how to proceed. It is a "conclusion or resolution reached after consideration" (*Oxford English Dictionary* 2020b). In naturalistic decision-making, making a decision is in fact defined as "committing oneself to a certain course of action" (Lipshitz et al. 2001).
2. A *choice* is a specific type of decision where enumerated options exist from which a selection has to be made.
3. *Judgment*, on the other hand, always carries a broader awareness and attention to sense-making, evaluation, and the formation of a subject's position toward an object of careful consideration, as in "the process of forming an opinion or evaluation by discerning and comparing" (Merriam-Webster 2020b), the "ability to make considered decisions or come to sensible conclusions" (*Oxford English Dictionary* 2020c) or "the capacity to recognize relationships, draw conclusions from evidence, and make critical evaluations of events and people" (APA 2020b). When you face a choice between two options but reflect on the boundaries of the presented decision, reject the framing, and instead pursue a third option, you exercise judgement. You make a different decision (commitment) as a result of your judgment. That is a complex human capacity that invokes reflection and situational awareness, qualities that are markedly absent in the machinery some describe today as "artificial intelligence" (cf. Crawford 2021; B. Smith 2019).

Imagine you are responsible for a software project. You are behind schedule by two weeks. A colleague suggests that instead of developing one of the main functional components, as initially planned, your team could spend this time integrating an open-source library. It would require some customization and there would still be some need for coding, but your colleague thinks it could save up to a month of effort. There is a chance of failure too, of course. What do you do?

In the rationalist tradition, you specify clear and unambiguous assumptions such as the time horizon and budget, enlist decision criteria—either in a list or in more elaborate structures such as utility trees—and then evaluate each alternative on each criterion to compare them and select the best. Cost-benefit analysis is a form of such multicriteria decision analysis (MCDA), architectural tradeoff decision support methods are another. MCDA requires clear assumptions and, at each point in time, treats two of the corners of the facts/value/scope triangle as fixed. It handles uncertainty via probabilities and the incommensurability of various scales by computing a unified value function, often expressed as utility.

Note that the situation has *become a decision* point because your colleague identified that your team could conceivably commit to a different action than initially planned. Otherwise, you may have simply proceeded. By enumerating two options, it has narrowed into a *choice*. But in the discussion that follows, your team may elect to redefine the framing of the decision. They may decide to widen the scope and inquire into other libraries, they may recognize that this library can also address another aspect of the system's functions, and they may choose to challenge the initial framing. For example, one could argue that the time horizon of this project is not a good scope for the decision since the effects, both positive and negative, play out over the entire lifetime of the system. Others may argue that integrating an open-source library has learning value and is fun. And yet others may argue that engaging and perhaps contributing back to the open-source community has altruistic and strategic value too. In other words, they will exercise *judgment*. In doing so, they will iteratively reflect on and discursively reposition claims about facts, values, and scope (see figure 6.2). The team's reasoning will often not fit easily into rigid MCDA frameworks. This is simply because the human capacity for judgment transcends the operations supported by MCDA. If we define decision-making via *only* the operations supplied by MCDA, we lose sight of the human ability to reflect, to judge the situation, and to generate different ways to act.

DECISION-MAKING IN THE RATIONALIST TRADITION

Winograd and Flores already argued that computing needs to "replace the rationalist[] orientation if we want to understand human thought, language, and action" (1986, 26).[3] Their focus was not on specific decisions *in* systems design, but on a general design orientation and paradigm. Despite their argument's influence in human–computer interaction (HCI), the rationalist tradition persists when it comes to questions of decision-making. Why is that the case? Let us take a closer look at the theory's appeal and origin.

According to this tradition, "intelligence is the work of symbol systems," and the brain is simply another example case of an intelligent symbol system (Simon 1996, 23). Judgment does not sit easily within this idea, but decision-making does. Keenly aware of the tension between normative and empirical accounts of rationality, Simon recognized that the rational choice model of *homo economicus*, which assumed perfect information about choices, criteria, and the environment, needed "drastic revision" to make it compatible with "the computational capacities that are actually possessed by organisms, including man, in the kinds of environments in which such organisms exist" (Simon 1955, 99). The main change Simon proposed was "taking into account the *simplifications* the choosing organism may deliberately introduce into its model of the situation in order to bring the model within the range of its *computing* capacity" (emphasis added). In other words, the focus is on human thought as a *subset* of computational processing, limited by assumed constraints. The underlying metaphor of the mind as computer is never questioned. Instead, based on admittedly "casual empiricism" (Simon 1955, 104), the rational model is modified to fit the idea of constrained computing capacity:

Because of the psychological limits of the organism (particularly with respect to computational and predictive ability), actual human rationality-striving can at best be an extremely crude and *simplified approximation* to the kind of global rationality that is implied, for example, by game-theoretical models. (Simon 1955, 101, emphasis added)

Simon's behavioral model of rationality evolves around a set of behavioral alternatives, some of which are considered by the organism, a set of possible outcomes, a subjective value function for each outcome, a set of consequences per alternative, and a probabilistic model of outcomes per alternative. He and his collaborators used this starting point to examine human problem-solving, organizational behavior, and administrative decision-making. The famous studies that laid empirical claims to Simon's theory of bounded rationality examined human problem-solving for highly constrained well-defined problems such as chess puzzles (Newell and Simon 1972). In contrast to the normative modeling exercises of game theory, these studies were descriptive and explanatory: they involved extensive think-aloud protocol analysis of human subjects. But when we read these protocols and analyses today with the benefit of hindsight, it is striking to see how strongly data collection and analysis itself were predicated on the preformed idea of the human mind as a computer. This reified metaphor made it appear natural that human problem-solving is a search in a defined problem space, performed in the mind by an algorithm that processes information obtained from perception and represented in the brain.

SIMON SAYS, OR: HOW REASON LOST ITS MIND

As a leading behavioral economist wrote, "50 years of dominance of the rational choice paradigm . . . has left most important questions unanswered" (Loewenstein, Rick, and Cohen 2008). Why? What happened during those fifty years? This section explores the legacy and deep influence of the rationalist paradigm, again centering on the work of Herbert Simon and his influences. This will allow us to more deeply understand why we should think of rational decision-making as a myth. I often focus in this book on Simon not because he single-handedly created the myth of rational decision-making (he didn't), but because his work is central to it and unique in its wide-ranging influence on the cognitive sciences, artificial intelligence, computer science, psychology, behavioral economics, design, and political science.[4] Each of these disciplines experienced for decades the torque of the computational model of mind, the framework of rational problem-solving as a spatial search in a bounded space, and the axioms of rationalist decision theory. In each discipline, careful work on the margins proved core assumptions wrong, but struggled to be recognized until the evidence became undeniable and the discomfort with existing paradigms too strong to ignore.

In the cognitive sciences, researchers recognized that cognition cannot be explained merely as information processing within the brain, nor is it reducible to computation—intelligence is *not* symbol processing (Maturana and Varela 1992; Varela, Thompson, and Rosch 1991). Objections came from multiple sides, reflecting the fragmented relationship between the disciplines involved in cognitive sciences, including psychology, biology, neuroscience, and sociology. From a sociological perspective, a well-known ethnographic study by Hutchins (1995), followed by others, argued for a shift in focus from the computational mind to a sociotechnical system involving external configurations such as the controls and displays in a cockpit interacting with the pilots. From a biological perspective, experiments in cognition demonstrated that the theory of cognition as representing external reality in the mind—as in a Von Neumann computer, and as a central tenet of the theory of intelligence as symbol manipulation—is empirically incorrect. The brain does not appear to store and manipulate representations of the outside world in neuronal registers through computational processes. The view of *enactive* cognition emphasizes instead how the structural evolution of the embodied mind and its historical coupling with an environment predispose it to act and interact with this environment, continually bringing forth what we experience as the present moment via a process of *autopoiesis*.[5]

In artificial intelligence (AI) research in computer science, the struggle is ongoing, despite early dissent (Weizenbaum 1976; Winograd and Flores 1986; Dreyfus 1972) and Lucy Suchman's influential refutation of the idea that *plans*, a central concept in AI work at the time, work like programs to be run by individual agents. In her detailed studies, it became clear that plans instead are weak resources used by reasonable humans to act meaningfully in their concrete situations. Her shift from plans to "situated action" resonates with Hutchins's studies (Suchman 2006). As late as 2019, Brian Cantwell Smith (2019) still had to explain that intelligence does not merely involve a procedural logic akin to a computer that operates within the given constraints, but also incorporates *judgment*, including the ability to reflect on the given framing as well as the agent's thought process and to make a conscious choice to transcend it, echoing much earlier calls (Weizenbaum 1976). This ability for judgment relies on an accountability to the real world that sets apart lived intelligence from computational machinery. Smith's reference to the continuous interaction of living beings with their environment mirrors the autopoietic theory arising from the biology of cognition and enactive cognitive science. But even today, the appeal of narrow, computationally understood AI remains strong. What so-called AI can do is not in a meaningful sense intelligent. Showcase examples for AI take place in well-defined domains to which computing is well-suited. These are impressive advances but not toward human reasoning. Tragicomical failures by self-driving cars, text generators, and image classifiers should remind us that these machines lack the human ability for reflective judgment.

In decision-making research, spanning behavioral economics, psychology, and later neuroscience, the rationalist research program was long dominant too, despite active discussions of descriptive, prescriptive, and normative perspectives (Bell, Raiffa, and Tversky 1989). While classical economics assumes that humans are rational, behavioral economics essentially assumed the opposite: they are irrational (Harper, Randall, and Sharrock 2016). From a popular perspective, behavioral economists like Kahneman are often portrayed as revolutionaries, but their work remained tethered to normative theories. "By emphasizing the divergence between actual human reasoning and standards of formal rationality such as logic and Bayesian statistics, they implicitly reinforced the normative authority of the latter" (Erickson et al. 2013, 24). But over time, the emphasis similarly shifted from an isolated individual with flawed computing powers to an appreciation of the macro-cognitive system that extends beyond the mind (G. Klein 1998; Hutchins 1995; Thaler, Sunstein, and Balz 2010). This involved important shifts in method too, which we discuss later.

Design-oriented disciplines such as design studies and later design research in HCI and information systems were also heavily influenced by Simon's work on the "sciences of the artificial," which he positioned as "sciences of design," again built on theories of rational problem-solving through satisficing search in defined problem spaces (Simon 1996; Rosner 2018; Dorst 2006). Here too, the burden of proof came to rest with those recognizing the limitations of such an appealing but narrow view. Here, too, the evidence was often of a qualitative, situated nature that was initially sidelined and ignored.[6] As a result, "much contemporary design research, in its pursuit of academic respectability, remains aligned to Simon's broader project, particularly in its definition of design as 'scientific' problem solving. However, the repression of judgment, intuition, experience, and social interaction in Simon's 'logic of design' has had, and continues to have, profound implications for design research and practice" (Huppatz 2015). There are many facets to this dominance, some of which are addressed in previous chapters (see also Rosner 2018). For example, detailed studies of design activity showed that Simon's core assumption of the problem space as existing prior to its exploration is misleading. Instead, designers co-construct and adapt their understanding of problem space and solution space simultaneously—the spaces co-evolve in mutual interaction (Dorst 1995; Dorst and Cross 2001). More generally, design as value-neutral scientific problem solving engenders a paternalistic attitude towards designers as well, as explored in chapter 4. It excludes both the notion of judgment as well as the idea of reflective practice, which evidently happens in design (Schön 1983). The poverty of this concept did not remain unnoticed even then. But Simon's reaction to such counterpoints as wicked problems in design (Buchanan 1992; Cross 1984) was to ignore the terminology and instead propose that "ill structured" problems were not fundamentally distinct from others (Simon 1973). Huppatz (2015) characterizes what I earlier called the evacuation of politics from design by emphasizing the disembodied nature of Simon's designer: "Freed of situated bodies, Simon's 'science of design' failed to engage with designing as a fundamentally social, political, cultural, and embodied activity." In the broader social context, this approach had lasting appeal: "The logic of optimization promises greater predictability and profit while rigorously stripping judgment, intuition, and experience from systems and service design" (Huppatz 2015).

Finally, in Simon's initial home discipline, political science, his work exerted significant influence too and helped shape what later historians called "Cold War Rationality." In *How Reason Almost Lost Its Mind*, Erickson et al. trace how the dominant vision about reason narrowed into a formalized, technical form of reasoning grounded in the formal abstractions of game theory and operations research, a view that explicitly banishes judgment. They describe the resulting view of rationality as a "distinctive combination of stripped-down formalism, economic calculation, optimization, analogical reasoning from experimental microcosms, and towering ambitions" (4) and write:

What was distinctive about Cold War rationality was the expansion of the domain of rationality at the expense of that of reason, asserting its claims in the loftiest realms of political decision making and scientific method—and sometimes not only in competition with but in downright opposition to reason, reasonableness, and common sense . . . what was Cold War rationality? . . . First of all, this rationality should be formal, and therefore largely independent of personality or context. It frequently took the form of algorithms . . . supposed to provide optimal solutions to given problems, or delineate the most efficient means toward certain given goals. (Erickson et al. 2013, 2–3)

In hindsight, the political forces that shaped this particular narrowing of view within the Cold War context have become much easier to discern, and the streams of ideas much easier to distinguish. Formal, mathematically grounded models of deductive reasoning became not only the normative ideal of how decisions should be made but were also used as the basis for models of how thinking actually works, where they underpinned a long stream of experimental and quasi-experimental research. This research was designed to understand *human* reasoning but anchored on ideas of formal and procedural rationality. Only late in this process did the objections to these flawed assumptions become so strong that they could no longer be ignored.

These accounts of rationality reveal important cross-disciplinary dynamics. There were significant objections within cognitive sciences, psychology, and behavioral economics to the dominant normative narrative and to the false dichotomy that anything not procedurally rational in the narrow sense was supposedly "irrational." But this dissent was hardly recognized and taken up in other fields such as computing or political science. Instead, "if we turn to the applications of the psychology of rationality to policy analysis, hints of dissent within the ranks are very rare. The monolithic definitions of rationality and irrationality inherent to the heuristics-and-biases school of thought still hold sway" (Erickson et al. 2013, 178). Evidently, and unsurprisingly, the same is true for the applications of the psychology of rationality in computing. What exactly is wrong with that?

WHY RATIONALIST DECISION-MAKING IS INADEQUATE

Rationalist models are appealing to researchers because their mathematical formulas promise a rigorous model of human behavior that supports the

collection of data, the detection of deviations, and the design of interventions. But it is important to understand that the axioms of rationalist decision theory were not empirical facts, but normative directives.
None of them had been established by observation. Empirically, the
rationalist theories of decision-making built on these axioms turn out
to be invalid in three important ways: assumptions, predictions, and
methods.[7]

INVALID ASSUMPTIONS

First, the rationalist theory assumes that preferences are fixed priors, that
options are independent from each other, that preference relations are
transitive, and that evaluation is independent of irrelevant alternatives.
These assumptions are empirically invalid: they do not describe how
humans reason (Beach and Lipshitz 1993; Tversky and Kahneman 1986).
The assumptions cannot be easily adjusted to fix the theories because
they are their foundational axioms without which the theories do not
work (Beach and Lipshitz 1993; Shafer 1986). Even more, the underlying
assumption that the computer is a reasonable metaphor for the mind has
never been verified either—on the contrary, research in the biology of cognition suggests that "the popular metaphor of calling the brain an 'information processing device' is not only ambiguous but patently wrong"
(Maturana and Varela 1992, 169). That the brain can *perform* information
processing does not reduce it to an information processor.

INCONSISTENT PREDICTIONS

Second, the rationalist predictions of the choices people will make are
inconsistent with human behavior. Empirical inconsistencies with the
original rational choice theory were known since the 1950s (Beach and
Lipshitz 1993; Shafer 1986). To economists, they were long not important
because their main focus is not the individual apparatus of cognition and
decision-making but the aggregate behavior of the economic system, so it
is to some degree defensible for them to abstract away from human reasoning. But for psychologists, the main focus is human reasoning, so the
emergence of cognitive psychology and the import of rational theory into
the field of psychology meant that the inconsistencies had to be addressed
(Harper, Randall, and Sharrock 2016, 15–43).

Three responses arose, and it remains important today to distinguish them. One dismisses deviations from rationalist theories as faulty and irrational behavior. The consequence is to set out and fix the behavior. "This view saves the theory and rejects the behavior" (Beach and Lipshitz 1993, 22). For obvious reasons, this reaction does not have a strong standing in psychology today—after all, the behavior did not go away, and this position does not explain why. It does, however, remain strong in studies of engineering and design behavior, where it is generally used to double down on formalization and method to bring behavior in line with theory.

Make no mistake, in its normative role, prescribing decisions for hypothetical Economic Man, classical theory is not subject to these criticisms. It is when behavioral scientists assume that these prescriptions apply to any and all human decisions that the mischief is done. Because its prescriptive role is assumed to have the same status as in its normative role, the legitimacy of the theory is not questioned when behavior does not conform to its prescriptions. Instead, it is concluded that the behavior, and thus the decision maker, is wrong or irrational and must be made to conform to the theory. (Beach and Lipshitz 1993, 29)

The second response was to adapt rationalist theory to correspond more closely to observed behavior. This was the path taken by Nobel-prize winner Kahneman and his colleague Amos Tversky. They did not drop rational choice theory but tuned its parameters to create prospect theory (Kahneman and Tversky 1979). In the place of absolute values attributed to enumerated alternatives, they proposed that each decision-maker has a subjective value function dependent on a current state and the losses and gains *relative to* that state. It is important to recognize that they retained all other normative assumptions (Beach and Lipshitz 1993), including the implicit metaphor of the brain as a computer. Instead of abandoning these orthodox assumptions, they merely applied the axioms differently, patching up the traditional model to extend its lifespan (Gigerenzer and Selten 2001, 13–37). On this basis, they developed an influential research program on *heuristics and biases* focused on showing, empirically, how human reasoning deviates from rational standards. Ironically, the normative standard of rationality is used to show that human beings are not rational. Thomas Sturm summarizes the normative aspect of their program succinctly in pointing out that it "claims, on empirical grounds, that human beings often and systematically violate *norms of rationality that derive from* formal logic, probability and decision theory" (2012, 66, emphasis added). The heuristics and biases view has dominated behavioral economics for decades (Kahneman 2011; Loewenstein, Rick, and Cohen 2008), and it had a marked influence on the way that engineering and design activities are understood in computer science (Mohanani et al. 2018). Let us add some entries to the dictionary:

> **rationality**, n.: that form of deductive reasoning which can be encoded and computed.
> **irrationality**, n.: those parts of human life that <u>rationality</u> has no access to.

The third response arose from the recognition of methodological flaws in rationalist research.

QUESTIONABLE METHODS

The methods used by behavioral researchers in this tradition, including the illustrious pair Tversky and Kahneman, designed simplified choice situations as context-free vignettes and prompted participants to select their preferred option. It is a very efficient approach to data collection. For example, in the famous Linda experiment, they described a person as follows:

Linda is 31 years old, single, outspoken, and very bright. She majored in philosophy. As a student, she was deeply concerned with issues of discrimination and social justice, and also participated in anti-nuclear demonstrations.

They then prompted participants to indicate whether it was more "probable" that Linda was (a) a bank teller or (b) a bank teller active in the feminist movement. The overwhelming majority of participants elected (b). The authors interpreted this as a "massive failure of the conjunction rule" (Kahneman et al. 1982, 94), which states that statistically speaking, the compound probability of (b) can never be higher than the probability of (a). In their view, cognitive heuristics led their participants to commit logical errors. From studies like this, they drew sweeping conclusions about the supposed universal presence of cognitive mechanisms and errors.

But there's a snag, and it has two parts. First, much of this award-winning work appears to have very questionable ecological validity (Harper,

Randall, and Sharrock 2016)—it describes the behavior of "people in the lab," not of people "in the wild" (G. Klein 1993, 36–50). In other words, the observed behaviors more often than not can just as well be explained (and manipulated) via the constellation of factors in the context, rather than the reasoning of the subjects themselves. For example, the effects described for the original Linda exam all but disappear with a slight reformulation (Sturm 2012; Gigerenzer 1996; Hertwig and Gigerenzer 1999, 291). Other experiments of similar structure could not be replicated at all (Harper, Randall, and Sharrock 2016). Major flaws have been pointed out in other similarly decorated studies (Harper, Randall, and Sharrock 2016, 44–81; Gigerenzer 1996; Sturm 2012).

Second, the participants' behavior can be interpreted very differently. The design of the Linda experiment started off with the mathematical theory of probability and the logical rule of conjunction. The authors operationalized these concepts loosely—very loosely—by formulating an English sentence containing the words "probable" and "and." This is a dubious operationalization of constructs. Probability theory in statistics is about repeated events, so it is not a valid theory to judge single events (Gigerenzer 1996, 593; Shafer 1986; Beach and Lipshitz 1993, 27). More generally, the logic of mathematics is not equivalent to the logic of language:

Sound reasoning begins by investigating the content of a problem to infer what terms such as probable mean. The meaning of probable is not reducible to the conjunction rule (Hertwig & Gigerenzer, 1995). For instance, the *Oxford English Dictionary* (1971, pp. 1400–1401) lists "plausible," "having an appearance of truth," and "that may in view of present evidence be reasonably expected to happen," among others. These legitimate meanings in natural language have little if anything to do with mathematical probability. Similarly, the meaning of *and* in natural language rarely matches that of logical AND. The phrase . . . can be understood as the conditional "If Linda is a bank teller, then she is active in the feminist movement." Note that this interpretation would not concern and therefore could not violate the conjunction rule. (Gigerenzer 1996, 593)

Tversky and Kahneman appear to fall prey to an *operationalism* in which their interpretation of *probable* was defined exclusively by the mathematical operations of probability theory. This would be perfectly appropriate within the confines of probability theory, but they were operating within a broader social context in which they asked people for an opinion. Participants were evaluated on the basis of a semantic interpretation they never agreed to. And this is not simply a matter of scientific rationality trumping common sense reason. On the contrary: Tversky and Kahneman explicitly justified the former with the latter (Gigerenzer 1996, 593). By denying the relevance of content and context, they were misled into an erroneous interpretation:

Recent studies using paraphrasing and protocols suggest that participants draw a variety of semantic inferences to make sense of the Linda problem. . . . Semantic inferences—how one infers the meaning of polysemous terms such as *probable* from the content of a sentence (or the broader context of communication) in practically no time—are extraordinarily intelligent processes. They are not reasoning fallacies. . . . Significant cognitive processes such as these will be overlooked and even misclassified as "cognitive illusions" by content-blind norms. (Gigerenzer 1996, 593)

Contradictions between rationalist theories and empirical data kept piling up, suggesting that "classical theory does not provide the conceptual depth that is needed to deal with real-world complexity; in some ways people seem far more capable than the theory" (Beach and Lipshitz 1993, 29). Harper et al. (2016, 61) conclude "we do not think the experiments these researchers undertake demonstrate these biases or aversions; in our judgment, they don't demonstrate very much at all."

There are two aspects that merit the use of rationalism as an -ism. First, note how these studies use normative assumptions as the basis for descriptive research. In linking normative theory and empirical observation, they have to face the question how to handle discrepancies between the two. We can recognize the rationalist tradition by the priority it places on theories that are normatively derived from logical principles but not empirically validated. In the face of conflict, the norms win.

Second, note how the abstraction of content and context from study design gives primacy to the normative framework's conceptual logic, while pushing situational awareness and social rationality aside. Critics advocate a broadening of the view to incorporate an awareness of the overall situation in which individuals reason and act, and to consider that, maybe, people are *reasonable* (Hertwig and Gigerenzer 1999, 300; Harper, Randall, and Sharrock 2016). I will return below in detail to the concrete implications this has on our understanding of *how system designers make decisions*, how they exercise judgment, and how we should approach studying this question. Table 7.1 briefly summarizes the mentioned rationalist studies and their

Table 7.1 Rationalist interpretations of empirical observations and alternatives to this view

	Rationalist interpretation	Alternative interpretation
The Linda experiment	The word "and" refers to the logical conjunction AND. Probability refers to the statistical likelihood of an event (%). The probability of the conjunction (b) must be lower than the probability of (a). Differing responses are false and reveal flaws in human reasoning. Humans use the "representativeness" heuristic in place of statistical reasoning. This leads them to commit errors. (Kahneman et al. 1982)	Probability theory is applicable to event series, not single events. The word "and" connects two parts of a sentence. Its meaning depends on the content and context. People are just as likely to interpret "probable" as plausible. The two statements are then not conjunctions. Instead, (b) indicates for many that despite her job, Linda remains committed to feminist principles. This is arguably more plausible than her having given up those ideals. Differing responses are reasonable and worth investigating because they may point to alternative views and interpretations. The study shows very little at all. "Instead of showing how people don't deploy logic, they show that they do deploy the logic of words." (Harper, Randall, and Sharrock 2016, 57)
Recent SE studies on managing technical debt	Decision-making follows rational choice, but imperfectly due to heuristics, biases, and cognitive limitations. Therefore, rational methods are better than whatever else practitioners are currently doing. When people do not follow rational methods, they behave irrationally.	When practitioners act in systems design, they exercise professional judgment in many ways. Methods are only one of many cognitive resources they rely on (Dittrich 2016; L. Suchman 2006). A plausible explanation for not using an available method is that according to the professional's judgment, the method is not suitable for the situation. This should prompt a reevaluation of the method just as much as a reevaluation of the professional's behavior. (Becker et al. 2018)

critique. It highlights that the rationalist interpretation is often question-able, that it has indeed been questioned at length, and that broadening our view beyond rationalist interpretations opens new perspectives on old questions.

THE EMERGENCE OF NATURALISTIC DECISION-MAKING

When it became undeniable that the assumptions, predictions, *and* meth-ods of rationalist decision theory as exposed in mainstream psychology and behavioral economics were questionable, some researchers concluded that they should be set aside. This involved three shifts. First, they aban-doned the "classical" rationalist *theory*: "classical theory cannot continue to be used as the standard for evaluating all decision behavior. . . . It is time to stop patching and propping an inappropriate theory. It is time to create a more useful theory" (Beach and Lipshitz 1993, 35). They started anew to develop new theories by focusing on behavior not norms (Beach and Lipshitz 1993; Zsambok and Klein 1997).

Second, they abandoned the *method* of lab experiments: "many com-promises have to be made to perform controlled experiments. The restric-tion on context, the absence of meaningful consequences, the use of tasks with well-defined goals, and particularly the elimination of expertise in studies presenting unfamiliar tasks, all raise doubts about whether the findings of these studies can be generalized to natural settings" (Klein and Wright 2016). Instead, they went into the field and studied professionals who were making important decisions, such as firefighters, military com-manders, surgeons, and engineering designers. They followed their day-to-day activities to observe and ask about the decisions they were making. This is why their research became known as naturalistic decision-making research (NDM).

Third, in contrast to the view on biases and shortcomings of human decisions, they were interested in how highly experienced individuals made decisions very well. Through this work, NDM researchers devel-oped a toolbox of research methods and guidance called *Cognitive Task Analysis* (Crandall, Klein, and Hoffman 2006; G. Klein 2000; Schraagen, Chipman, and Shalin 2000). A central component is that they expanded the horizon of the system of interest from the individual mind to the situation and its factors. This view of *macrocognition* (Klein et al. 2003; Klein and Wright 2016) incorporates an appreciation of the relevant factors that combine in each situation differently to shape the outcomes of decision-making. It places the individuals and groups engaging in decision-making into a concrete situation and emphasizes the content and texture of that situation. This formed an entirely new paradigm distinct from rationalist research, with different base metaphors, methods, and evaluation of validity.

Whereas the behavioral decision-making community focuses on human limitations and seeks ways to reduce biases and mistakes, the NDM community, as it performs macrocognitive research, focuses on human capabilities and regards good performance as much more than the absence of mistakes. Good performance is also about discoveries and insights; it is about the strengths of decision makers, and the importance of experience. Experience serves a variety of functions including a larger repertoire of patterns and associated actions, a richer mental model of how things work to support inferential reasoning and sense-making for diagnosis and anticipation. (Klein and Wright 2016, 3)

What they found was remarkable (G. Klein 1998; Zsambok and Klein 1997). Initially, they found no "decisions" at all: people denied making choices among options. This made little sense at first. It was only when they broadened their search beyond "choice" that *decision-making* appeared: people did commit to actions, but not by evaluating options against criteria. Importantly, even professionals trained in MCDA, and who believe in its value, don't *use* it much (Isenberg 1984). Instead, in recognition-primed decision-making, a prominent model that has frequently been documented, people rapidly identify and process cues in the environment that allow them to match salient features of a situation against their experience, and they use their experience and technical knowledge to generate one plausible course of action. They then use various techniques, above all mental simulation, to evaluate how well this option would work. If they are satisfied, they proceed; if not, they adapt the option or drop it to generate a new path of action. They never perform pairwise comparison, and they have no need to explicitly articulate criteria for evaluation (G. Klein 1997).

DEFENSES OF RATIONALIST THEORIES

Two explicit defenses for the rationalist paradigm are typically brought forward against its critiques. One defends the rational process as a regulative ideal worth striving for (Baron 2012), even if we "fake it" in design (Parnas and Clements 1986). This defense has normative validity—it is still often useful and valuable to justify decisions in terms of systematically specified criteria and to articulate which alternative options were considered. (I did a PhD on that and have few regrets.) But this argument does nothing to justify the attribution of *descriptive* validity. Normative value does not beget descriptive validity (Beach and Lipshitz 1993). There *is* to a limited degree a *performativity* in normative models: for example, to some degree normative economic models *produce* the behavior they describe (MacKenzie and Millo 2003). But this aggregate observation does not imply individual behavior and is certainly not valid for engineering and design methods (Dittrich 2016). It is naïve to assume that teaching rationalist methods will simply produce behavior that corresponds to them. In addition, this defense fails to address the deeper concerns about the issues raised by the reified, operationalist frame of thought that is introduced to the discussion and study of human behavior by the metaphor of thought as information processing. Methods are not programs to be run on the computing hardware of team member's minds, and computing education does not install an operating system in students' heads.

The other defense asks, "what's the alternative?," following a long-standing tradition that relegates anything outside narrow rational behavior into the realm of irrational behavior (Erickson et al. 2013; Sturm 2012). It is important to recognize that this argument is operationalist itself: it fails to recognize anything not defined by its own specification as a valid form of reasoning and leads, ironically, to a false dichotomy that considers the opposite of rationalist to be irrational behavior.

This is not to deny that people are also perfectly capable of using rationalist methods as part of their cognitive toolbox. These methods are simply not always appropriate, and the heuristics people use instead often outperform rationalist methods (Gigerenzer and Selten 2001). Unsurprisingly, it proved difficult to convince the rationalist camp to change course, despite conciliatory voices emphasizing commonalities across the schools of thought (Kahneman and Klein 2009). The relationship between descriptive and prescriptive research on decision-making is complicated (Bell, Raiffa, and Tversky 1989) and remains so. Advances in neuroscience suggest that multiple modes and systems of decision-making seem to coexist in the brain. Rational models do get used for clearly circumscribed tasks, while methods such as recognition-primed decision-making and other strategies identified by naturalistic research are used in many less circumscribed situations, for example to structure problems (Loewenstein, Rick, and Cohen 2008; Kahneman and Klein 2009).

Systems design research needs to take both views and understand their relationships. We need to question whether the boundary of decision-making supposed by rationalist theories are meaningful and helpful for us in understanding design for sustainability and justice, or whether it would not be smart to shift our focus: from choice to judgment, from rationalist theories to naturalist theories, and from the mind as a computer to the social situation of decision-making.

BEWARE THE NORMATIVE FALLACY

Having explored how other fields struggled to extricate themselves from the grip of an obsession with a limited theory, let us take another close look at the myth of rational decision-making (RDM). RDM manifests as myth when its scope of relevance is overextended. In computing research, RDM has widely been uncritically accepted as "the" theory of decision-making, and the nuances explored above often collapse into one "decision-making" concept that is defined as choice between enumerated options.

What are the consequences? Here I will explore one in more depth: RDM has socially preformed empirical software engineering (SE) research and misled it to severely misinterpret how people use methods, make judgments, and arrive at decisions. For good reasons, MCDA is central to SE methods for making tradeoff decisions: It offers a rigorous, systematic, repeatable and teachable approach to making decisions based on solid evidence, and it facilitates the review of that evidence. For decision situations that meet specific criteria, there is little doubt that these methods should be applied correctly. But how people actually use these methods, and what it means to "use a method," differs from what RDM assumes, and that places empirical research on practice in tension with the prescriptive nature of method development (Dittrich 2016). As mentioned earlier, the rationalist tradition sees methods as programs, and this is unsurprising given the role of concept-mapping. When computer science curricula focus almost completely on how programs work, it seems natural to rely on explanatory schemas from the domain of programming to understand what *programmers* do. To those captive to RDM, any deviations appear as irrational mistakes. In this clash between theory and practice, practice loses, but so does theory. "Divergence from what method developers and research community recommend and diversity of practices should not by default be regarded as a problem but as source of understanding the rationalities of practice that then can inform method development and appropriation" (Dittrich 2016, 228).

The orthodox view of "methods as programs" is based on an operationalism that thinks of the elements of design and development activity as clearly definable discrete atomic operations linked by specified relationships. Process modeling formalizes these elements and relationships. A method is defined by these operations, and deviations from the operations in practice are either invisible or seen as defects. But this operationalism does not merely mean that everything that matters can be operationalized—it also defines, reversely, what gets recognized as belonging to a class of things. The operationalism often structures entire research projects, preconfiguring behavioral studies of engineering activity to only recognize those parts of practice that fit the decision-making operations defined by MCDA.

Behavioral studies of systems design matter because understanding how design decisions go right or wrong is a key step to doing better. But the operationalist understanding of RDM leads many studies of engineering and design practice to commit a normative fallacy (Campbell 1970)—they misappropriate normative theories and prescriptive models for descriptive purposes.

Suppose we study the behavior of participants facing a risky software project situation. A series of twenty bugs were identified in short sequence in a new system under development. They appear somehow related, but it is unclear how. One is critical, and nineteen are severe. The team considers the business value of fixing one critical bug equivalent to fixing four severe bugs. Most team members want to focus all attention on the critical bug first (strategy a), but one claims to know how to fix all twenty and wants them to pursue their strategy (b). Our study participants must choose between two competing strategies: In their judgment, strategy (a) is almost certain to resolve the critical bug in the system, strategy (b) has a decent shot at fixing all bugs.[8] Many participants readily choose (a).

According to the normative model, this is, strictly speaking, an "error" because in its calculation, the expected value of option (b) is much higher. For example, if we set (a) at 90 percent and (b) at 33 percent, we arrive at expected values of 0.9 against 1.9. A sensitivity analysis may confirm that the two options' ranks are very robust against estimation errors.[9]

The question is not whether we agree with this assessment. The normative fallacy comes into play when we frame research questions and collect data *about* this situation. If we set out to describe how participants make their choice, should we ask them "how did you compute the value of each option?" and "how did you weigh your factors?" Doing so would mean committing the normative fallacy: It would assume that they decide by applying numeric operations to abstract concepts without further considering the content and context of the situation they are in to reflect on the meaning of these numbers. Instead, we could ask an open-ended question: "How did you make your choice?" If given a chance, participants may rightfully introduce concerns outside the framing of the gamble which make it perfectly reasonable for them to prefer option (a). They might say that it can be communicated more effectively to the stakeholders, hope that fixing the critical bug produces insights that help fixing the others later, or believe that resolving the critical bug first will reduce the stress on the team. They may also emphasize the nature of ambiguity: After all, the percentage estimates are in truth based on ambiguous information. Their concerns and judgments transcend the original framing of the gamble and situate it in a broader context. They seem much more reasonable than the normative assumptions of rational choice, but they do not fit into the confines of "rational decision-making."

THE NORMATIVE FALLACY IS COMMON

In behavioral SE studies, unfortunately, this normative fallacy is common. Our systematic literature reviews in key areas found that many studies use prescriptive theories to collect and interpret data about behavior for descriptive and explanatory purposes (Becker et al. 2018; Becker, Walker, and McCord 2017). Most importantly, the normative fallacy manifests as studies in which data collection and analysis are incorrectly predicated on the narrowly framed concepts of factors, weights, ranking, and choice.

Broader aspects of reasoning including the role of expertise, experience, cognition, incentives, mental simulation, judgment, and perception are never considered. For example, one study about technical debt decisions asked participants these questions:

1. What factors are considered when you make a decision about when to fix a defect?
2. How are these factors weighted? (Snipes et al. 2012)

Responses to these questions were taken at face value. This may appear normal, and it is normal in SE research. But accepting the findings at face values means accepting that the actual decisions of these participants are *adequately described* by the operations specified in MCDA: criteria specification, weighting, evaluation, ranking, and choice. That is, these questions make no sense unless we take the normative theories of rational choice as descriptively valid. That is a mistake, so the questions above simply failed to capture how the participants really made their decisions. Instead, they nudged them to describe their reasoning *as if* they had considered weighted factors.

Even though most participants in these studies deviate from the proposed norms, the norms and assumptions are never questioned. Instead, the deviations are considered deficiencies: errors that need to be fixed. For example, in another study in which the observed practices did not correspond to the proposed method, the researchers developed a data categorization scheme predicated on MCDA to make sense of it. We are left to wonder what they would have found had they looked beyond it. Similar bias pervades many discussions in which numerical optimization across multiple objectives is depicted as automatically superior to individual expertise and team knowledge, as in this case: "Technical debt . . . is currently managed in an implicit way, if at all. Decisions are largely based on a manager's experience, or even gut feeling, rather than hard data gathered through proper measurement" (Guo, Spínola, and Seaman 2016, 160). But what is called *gut feeling* here has long been relocated to the brain and reinstated as social rationality, experience, and judgment.

This is not to say that participants in this example *should not* consider a set of factors and gather evidence: simply that their reasoning will remain invisible to those who perform the study. Crucially however, because these questions are posed in the context of a scientific study by an academic research team with scientific credentials, the participants certainly *will* provide answers. In doing so, they will retroactively construct plausible factors and weights. Papers that uncritically report these answers, though not empirically valid, often pass peer review because reviewers in SE are also not trained in behavioral research or psychology. The findings are then cited to support further research, most of which is again prescriptive. In this way, misunderstood empiricism reinforces a misleading narrative. The myth of rational decision-making is reproduced via its retelling.

Imagine if the study above had instead asked an open-ended question: "How did you decide when to fix a defect?" This simple reorientation would already avoid two important misunderstandings produced by the torque of the RDM myth. First, it sets up the expectation that the participants' reasoning will be described on their own terms and leaves open in what way the result will be compared and contrasted with various theories of decision-making. Second, when and if such comparisons happen, if the collected responses indicate a divergence between rationalist models and participants responses, the open-ended structure of their responses will make it easier to see when their reasoning is different. It will make it easier to recognize that their decision-making is not an impoverished *version of* MCDA but could instead be interpreted as balancing competing priorities in a complex situation. In other words, we should get ready to entertain not only the possibility that our theory is wrong but question how it prestructures our investigations. Table 7.2 summarizes this

Table 7.2 The normative fallacy in empirical research paradigms

		Normative research (what should be?)	Empirical research (what is?)
Theoretical framework	Descriptive and explanatory	OK: normative recommendations grounded in empirical findings	OK: empirical research
	Prescriptive	OK: deductive or constructive research oriented towards design or modeling	NOT OK: normative fallacy[a]

Note: a. Testing theoretical assumptions of prescriptive theories *as hypotheses* in empirical research is of course fine, but that is not a case of this quadrant: In this case, the theory takes the role of research object, not of theoretical framework.

discussion by illustrating, in simplified form as "ideal types," the possible combinations that arise from the use of descriptive or prescriptive theories in normative or empirical research.[10]

The normative fallacy carries important consequences:

1. Participants *will* respond to the questions above, and those responses will reinforce the empirical basis for normative theory, even when this empiricism is flawed and tautological.

2. Because of the unquestioned acceptance of RDM as normative *and* descriptive standard, any deviation will be treated as a defect and labeled as bias or error.

3. Participants' rejection of decision-making methods is often explained away with the naïve suggestion that they just need better training or more time so they will see the right way.

4. In all this, the uncritical empiricism built on the normative fallacy ultimately turns into ideology. And it keeps us from asking the important questions: How *do* people make tradeoff decisions when the outcomes are at a distance? What can we learn from those who act with long-term vision? Do our methods inadvertently tilt the field against sustainability and justice? How can we do better?

CONCLUSIONS

The myth of rational decision-making tells us a story of criteria, options, weights, ranking, and choice. This is not what happens, even for those people who are trained in and believe in these methods. In the end, it appears that "humans are not two eyeballs attached by stalks to a brain computer" (D'Ignazio and Klein 2020, 85). The rationalist logic of choice is one tool in the toolbox of human reasoning. When applied as a description, it is narrow, flawed, and misleading. People are perfectly capable of deploying rationalist models and methods, but they often choose not to, and they often have good reasons.

It is hard to get over the myth, because the reluctance to abandon cherished paradigms is always strong (Kuhn 1962; Ralph and Oates 2018; Ralph 2018). When theory and data collide, "the possibility that the theory is inappropriate is seldom entertained" (Beach and Lipshitz 1993, 28). The unquestioned adoption of rationalist theory unfortunately continues in some computing fields despite the indisputable fact that the rationalist model of decision-making misses countless ways in which smart humans reason intelligently and function effectively in complex open-ended environments like systems design. It misses situated and experiential knowledges, moral reasoning, judgment, and the nuanced nature of intuition, recognition, and expertise (G. Klein 1998). How is a theory of choice based on the human mind as a defective calculating machine supposed to lead us into a future in which everyone designs responsibly, sustainably, and justly? It won't.

But if that is not how it happens, what else is happening? How do people take all these decisions in systems design that affect, at a distance, those in the future and far away? Without understanding that, we cannot possibly hope to design more sustainably, and more justly.

Moving beyond the myth of rational decision-making, and avoiding the normative fallacy, opens new opportunities. We can reorient ourselves: from methods as program to methods as resources in situated action; from people as information processors to people as reasonable and purposeful social beings; from engineering practice as applied science to engineering practice as a social epistemic practice on its own ground; from rational choice to reasonable decisions and wise judgments; and from rationalist quasi-experiments to naturalistic studies. We can focus on identifying wise judgments and supporting reasonable decisions based on macro-cognitive perspectives. We can study judgment *and* decision-making accounting for reflection, critical awareness, the critique of boundaries and value systems, and the composition of the macro-cognitive environment in a decision situation. We can aim to understand which constellations are more likely to produce short-sighted decisions, and we can support design teams in redesigning the architecture of their situations, so they are better able to take a long-term view and consider distant impacts and outcomes of their decisions. Chapter 11 explores this path.

8

PROBLEMS ARE FRAMINGS
THE DISCORDANT PLURALISM OF JUST
SUSTAINABILITY DESIGN

The language we use[] is not neutral and can even become a trap, confusing our discourse on reality with reality itself . . . several authors in the objectivist tradition have not been able to avoid this trap, with the result that they have become prisoners of their own language: they have confused the way they are talking about unsatisfactory situations with reality itself.
—Landry (1995)

It becomes morally objectionable for the planner to treat a wicked problem as though it were a tame one.
—Rittel and Webber (1973)

Can design solve the wicked problems of sustainability and justice? From a problemist standpoint, absolutely. The famous *wicked problems* concept, initially coined in the context of urban planning and policy, has influenced thought in design widely (Buchanan 1992). But the unfortunate name seems to have led many astray. This chapter will address three questions in search of a meaningful position.

1. Can design in principle solve wicked problems involving sustainability or justice concerns?
2. What does it take to intervene ethically in such a situation?
3. How can critical systems thinking (CST) help us to position systems design for meaningful action in this context?

CAN WICKED PROBLEMS BE SOLVED THROUGH DESIGN?

What happens when designers solve a problem? First, the problem needs to be stated, then design activity results in some form of intervention aiming to "chang[e] existing situations into preferred ones" (Simon 1996, 111). Problem framing is so important in design because it directs the work that results in change. Recall that soft systems thinkers showed that any articulation of a problem relevant to a situation is made by someone, implies a worldview, and represents a framing that works like a lens placed in front of the situation. Even if a problem statement does not directly name the objective or satisfaction criterion—or if the criterion later gets restated—how the statement frames the problem profoundly shapes the purpose and direction of the design activity. In a sense, stating a problem cuts a slice out of a problem situation to frame that slice as something worth solving: "problems are not given, nor are they reducible to arbitrary choices which lie beyond inquiry. We set social problems through the stories we tell" (Schön 1979, 150). To frame a problem is to tell a story:

Each story selects and names different features and relations which become the "things" of the story—what the story is about. . . . Each story constructs its view of social reality through a complementary process of naming and framing. Things are selected for attention and named in such a way as to fit the frame constructed for the situation. Together, the two processes construct a problem out of the vague and indeterminate reality which John Dewey (1938) called the "problematic situation." They carry out the essential problem-setting functions. They select for attention a few salient features and relations from what would otherwise be an overwhelmingly complex reality. They give these elements a coherent organization, and they describe what is wrong with the present situation . . . to set the direction for its future transformation. (Schön 1979, 146)

The telling of any story is socially preformed by metaphors and frames. "Frames structure the way we think, the way we define problems, the values behind the definitions of those problems, and what counts as 'solutions' to those frame-defined problems . . . other frames allow us to see other problems, other causes, and other solutions" (Lakoff and Ferguson 2009). As Schön writes, the "participants in the debate bring different and conflicting frames, generated by different and conflicting metaphors. Such conflicts are often not resolvable by recourse to the facts—by technological fixes, by trade-off analyses, or by reliance on institutionalized forms of social choice" (Schön 1979, 139).

When Rittel and Weber articulated the *wicked problems* concept in the 1970s (Rittel and Webber 1973; Churchman 1967), the limitations of a reductive problem-solving approach had become apparent to those in social policy. An obvious immediate challenge is the complexity of real-world situations. Side effects and ripple effects urge us to expand the considered system boundary (Rittel and Webber 1973, 159). This need to "sweep in" more aspects of the environment to gain a more holistic understanding of the problem situation is potentially never-ending (Churchman 1971, 197; 1979a; Ulrich 1985). Because this difficulty applies to any social problem situation of relevance, "the classical paradigm of science and engineering—the paradigm that has underlain modern professionalism—is not applicable to the problems of open societal systems" (Rittel and Webber 1973, 160).

Wicked problems are often characterized by their original ten properties:

1. There is no definitive formulation of a wicked problem.
2. Wicked problems have no stopping rule.
3. Solutions to wicked problems are not true or false, but good or bad.
4. There is no immediate and no ultimate test of a solution to a wicked problem.
5. Every solution to a wicked problem is a "one-shot operation"; because there is no opportunity to learn by trial and error, every attempt counts significantly.
6. Wicked problems do not have an enumerable (or an exhaustively describable) set of potential solutions, nor is there a well-described set of permissible operations that may be incorporated into the plan.
7. Every wicked problem is essentially unique.
8. Every wicked problem can be considered to be a symptom of another problem.
9. The existence of a discrepancy representing a wicked problem can be explained in numerous ways. The choice of explanation determines the nature of the problem's resolution.
10. The planner has no right to be wrong. (Rittel and Webber 1973)[1]

But skimming this concise list does not do justice to a central difficulty behind the ill-named concept. The complexity of the situation is often considered the primary challenge, but following the ninth criterion leads us to another one that is more profound.

In dealing with wicked problems, the modes of reasoning used in the argument are much richer than those permissible in the scientific discourse. Because of the essential uniqueness of the problem (see Proposition 7) and lacking opportunity for rigorous experimentation (see Proposition 5), it is not possible to put [a hypothesis for a solution] to a crucial test. That is to say, the choice of explanation is arbitrary in the logical sense. . . . The analyst's "world view" is the strongest determining factor in explaining a discrepancy and, therefore, in resolving a wicked problem. (Rittel and Webber 1973, 166)

The divergent views of stakeholders are based on discrepancies in their worldviews and interests. There is no independent criterion, no "position from nowhere" by which to decide which worldview to prioritize a priori. That is why such situations are "wicked." Many different problem formulations can be given, so a mechanism is needed to evaluate and distinguish these formulations.

The scientific discourse lacks the modes of reasoning—the social rationality—to address this crucial character of wickedness. First, in the positivist tradition, a problem involving many stakeholders is no different from a problem posed by one. There are simply more data, a reason used by experts to marginalize other views (Landry 1995) in the name of scientific expertise. This is where the myths of rational decision-making, value-neutral technology, objective problems, and solvency exert the strongest grip.

Second, some contradictions in conflicting problem statements cannot be resolved via logical means alone because they arise from contradictions in the underlying frames and metaphors used to tell the story about which problems should be solved. These are not simply factual disagreements that could be resolved logically, because they form the conceptual structures and metaphors through which participants see the world and articulate their understanding of it. Problem framing "often depends upon metaphors underlying the stories which generate problem setting and set the directions of problem-solving," and conflicts between the frames used by different stakeholders "are not problems. They do not lend themselves to problem-solving inquiry" (Schön 1979, 150). Instead, many of these contradictions are value conflicts between incommensurable ends. "Ends are incommensurable because they are embedded in conflicting frames that lead us to construct incompatible meanings for the situation" (Schön 1979, 150).

There is no legitimate substitute for democratic reasoning in such situations, and that is why "it becomes morally objectionable for the planner to treat a wicked problem as though it were a tame one, or to tame a wicked problem prematurely, or to refuse to recognize the inherent wickedness of social problems" (Rittel and Webber 1973). Not only does scientific reasoning fail to provide the conceptual framework required to handle this situation, but it hides its own insufficiency by masking the inevitable role of values and politics in the design process. The myths of systems design evacuate the appreciation of the political nature of design and leave behind an impoverished and inadequate model of rationalist problem-solving.

But if there cannot be a singular objective articulation of a wicked problem, then there is nothing that could be *solved*. What exists is a *predicament* offering a space for meaningful action: a messy problem situation that sets a context for a discourse characterized by multiple perspectives that can be mutually contradictory and even irreconcilable. If wicked problems do not have solutions, it is misleading to call them problems. They are quagmires, predicaments, difficult situations. And indeed, when we consider what the ten characteristics are *about*, we notice that they don't describe problems but properties of situations in which problems are formulated by people. The term has misled many scientists, engineers, and researchers into referencing the concept in their exposition of a real-world problem situation, only to advance straight to their proposed "solution," prematurely "taming the wicked problem." In doing so, "they have confused the way they are talking about unsatisfactory situations with reality itself" (Landry 1995, 339). A module for an interaction design course (Wong 2020) is a paradigmatic example of a well-established perspective on design positioned to solve wicked problems. "Have you ever come across a problem so complex that you struggled to know where to start? Then you have stumbled upon a wicked problem." The piece suggests that systems thinking begins with decomposition—"You can utilize systems thinking if you break the information down into nodes (chunks of information such as objects, people or concepts) and links (the connections and relationships between the nodes)" (Wong 2020)—in a paradigmatic uncritical interpretation of the systems idea through an objectivist lens. The module proceeds to advocate for solving wicked problems without recognizing the crucial issue: Divergent worldviews imply conflicting reference systems that require problem negotiation. Instead it assumes that the problem has an existence independent of its formulation. The author interprets the denominating term "problem" in operationalist terms. I have documented the collateral damages straddling the path of this approach in earlier chapters. Even progressive voices seem to fall into the same trap, taking the noun "problem" as the operating concept, merely modified by the attribute "wicked," when they write about "solutions" for wicked problems (e.g., Steel, Lach, and Weber 2017; Wiek, Withycombe, and Redman 2011).

That is why the concept's name is so unfortunate: The name suggests that the concept still fits into the conceptual frame of problem-solving, when its characteristics emphasize the need to transcend it. To see wicked problems as solvable, as many do (Kasser and Zhao 2016; Steel, Lach, and Weber 2017; Wong 2020; Lindberg et al. 2012; von Thienen, Meinel, and Nicolai 2014), is to fall prey to a positivist misunderstanding of the problem concept (Landry 1995). This convenient misunderstanding fits the operationalist mindset perfectly and reinforces the false consciousness of problemism. But wicked problems cannot be solved, neither through design nor through any other means. It is more constructive to focus on *wicked problem situations*—complex, conflict-heavy situations characterized by the ten characteristics, in which different problem frames suggest different interventions.

> **conflict**, n.: Something that does not match up and needs to be fixed. Typically identified between (a) alternative design solutions for a given problem, easily addressed by identifying and modeling the costs, benefits and risks and then trading them off rationally to select the optimum choice (that's what engineers do); or (b) between development branches. (May the wrath of the Gods be upon you in eternity while you slowly roast in hell.) Other forms of conflict are invalid and irrational.
>
> **problem**, n.: something that can be fixed or solved.
>
> **problem-solving**, n.: the process of fixing things that aren't broken (because they don't exist) and thereby creating new problems.

EPISTEMOLOGICAL RESTRUCTURING: WICKED PROBLEMS
IN CRITICAL SYSTEMS THINKING

When we design in wicked problem situations, we intervene to cause improvement. Dorst (2006) regards "design as the resolution of paradoxes between discourses in a design situation" (17). This primary attention to the discourse has to be a central feature of just sustainability design. But who or what is to be the arbiter of what counts as improvement? Whether something is genuine improvement is a question of legitimacy as much as scientific and other knowledges. What does it take then to design ethically? Critical systems thinking offers a few lessons.

Problems are like framed lenses through which we view a situation. When a *wicked* problem situation invokes issues of sustainability or justice, asymmetric vulnerability shapes the discourse, introduces distortion into the lenses, and thus affects problem framing. Because the conflict-rich nature of the situation brings the political nature of discourse to the forefront, we cannot legitimately assume consensus, but need to appreciate the politics of problem framing. We prepare to do this by examining the epistemic restructuring that resulted in CST.

Proponents of CST recognized the inadequacy of rationalist problem-solving and the illegitimacy of applying physical science methods to social and sociotechnical situations. On the other hand, they also knew first-hand of the technical effectiveness of "hard" problem-solving methods in appropriate contexts. After all, they themselves had developed some of those methods.

To someone subscribed to the ideology of problemism, the shift from problem to situation, from hard to soft systems thinking, and from verification of correspondence to a reliance on consensus, may appear paralyzing. "If we need consensus on everything, we'll never achieve anything. Is that what you want?" But the choice we face is not one between problemism and paralysis. Instead, CST developed an epistemology and methodology to underpin ethical interventions in complex problem situations that makes appropriate use of rationalist methods. In doing so, its proponents had to navigate between the Scylla of positivism and the Charybdis of relativism. They adopted Checkland's shift away from a naive focus on "problems" to an appreciation of the "problem situation." Contrary to Checkland, however, they emphasized how material differences and ideologies shape social reality, so that interpretive perspectives such as soft systems methodology (SSM) are not considered adequate on their own. After all, "one of the most far-reaching exercises of power is in the structuring of the world-views of others, which in turn will be reflected in the definition of a problem" (Thomas and Lockett 1991, 93).

Importantly, Flood and Jackson positioned existing methods and approaches along two axes: the complexity of situations and the complexity of social relations in those situations (Flood and Jackson 1991a; M. Jackson 2003). This allowed them to map out for which kinds of situations existing systems thinking approaches were suitable and to develop multi-methodology frameworks for combining suitable approaches, including *Critical Systems Practice* (M. Jackson 2019). They recognized that the social complexity of problem situations must have primacy in guiding what is an appropriate methodology for intervening in situations. Table 8.1 groups some approaches we encounter in this book in terms of this framework.

Perhaps the hardest lesson from CST is that the critical turn is unwelcome. In facing the difficult task of writing back to the rationalists, soft and critical systems thinkers ultimately failed to convince the mainstream to change direction (Ulrich 2004; M. Jackson 2019; Kirby 2003). But they created organizational and institutional change and significant new streams of research, practice, and education that can help us restructure and reorient systems design in computing.

Structurally and historically, there is a striking parallel and a direct lineage between the rationalist mainstream in operations research (OR) that CST challenged and the rationalist design mainstream that still dominates computing. The myths of computing originate in the sphere of OR with which it shares its roots. The ambition to intervene in the world, which emerged so boastfully in OR, has only grown in computing. What is different today? The effects of computing have scaled up and now have significant aggregate structural impact (recall the classification from chapter 1). Rationalist theories and methods have absorbed and assimilated some of the critical terminology (without adopting its intent). Tech companies are much more powerful than before. And just sustainability raises the additional challenges of asymmetric vulnerability and moral corruption.

Table 8.1 Classifying systems approaches (adapted from Jackson 2003)[a]

		Stakeholders (*social complexity*)		
		Unitary	**Pluralist**	**Coercive**
Situation (*domain complexity*)	**Simple**	General systems theory	Soft systems methodology	Critical systems heuristics
	Complex	System dynamics, cybernetics, climate models, systemic effects of ICT for sustainability: SusAF[b]		Systemic intervention, critical systems practice, autonomous design,[c] just sustainability design

[a] Jackson (2003) adapts this from the original system of systems methodology (Flood and Jackson 1991a). In my version, the axes are renamed for consistency— "Participants" have become "stakeholders" to include those affected but not involved, and "systems" (which originally was called "problem context") has become "situation." Note that "coercive" situations include situations in which coercion is "mild," while "pluralist" refers to approaches that do not come equipped to recognize power and coercion.

[b] SusAF, introduced in chapter 1, remains silent about social complexity, and its language suggests that it models (corresponds to) real effects in the same way as system dynamics and other hard systems approaches do.

[c] I place autonomous design here, rather than in the center, although the suggested design process or approach itself does not appear to incorporate ways to handle the nature of situations characterized by the "coercive" label. There are two reasons. First, Escobar's critical intent lies in the ontological reorientation and in creating a space in which autonomous design can flourish, protected from the coercive nature of the patriarchal, rationalist design paradigm. Second, the principles of autonomous design are decidedly pluralist and reminiscent of soft systems thinking (Escobar 2018, 184–85), but it is worth noting that Escobar speaks of "building *a* model of *the* system that generates *the* problem" (Escobar 2018, 185; emphasis added), which runs counter to soft systems epistemology.

To *do* just sustainability design, we need a design methodology that is robust, critical, multi-methodological, able to handle both domain complexity and social complexity, *and* is able to incorporate and justify rational methods for their appropriate use. Rationalist methods must become components of reflective methodologies with a critical orientation. CST provides the theory and methodologies for doing just that:

for embedding computational problem-solving into a critically systemic design approach. In the next section, we explore how it can help us organize the social life of systems design to be critically appreciative and technically effective.

TOOLS FOR THE POLITICS OF FRAMING A PROBLEM

CST resulted in important conceptual frameworks, methodologies, and methods. Critical systems heuristics, introduced in chapter 5, may be the most robust and well-defined framework. Jackson positions it on the "simple" range of domain complexity, and it is true that it does not in itself provide concepts to address domain complexity, but it has been applied in highly complex situations, including the evaluation of sustainable development (Reynolds 2007). We will see more of it soon.

Recall that CST aims for (1) critical reflection on normative claims, (2) emancipation, and (3) methodological pluralism. In Ulrich's words,

> It is only by giving an equal status to the rationalities of the involved and the affected that we can prevent the former from making themselves the judges who define the measure of improvement. In short, the affected must be helped to emancipate themselves from the rationality of the involved, from the premises and promises of the "experts." [so we need] a *dialectic of expertise and emancipation*, of professional competence and democratic participation of citizens. (Ulrich 1983, 290)[2]

Midgley built on CSH to advocate a shift to process philosophy to overcome some of the inherent dualist challenges central to twentieth-century philosophy. His process philosophy centers on the activity of making boundary judgments, which simultaneously gives rise to the object and to the subject making the boundary judgment. In his methodology, systemic intervention, first-order boundary critique reflects outwards on the boundaries of the object, while second-order boundary critique reflects back on the boundaries of the "knowledge generating system" making the first boundary judgment (Midgley 2000), say a team of operations research practitioners in a social housing project (Midgley, Munlo, and Brown 1998). Here, boundary critique enables a structured reflection, using CSH, on positionality, situated knowledges, privilege, marginalization, and emancipation. Through such reflection, Midgley (1992) recognizes and addresses ethical dilemmas, and he shows how power asymmetry in systems design leads to the marginalization and devaluation of some stakeholders and their views to a profane status while others are elevated to a sacred status beyond debate.

We can observe this tendency in much of the discourse about smart cities. Shannon Mattern (2017) quotes and criticizes a tech writer's view of the smart city: "The city is a computer, the streetscape is the interface, you are the cursor, and your smartphone is the input device" (Mcfedries 2014, 36). Note how the metaphor relocates city life into the domain of computing, where the speaker is the expert. What role do participants have in this approach to design their urban future after this relocation? What rights do they have? In many smart city approaches, professional knowledges and technological expertise have served to marginalize the interests, perspectives, and values of those who have to live with the outcomes of technology-driven urban planning (McCord and Becker 2019).

This is often done under the guise of participation itself and can be quite insidious. In the Sidewalk Labs Toronto project, for example, which was ultimately abandoned after significant pushback from organized civic rights advocacy groups (Wylie 2020), public townhalls called for broad participation. Those who attended were assigned to small groups. Each group was to respond to a predefined question. No opportunity was included to speak up, voice concerns, and create a conversational space among the attendants other than that predesigned by the organizers (McCord and Becker 2019). Effectively, those affected are configured *by design* into the role of a data source or data collector rather than a designer or decision-maker (Palacin et al. 2020). Like a cursor, indeed, they are moved around by someone else.

CST provides the concepts to make this visible and the methodologies to do better. In urban life, and today in the smart city debate, what is sacred is often the professional expertise of the technologists and the market logic of neoliberal capitalism, while those who must live the consequences of smart city developments are considered mere service consumers. As Ulrich (1983) asserts: On the topic of their lived experience, only the residents are the true experts, and ultimately, only those who are in some form affected by a development can lend legitimacy to it. The proposals of professional experts are of course often informed by a profound understanding of the pertinent challenges and opportunities of domain complexity. But their legitimacy ultimately rests on democratic grounds: involvement of, and power to, those affected. CSH supports us in making visible how expert knowledge and lived experience are situated. By using it to question the stated purposes and their justifications, we can map the reference system of assumptions and values on which a project rests. In the less-than-ideal case where those in "sacred" positions refuse to open up their claims for scrutiny, CSH at a minimum allows us to make visible that the situated knowledges of the experts cannot legitimately be justified (McCord and Becker 2019). In an ideal case, CSH allows either for the emergence of a richer shared understanding or a *discordant pluralism* in which multiple worldviews coexist without being forced to reconcile all disagreements. The latter is a CST concept well suited to what Escobar (2018) later calls a pluriverse—it is "local, contingent and historically situated" (Gregory 1996b, 52), it brings radically contradicting perspectives into dialogue without assuming reconciliation and closure, and it nevertheless maintains the need for a normative stance to anchor ethical decision-making.

Through these and other arguments, proponents of CST made important methodological contributions to social science, most importantly on action research, multi-methodology, reflexivity, participation, and socio-technical systems. The restructuring performed in the critical turn in systems thinking engaged deeply with social theory; widened the epistemic horizon beyond the hard and soft systems paradigms; and developed a critical systems theory that can underpin multi-methodological practice.[3]

The differentiation in domain complexity and social complexity helps us see what went wrong with the *wicked problem* concept. From an orthodox rationalist viewpoint, the salient characteristics of wicked problems appear as features of *domain complexity*. They characterize hard, complex problems. There is nothing more appetite-inducing for those of us trained in science and engineering! The interaction design course mentioned previously is one of countless examples dominated by questions focused on that one dimension. The well-meaning sustainability engineer I mentioned in the Introduction who wanted an approach to solve the wicked problem of sustainability is one of many educated in such thinking. Meanwhile, this viewpoint easily overlooks that what truly makes those situations wicked is the *social complexity* they entail: conflicting worldviews, incommensurability, and the effects of power, marginalization, false consensus, asymmetric vulnerability, and moral corruption. And to make it worse, this viewpoint lacks the reflective awareness to recognize its own false consciousness.

To summarize: any articulation of a problem relevant to a situation is made by someone, implies a worldview, and represents a framing that works like a lens placed in front of the situation. Problem frames can be traced to their assumptions and worldviews. Schön and Lakoff suggest that we examine deep frames, frame conflicts, and value conflicts among stakeholders. CST provides the conceptual tools to do so. We can use CSH to make visible the reference system that underpins each problem frame. Ulrich also points us to the interlinking between claims about system boundary, facts, and values. Following Midgley, we can engage in first- and second-order boundary critique to reflect on the boundaries of the frames and the knowledge generating system producing those frames simultaneously. Boundary critique and CSH will also help us detect whether the politics of a situation provide conditions for genuine participation.

As Gardiner writes, "[t]he dominant discourses about the nature of the climate threat are scientific and economic. But the deepest challenge is ethical" (2014, xii). Just sustainability design must face this challenge, step by step. Chapter 10 demonstrates how CSH can be introduced into an engineering process to explicitly guide the participants' attention toward a reflexive understanding, facilitate first- and second-order boundary critique, support the development of critical appreciation, and meaningfully orient the focus of design toward the emancipation of those most affected. But before that, I will illustrate what the epistemological shifts in systems thinking mean for the evaluation of sustainability and sustainable development.

EPISTEMIC RESTRUCTURING IN SUSTAINABILITY

The soft and the critical turns in systems thinking have manifested in generations of systemic frameworks for evaluating sustainability. Initial frameworks such as the World Model underpinning the Club of Rome report were based on a hard systems worldview taking the correspondence of models to the real world as a starting point. Variables such as the world population, the degree of industrialization, the average temperature, and the amount of CO_2 emitted yearly are assumed, and their dynamic causal relationships are modeled. These models produce high-level sustainability indicators (biomass, warming, etc.). These are crucial to understand aggregate impact on a planetary scale and assess to what degree humanity is exceeding planetary boundaries. The underlying epistemology is reflected in small-scale frameworks such as the SusAF model introduced in chapter 1 as well, even if the models are declared speculative.

Later generations of sustainability evaluation frameworks took a more situated, localized standpoint, often aiming to evaluate regional efforts. This is not in conflict: they address different problem situations. But when bringing the evaluation of sustainability into local contexts, Bell and Morse (2008) learned that hard systems approaches were simply not up to the task. They are perfectly capable of addressing domain complexity but struggle on other fronts. A crucial issue for a localized application of a sustainability evaluation framework is the definition of boundaries in time, space, and social sphere. These issues can be sidestepped and avoided in large-scale global models because the planetary boundary makes for easy consensus. But where exactly is the boundary to set when evaluating the effects of a new airport runway? A highway expansion? A train line? Geographic categories soon turn out to be problematic for many reasons. With each shifting of a boundary, different facts become relevant, different stakeholders are identified, and different value judgments are brought to the forefront. Different evaluation frames are sometimes commensurable and reconcilable, sometimes not.

Bell and Morse recognized that the hard systems view is inadequate for this social complexity and turned to soft systems thinking in the form of SSM. Their systemic sustainability analysis (SSA) is "the participatory deconstruction and negotiation of what sustainability means to a group of people, along with the identification and method of assessment of indicators to assess that vision of sustainability" (Bell and Morse 2008, 147). Based on the pluralist belief that divergent opinions can be negotiated, SSA specifies sustainability as a subjective evaluation of a system. The Imagine approach to perform SSA is decidedly a soft systems approach built on reflective action research and subjectivist epistemology. Based on "understanding the context" through a wide range of techniques, including visual tools and system definitions adopted from SSM, Imagine aims to identify a wide range of stakeholders. The approach explicitly embraces their multiple views of sustainability, but it remains silent on how real conflict can be detected and resolved. The inclusion of stakeholders in the process is intended to guarantee that a comprehensive view is represented once agreement is reached (Bell and Morse 2008, 170), but it should be clear by now that this is far from guaranteed. The authors acknowledge the problematic imbalance of expertise and power inherent in the colonial nature of "sustainable development," in which supposed experts from developed countries define sustainability for developing countries (Chambers 1997; Escobar 2011; Gómez-Baggethun 2019), but they do not perform the critical turn. Instead, "participation and inclusion" is labeled an "outstanding issue" (Bell and Morse 2008, 196). This definitely seems a missed opportunity. In contrast, Martin Reynolds's (2007) work on evaluating sustainable development projects applies CSH. This produces a nuanced view of the *social complexity* dimension but remains comparably silent about domain complexity: that is, the content of sustainability evaluation itself.

These three approaches illustrate how each systems paradigm orients evaluation. Table 8.2 summarizes the viewpoints. Clearly, a critically systemic approach for evaluation will need to reintegrate hard systems approaches for handling domain complexity on appropriate and legitimate social foundations. This is a core challenge that just sustainability design needs to work toward. Recent CST work can provide the methodological basis for that, because this is precisely what it grappled with (Midgley 2000; Jackson 2019).

ENCOUNTERING SUSTAINABLE DEVELOPMENT

With all this in mind, let us briefly revisit the nature of "sustainable development" and its predecessor, development. In *Encountering Development*, Arturo Escobar retraces the invention of "underdevelopment" in the late 1940s and writes: "Development fostered a way of conceiving of social life as a technical problem, as a matter of rational decision and management to be entrusted to that group of people—the development professionals—whose specialized knowledge allegedly qualified them for the task" (Escobar 2011, 52). World Bank economists rejoiced in the "marvellous

Table 8.2 Sustainability evaluation on hard, soft, and critical systems terms

Sustainability	. . . on hard systems terms	. . . on soft systems terms	. . . on critical systems terms
Examples	World Model, IPCC reports	SSA/Imagine	Evaluation based on CSH
Strengths	The integration of scientific evidence across a wide range of sciences using the full force of modeling techniques produces the most reliable assessment we know of aggregate human impact on life on this planet.	The participatory, interpretive approach introduces pluralist perspectives, addresses difficulties of incommensurability, and assures broad participation from stakeholders.	The approach addresses social complexity and power by making visible the sources of motivation, control, expertise and legitimation. By addressing the social complexity dimension, the approach has the best claim to legitimacy. This is also the only framework that directly speaks to the ethical dimension of asymmetric vulnerability.[a]
Gaps	Because of the absence of a conceptual framework to address social complexity, silent exercises of power determine the terms and boundaries of debate. Divergent viewpoints are marginalized to "profane" status. The evaluation is profoundly political, but its political nature is not subject to an open and equitable debate.	Because of the absence of social theory to recognize and handle the exercise of power, dominant viewpoints will still determine the terms and boundaries of evaluation and marginalize divergent views.	CSH does not provide *content* for dealing with domain complexity. Instead, it relies on existing hard and soft approaches as appropriate. This means it requires its practitioners to understand all three and navigate their epistemic differences.

Note: a. CST also supplies ample conceptual arguments to extend the consideration of stakeholders beyond the human (Ulrich 1983; Midgley 2000; Stephens, Taket, and Gagliano 2019).

number of practically insoluble problems" they found in countries like Colombia (Escobar 2011, 55) after the World Bank itself "defined as poor those countries with an annual per capita income below \$100 . . . if the problem was one of insufficient income, the solution was clearly economic growth" (Escobar 2011, 24). Here too on this grand historic scale, the specification of the problem had left only room for one solution.

When it became clear that there was no way the planet could sustain eight billion people *at the level of material activity of the Global North*, two responses dominated. Both were colonial. One blamed "overpopulation" (again, the problem implies its solution). The other buried the concept of nature and elevated in its place the "management of natural resources"— "the twin of gluttonous vision. . . . What is at stake for these groups of scientists and businessmen . . . is the continuation of the models of growth and development" (Escobar 2011, 193). While earlier, "redistribution of wealth was the favored option to harmonize environmental protection and social justice" (Gómez-Baggethun 2019, 71), with sustainable development, "growth was no longer presented as the cause of environmental problems but as the remedy" (Gómez-Baggethun 2019, 71). The primacy of development was reified by qualifying it as "sustainable" under certain conditions:

two old enemies, growth and the environment, are reconciled (Redclift 1987). The [Brundlandt] report, after all, focuses less on the negative consequences of economic growth on the environment than on the effects of environmental degradation on growth and potential for growth. It is growth (read: capitalist market expansion), and not the environment, that has to be sustained. (Escobar 2011, 195)

In other words, the aim of sustainable development was "to save 'the religion' of economic growth and to deny ecological breakdown" (DeMaria and Latouche 2019, 149). In doing so, sustainable development framed the relationship between current and future generations as a conflict and manifests a colonial ideology reinforcing institutions dominated by the Global North.[4] This is not to say that the pursuit of most UN Sustainable Development Goals is unworthy, but that "sustainable development" as a framework cannot provide an ethical foundation for just sustainability design.

CONCLUSIONS

Wicked problems are not problems; they are situations characterized by conflicting goals and irreconcilable worldviews. Wicked problems cannot be solved as such because their existence is neither singular nor objective. To see wicked problems as solvable is to fall prey to an operationalist and positivist misunderstanding of the problem concept. Recognizing this trap allows us to pass it by. "The kinds of . . . activities that seem to meaningfully contribute toward sustainability are not those that solve well-defined problems, but rather those that contribute more subtly to a shift in culture or power. While this leaves us in the uncomfortable position of not necessarily knowing what to design . . . it does at least mean that we are looking in the right place for inspiration to strike" (Knowles, Bates, and Håkansson 2018, 7).

This chapter disentangled how domain complexity in wicked problem situations relates to social complexity. This helps us to better understand why rationalist approaches in design are so ill-equipped to design for sustainability and justice: When they handle social complexity at all, they handle it *as a case of domain complexity*. They treat the divergence of stakeholder views and the incommensurability of underlying worldviews as solvable problems of known complexity rather than as challenges to their own worldview. That usually serves those who design rather well. In problemism, the privilege hazard meets Gardiner's danger of moral corruption.

In situations where sustainability or justice are central concerns, irreconcilable worldviews meet asymmetric vulnerability. This creates a strong distorting force for marginalization and moral corruption. To ethically address real-world complexity in wicked problem situations and develop effective interventions, just sustainability design must first address the *ethical* complexity in the situation. This means to prioritize legitimacy over technology, which requires us to focus on establishing a fair discourse. Because the ideal speech situation can never fully be established, we turned to the dialectical tools of critical systems thinking to complete the proposed restructuring of systems design away from problemism toward just sustainability design.

III

REORIENTING SYSTEMS DESIGN

How will the rest of the story unfold? It is within our power to write a different future, if we can summon the courage to do so.
—Hickel (2020, 290)

So, I think my problem, and "our" problem, is how to have simultaneously an account of radical historical contingency for all knowledge claims and knowing subjects, a critical practice for recognizing our own "semiotic technologies" for making meaning, and a no-nonsense commitment to faithful accounts of a "real" world, one that can be partially shared and that is friendly to earthwide projects of finite freedom, adequate material abundance, modest meaning in suffering, and limited silliness.
—Haraway (1988, 579)

9

LEVERAGE POINTS FOR CHANGE
FROM INSOLVENT COMPUTING TO JUST SUSTAINABILITY DESIGN

To critique a particular normative regime is not to reject or condemn it; rather, by analyzing its regulatory and productive dimensions, one only deprives it of innocence and neutrality so as to craft, perhaps, a different future.
—Mahmood (2015)

So how do you change paradigms? . . . you keep pointing at the anomalies and failures in the old paradigm, you keep speaking louder and with assurance from the new one, you insert people with the new paradigm in places of public visibility and power. You don't waste time with reactionaries; rather you work with active change agents and with the vast middle ground of people who are open-minded.
—D. H. Meadows (1999, 18)

Just sustainability really is a perfect storm: the immense complexity of interacting spheres of human artifacts, societies, and the rest of nature meets the social complexity of designing in wicked problem situations with asymmetrically vulnerable stakeholders. Its characteristics mean that the computing field urgently needs to restructure the way it conceptualizes systems design to address its insolvency. Restructuring reorganizes the elements of a system so that it becomes more capable of fulfilling its obligations. Part II reorganized the narrative of what happens in systems design for sustainability and justice so that systems design research, education, and practice can improve. The conceptual restructuring of theories and methods completed in part II provides a basis for *reorienting* systems design. With the help of our critical friends, I separated computer science from its dominant myths and offered an alternative set of foundational principles. This might entail a sense of loss. In his book on "life after capitalism," Jackson writes that the role of myths is "to furnish us with a sense of meaning and to provide a sense of continuity in our lives. That need is a perennial one. The loss of a sustaining myth undermines our sense of meaning and threatens our collective wellbeing. Developing new myths, better stories and clearer visions is as essential as understanding the dynamics of collapse. Perhaps more so" (T. Jackson 2021, 56). In displacing the rationalist myths of systems design, the critical turn in computing elevates robust counternarratives:[1]

Software is never neutral: Recognize that computer science is not merely the systematic application of abstract value-neutral scientific methods to the real world. Values, ethics, and politics are not separate issues to be treated as "additional considerations" in computing. They are constitutive of the foundations of what computing has become. As a socially entangled discipline, it can no longer see itself as purely technical but must recognize its sociotechnical nature. The foundations of computing must include those social and humanistic perspectives that help us understand that nature. This shift addresses the myth of value-neutral technology (VNT).

People are more-than-rational: Reevaluate the role of rational decision-making and place much more attention on the difference between prescriptive approaches to decision-making and descriptive studies of practice. Critical reflection is the link between these two. This shift addresses the myth of rational decision-making.

Problems are framings: Because framings are inevitably made from partial perspectives, we must recognize the inability of computational thought to construct valid problem framings *on its own*. To address the myth of objective problems, we prioritize dialectical problem framing.

Design must be critically systemic: As a consequence, in systems design, rationalist computational methods must be placed within reflective methodologies using critically systemic frameworks. This methodological shift addresses problemism in systems design.

The concept *just sustainability design* (JSD) encapsulates the orientation away from a problem-solving paradigm to a critically systemic engagement with wicked problem situations. I introduced it in the Introduction to describe a systems design paradigm that aims to address the challenges of just sustainability (dispersal; uncertainty/ambiguity; fragmentation; power imbalance and asymmetric vulnerability; incommensurability). To do so, JSD needs to fulfill at least the following criteria:

1. *Constructive and critical*: Critique is an essential element of change. To reorient systems design toward sustainability and justice, critique must also examine the norms themselves that govern how computational systems are designed today. In doing so, JSD must be critical without abandoning the generative aspects of engineering and design. The idea of critical friendship is central to achieving this.

2. *Systemic*: Because the social, the technical, and the natural are entangled with political, cultural, and economic dimensions, JSD must take a systemic perspective—a perspective that prioritizes the consideration of wholes and their relationships over the isolated analysis of individual components and their properties. The boundaries of meaningful "wholes" rarely align with organizational and technical boundaries, so the commitment to systemic thought implies the acknowledgement that technology design always designs sociotechnical rather than purely technical systems.

3. *Dialectic*: The climate crisis has brought to the forefront central challenges for epistemology and collective action that sustainability and justice advocates have long grappled with, including the incommensurability of conflicting worldviews, the inevitable selectivity of each, and the need to nevertheless find common grounds for collective action. Just sustainability design—and in fact any systems design for the twenty-first century that tackles relevant social issues—must transcend the monological forms of reasoning expressed by traditional science-driven approaches to "solving social problems" through deductive means alone, in favor of a pluralist dialectics of design in which multiple worldviews can meaningfully engage.

4. *Diachronic*: The delayed temporal nature, and the path-dependent nature of design decisions and ecosystems, require a diachronic design

perspective— that is, a perspective aware of temporal scales and dynamics—studying its phenomenon as it evolves over time (Merriam-Webster 2020a). JSD must account for the historical profile of the processes that led to the system design, the life cycle of the system itself, the downstream and long-term impacts and consequences, and the temporal dynamics of design itself, rather than taking an atemporal focus on the "now" of design.

5. *Contingent*: The universalist approaches touted by science-driven methodologies are in fact neither universal nor appropriate to the case. Their "unmarked" worldview is partial, resting on unspoken assumptions that are Western and colonial rather than truly global. Far from being "independent of culture," as some would like sustainability to be, universalist approaches express an unacknowledged ideology built on a colonial and racist legacy. Just sustainability design must be aware of its contingency and the partiality of its perspective and equipped to reflect on its already given context, boundaries, and assumptions. I use contingency rather than situatedness or located accountability (Suchman 2002) to emphasize the importance of attending to the set of assumptions that are implicitly mobilized to justify systems design choices. But contingency also means that rather than being declared complete and optimal, JSD must be thought of as a proudly incomplete project, fashioned to learn and evolve.

6. *Legitimate*: The orientation toward sustainability and justice highlights the asymmetry of distant effects of design choices. Because full participation of those affected in design is not possible, JSD must grapple seriously with the question of justification. Rather than seeking a technical optimization or abandoning the generative orientation of design and engineering, the approach must aim to prioritize questions of legitimacy when systems design choices are to be justified based on their potential effects on those not involved in design.

7. *Reasonable, rather than rationalist*: In justifying critically and systemically the choices made in systems design with regards to their uncertain and distant effects, JSD cannot rely solely on rationalist modes of deduction. They would lead us right back into the traps of universalist frameworks of an ideological nature. Instead, it must be built on reasonable arguments. This is important, and it applies both to the

level of argumentation needed to develop and justify methodological commitments and principles and to the discursive level of reasoning through systems design choices. This does not preclude the application of scientific and rationalist frameworks for those issues and contexts to which they are appropriate in situations in which that is legitimate.

8. *Replicable, rather than repeatable*: Because of its emphasis on contingency and legitimacy, JSD will not produce one absolute method for design that can supposedly be repeated to yield optimal outcomes. Instead, it asks for replicability across different contexts, building and growing our understanding of how it can work in heterogeneous ways across diverse contexts.

These principles have led me to conduct research that critically evaluates proposed systems designs for sustainability and/or justice and the processes by which they come about (McCord and Becker 2019), develops design tools that support the collaborative exploration of distant effects in systems design among heterogenous stakeholders (Penzenstadler et al. 2018; Becker et al. 2016), studies the macro-cognitive systems of judgment and decision-making in systems design (Fagerholm et al. 2019), examines genuine participation vs pseudoparticipation in and by design (Palacin et al. 2020), and develops systems design methods that embody the criteria listed above (Duboc, McCord, et al. 2020). In my academic organizing, the just sustainability design framework has motivated the facilitation of conversations about the role of human values in computing (Becker, Engels, et al. 2019), the collective social responsibility of computing professionals and academics (Becker, Light, et al. 2020; Saxena et al. 2020), the ethical tensions in requirements engineering work (Becker, Betz, et al. 2020), and the leverage points available to our societies to abolish the conditions that allow Big Tech to have such excessive power (Barendregt et al. 2021).

I think of just sustainability design as a region and a mindset, rather than a fixed point, theory, or singular method. Its scope is defined by the principles above (critical and constructive, systemic, dialectic, diachronic, contingent, reasonable, legitimate, and replicable). Its origin lies in its aim to transcend the monological forms of reasoning expressed by traditional science-driven approaches to "solving social problems" through deductive means alone in favor of a dialectics of design in which multiple worldviews can meaningfully engage. I do not offer a comprehensive design framework here, but a roadmap that can be used to chart diverse paths across this region. I offer principles and high-level goals, chart tentative pathways and their challenges, and explore leverage points for change. One can argue that this framework amounts to a paradigm shift, and it would be naïve to assume that presenting eight principles is enough to make that happen. Rather, we will need to shift perspectives: from a focus on the narratives and assumptions of systems design to considering the social structures and forces that constitute and shape systems design practice. But change can start anywhere.

LEVERAGE POINTS FOR CHANGE

I am far from the first to suggest the need to reorient computing, or design, away from rationalist, modernist, patriarchal, capitalist foundations (Winograd and Flores 1986; Escobar 2018; Knowles 2013; Costanza-Chock 2020; Benjamin 2019). Many have recognized that such a reorientation is an indispensable step toward bringing human activity into harmony with the rest of nature (Hickel 2020; Midgley 2000; Kimmerer 2013; Escobar 2018). As Escobar (2018) writes, "the practice of transformation really takes place in the process of enacting other worlds/practices" (99). In this third part, I show how I enact this reorientation in my practice to illustrate the opportunities, difficulties, and consequences of this book's argument and to provide starting points and suggestions for others interested in enacting similar transformations.

The concept of *leverage points* (D. H. Meadows 1999) proposes that in a complex system, we can understand and evaluate possible interventions by considering how they address the structural dynamics of the system's organization. Some interventions may be easy to implement but have only localized or fleeting impact, because the dynamic behavior of the system is bound to erase any positive change over time. For example, pouring water on a hot stove will not permanently cool it down while it's on, but might well lead to steam burns and electrical shorts. Other interventions carry leverage: they shift the structure of the system's elements such that the newly changed system begins to exhibit new behaviors or initiate processes that lead to large-scale change. Disconnecting the stove from electricity might be an example. (I will leave it to you to draw parallels to geo-engineering.) Typically, interventions with small leverage are easier to implement, but the relationship between leverage and difficulty is not simple. By analyzing a system, we can evaluate which interventions may be most promising: feasible, yet powerful.

Meadow's classification of leverage points is built atop an understanding of the existing system from the viewpoint of system dynamics, so I will not reproduce it here in full. In ascending order of their power, Meadows proposed twelve intervention points. The weaker six are specific to system dynamics. They provide highly visible points of intervention such as tax percentages or efficiency measures, and they are often targeted by managers and policy makers. But they rarely shift the behavioral patterns of the system that produces them because these patterns emerge from the deeper structure of the system's organization. The highest leverage points within this group are the strength of balancing and reinforcing feedback loops, whose influence is often underestimated. The second half of Meadows's leverage points can be interpreted outside of system dynamics. In ascending order of power, these are:

- The structure of *information flows* determines who, and which part of the system, has access to what kinds of information.
- The *rules* of the system include incentives, punishments, and constraints.
- The power to add, change, evolve, or *self-organize* the structure of the system implies the ability to change any of the prior aspects, such as information flows.
- The *goals* of the system determine what behavior it will strive for.
- The *mindset* or paradigm out of which the system arises will shape its goals, structure, rules, delays, and parameters.
- The power to *transcend paradigms* involves difficult shifts in mindset. (adapted from D. H. Meadows 1999, 3)

The first three refer to the *design* of structural conditions that regulate how the system is coordinated; the final and strongest three, to its underlying *intent*: that is, "the norms, values and goals embodied within the system of interest" or "the emergent direction to which a system of interest is oriented" (Abson et al. 2017, 32–33). Within the world of just sustainability design, these leverage points might direct designers to act on the following:

Information flows: Researchers make publicly visible the sustainability debt of a large information system so that it can be evaluated by other stakeholders.

Rules: Lawmakers extend the legal frameworks of "extended producer
responsibilities" mentioned in chapter 1 to increase the extent to
which IT companies are liable for the environmental degradation and
health implications caused by designed obsolescence.

Self-organization: Local communities organize to form permanent groups
that evaluate a given system's sustainability debt or work to persuade
lawmakers to extend the responsibilities of producers, ban facial recognition technology, or regulate consumer data collection.

Goals: A start-up constitutes itself as a cooperative designed not to maximize shareholder return but to best serve its constituents.

Mindset: Increasing numbers of aspiring computing professionals commit
to, organize as, and practice systems design in a manner committed to
principles of design justice, data feminism, autonomous design, or just
sustainability design.

Beyond one paradigm: Powerful narratives of change sometimes hit home
across diverse constituents. From a critically systemic view, this reminds
us all of the need for epistemic pluralism. We want to build "a world
where many worlds fit," as the Zapatista slogan at the heart of Escobar's
(2018) work on *pluriversal design* suggests.

This list illustrates the range of possible interventions that can move
computing design from insolvency to just sustainability. It further illustrates that some leverage points have the power to cause interventions
on other leverage points. Each of these interventions implicitly assumes
a system boundary. The system in *goals* is the social organization of the
start-up. The system in *information flows* includes the object of evaluation,
the researchers, and others for whom this information serves as input for
some decisions (perhaps the *rules* regulators). Reflecting on these boundaries is crucial for evaluating possible interventions.

Meadows cautioned that her list is "not a recipe for finding leverage points. Rather it's an invitation to think more broadly about system
change" (D. H. Meadows 1999). Some leverage points relate to mindsets
and paradigms, and I make no secret of the fact that this book argues for
a shift in mindset. I harbor few illusions about the ease of doing so. But
smaller change can trigger larger shifts. We really do need to *act locally*
where we can, while *thinking globally* and keeping larger contexts in mind.

MY LEVERAGE POINTS

The *leverage points* concept suggests that the analysis of complex systems can help us to identify interventions that exert leverage beyond localized change, but the chapters in part III do not provide a comprehensive account and assessment of all the actions we may take to shift our practices of designing systems to be more sustainable and just. That is likely a good subject for a separate book beyond my own ability. Instead, I discuss how I have reoriented my own work, reflecting on the margins of maneuver that I see for myself, I survey some initiatives for change, and I consider some proposals worth exploring.

As an academic, I am asked to describe, measure, and evaluate my impact in terms of teaching, research, and "service," that is the organizing, coordinating, and administrative work that helps keep everything else running at my university and in my research communities. Since these are distinct categories of work in my employment contract, it makes sense to look for leverage points in each. What can I do within each to enact change, how significant is that change, and can I identify changes in each that can trigger larger shifts? I can try, aiming to be mindful of my privilege, humble about my abilities, and courageous in my ambition.

Research is at once the most obvious leverage point and famously slow. Academic research can produce insights that change public knowledge, influence public perception, and shift design practice. Even when it does, however, it often takes a very long time for new insights, methods, and frameworks to be adopted in practice. And yet not all research is slow to make an impact, as we can see, for example, in critical research about machine learning algorithms and their role in our societies. It is important to recognize the role of advocacy and organizing in leveraging research insights for policy change. For example, substantive research about facial recognition was used by organizers to advocate for banning it. Personally, I am in a privileged position, able to refocus my research where I see leverage. After all, addressing the *mindset* of problemism is a very difficult leverage point, but it is one with very strong leverage. Within the scope of the paradigm change this book argues for, I believe that the myths present us with leverage points we can tackle to move minds and views, and reframe issues. If the myths of computing depoliticize the discourse of systems design, then we can restore the acceptance that computing already is political by retelling the story differently, debunking false claims, and reframing the perspective. This is slow, and the work of persuasion takes time, but I would not be writing this book if I did not consider them worthwhile, and dare I say, you would not be still reading if you did not at least partially agree. I have not always followed Meadows's suggestion not to "waste time with reactionaries" highlighted in the epigraph, but I have eventually learned to "work with active change agents and with the vast middle ground of people who are open-minded." Chapters 10 and 11 speak to this.

Teaching is a notoriously slow leverage point too, but it is often said that the biggest impact most academics have at the end of their career is to be found in their students. And it does matter what we teach and how. "After all, educators hold the power to shape public perception of computing. We do this through the problems we focus on in our classrooms; through who we choose to teach; in how we shape students' career choices; and in how we conceptualize computing to journalists, social scientists, and society. The world has critical questions about computing and it is time we started teaching more critical answers" (Ko et al. 2020, 32). I have been fortunate to be able to reshape what I teach, and how I teach it. Many of my students are deeply concerned about the state of the world in terms of sustainability and justice. They are dissatisfied about the role of technology in our societies and deeply uncomfortable with the ethical dilemmas they anticipate facing in their future professional roles. They want to make a difference. Equipping them with critically systemic thinking and an appreciation for the other critical friends of computing has been a deeply rewarding experience. As one reflected in 2021:

I know that once I leave the academy, ethics and professional responsibility will be ignored in the workplace unless I make a point of bringing it up myself. I am comforted by the fact that through this course, I've built a toolkit of critical thinking frameworks and principles for organizing against harm that prepares me for the rocky road ahead. When I graduate, I will be a tech worker but I will also be a co-conspirator working in solidarity with my "users" to fight back against injustice.

Finally, "the traditional academic roles of research and teaching are not sufficient to drive transformative change in a time of rapidly accelerating global crises, so those with the greatest knowledge and understanding of these crises have a moral obligation to provide leadership, and engage in advocacy and activism" (Gardner et al. 2021, 5). *Service* too presents very specific and largely bureaucratic constraints, as well as the entrenched hierarchies of academic conference organizing, disciplinary communities, university departments, and funding policies. It is also a place where the matrix of domination profoundly influences what can be done by whom. Many academics cannot afford to withdraw their labor from departments where they are marginalized or from reviewing at conferences where their research is not valued. But whatever the constraints, we can prioritize and try to apply our labor and skills to elevate marginalized voices, to support those who have more profound insights than us and less visibility, and to help organize. That is what I have been trying to do. Chapter 12 will speak to the work of organizing collective action.

RESTRUCTURING AND REORIENTING

To flourish, just sustainability design must be critical without abandoning the generative aspects of engineering and design, as captured in its first principle. Embedding computational thinking and problem-solving into critically systemic frameworks retains their analytic strengths, allowing us to deploy them ethically and productively. Part II provided the conceptual and theoretical underpinning for this restructuring. It focused on the elements and relationships of key assumptions in the theories, methods, and practices of systems design. Technically generative features of design practice currently built on the myths of computing can be recast into new ways of thinking and working. Some aspects are worth salvaging, after all: We need problem-solving abilities; we need computational thinking; and we need some rational decision-making. The proposed restructuring alters layers of knowledge and practice of established computing disciplines.

The next chapters explore what that means. Chapter 10 illustrates a critically systemic shift in requirements engineering, and chapter 11 presents a shift in decision-making studies. Both illustrate how the principles of JSD reorient my own work. Other critical friends have developed different perspectives and themselves present important critical friends that are not all shown here, but from whom and with whom JSD should expect to learn and grow.[2]

10

CRITICAL REQUIREMENTS PRACTICE

A critical technical practice will . . . require a split identity—one foot planted in the craft work of design and the other foot planted in the reflexive work of critique. . . . This strangeness will not always be comfortable, but it will be productive nonetheless, both in the esoteric terms of the technical field itself and in the exoteric terms by which we ultimately evaluate a technical field's contribution to society.
—Agre (1997b)

If I ask you to design a new sustainable digital payment system, where do you start? Which questions do you have?

You might ask about existing infrastructure and systems (context), about who wants to make and accept payments (users), for what and how much and when (scenarios), or where and under which conditions (interfacing systems). You will have questions about existing standards (constraints). You will ask to speak to the users and other operators of this future payment system to identify what they want to achieve by using the system (stakeholders and goals). You will use these insights to figure out what the system has to do (functions) and how it should behave (qualities). And you will hopefully consider the direct and indirect effects of the system, including its environmental cost (sustainability debt). You may create mock-ups and prototypes to elicit reactions from stakeholders. The conceptual space you are traversing is inhabited by multiple fields and their frameworks, which have grown up in different regions. From Design Land, the UX clan and the design methods family send their regards. From Engineering Land, we welcome requirements engineering and the engineering methods family.

I positioned requirements as a "key to sustainability" in chapter 1 because requirements activities reconcile the social with the technical. In just sustainability design (JSD), those activities reconcile the social, the technical, and the environmental: We move between what Goguen (1992; 1994; 1996) called the "wet" world of living experience, characterized by its situated, emergent, embodied, indeterminate, contingent, ambiguous, and open-ended nature, and the "dry" world of abstract models. Ideally, we make each legible to the other. The term *requirements* carries three meanings: First, it can refer to stakeholder needs and aspirations; second, to *statements about* stakeholder needs and aspirations, and third, requirements are understood as the web of concepts that are needed to reconcile the social and technical when making choices in systems design. In this latter interpretation, requirements encompass and connect stakeholders, their concerns and goals, scenarios of system use and operations, as well as the features and qualities expected of a system under development and the constraints it must adhere to. It is the broad understanding in the third sense that is captured here and represented in table 10.1.[1]

Requirements thus provide the frame of reasoning that ultimately justifies technical design decisions by reference to the social world. To play a meaningful role in JSD, requirements engineering (RE) cannot pretend to be neutral. It is not hard to see that the shift advocated in this book challenges existing social arrangements, values, and power relationships. No engineering or design method or tool can on its own change these power relationships or magically establish ideal speech situations in which all stakeholders are given equal footing. But if we agree that the practice of RE is political, and that the margin of maneuver available to RE professionals in systems design is shaped by current frameworks and ways of thinking, then the question arises: How do we reorient RE in practice? What could a *critical technical practice* for RE look like? How could it widen the margin of maneuver and carve out a space for enacting social change in systems design? These are the questions addressed in this chapter.

The field's name is unfortunate. Requirements are not so much engineered as they are constructed, elicited, developed, and negotiated (Alexander and

Beus-Dukic 2009, 7–15; Nuseibeh and Easterbrook 2000). They belong to the "discursive" sphere of systems design more than to the material structures created by it. But the naming reflects that the field's preferred approach to reconciling social and technical issues is to apply systematic, measurable techniques to model and specify requirements. This chapter will take a deep look into the relationship between RE's "dry" frameworks and the implications just sustainability concerns raise in the "wet" world of human experience.

> **requirements engineering**, n.: the social practice of turning wet, interesting issues such as human values, politics and moral decisions into dry, complicated diagrams (models) that create the illusion that the work to be done is solidly understood.

To address sustainability, requirements activities must undergo a paradigm shift, as illustrated in table 10.1. Key tasks are listed in logical sequence on the left. Concerns of effectiveness and efficiency dominate current practice. JSD requires a shift, articulated on the right. These activities do not only take place when they are explicitly attended to through formal models and explicit methods. The conceptual linkage between these steps remains relevant when we leave the RE worldview. Any systems design effort will undergo almost each step listed in table 10.1, whether explicitly or implicitly. In many cases, these steps remain silent and undocumented. For example, agile projects typically will not conceptually separate the user stories that describe prospective system use from the described system's scope, features, and qualities. In other cases, these aspects are documented in excruciating detail. Regardless, they form a path-dependent set of choices that systems design passes through like gates in a maze. In any given project, the series of actual choices will rarely happen linearly according to their dependencies. But choices listed earlier in the table cause ripple effects through subsequent decisions. For example, the scoping of a project influences which stakeholders should be involved, which in turn shapes what voices will have a seat at the table when success criteria are determined.

Accounting for asymmetric vulnerability in just sustainability has profound implications on how we conduct these tasks. For example, it demands a shift in how we identify stakeholders, classify them, and incorporate their views. Current practice often involves a stakeholder influence matrix that maps stakeholders on two dimensions: influence over the project and interest in the project. This is often used to ensure that those with the most power remain supportive of the project. JSD will at least replace this with a matrix that prioritizes not those involved but those affected: by replacing the interest axis with an axis describing the expected impact of the system on their well-being, whether or not they are aware of it. This stakeholder map should then be used to prioritize those with the least influence who are most affected throughout the project stages, including all tasks listed in table 10.1.

It also shifts how we identify and account for risks. While current risk management practice often prioritizes the success of the project and its timely completion, JSD risk assessment may prioritize the identification of sustainability debt and the risk the project poses to its future environment, using tools such as the sustainability awareness framework (Duboc, et al. 2020) and long-term scenarios (Nathan et al. 2008) to identify systemic effects. When a project identifies significant risks of externalized sustainability debt but aims to proceed nevertheless, it ought to have an ethically sound justification, and ideally, that justification can be audited.

These kinds of concrete actions, however, can only take place successfully when accompanied by, and based on, more profound reorientations. For example, the shift in stakeholder matrix is easier said than done: those with influence may not exactly approve of it. The shift in risk assessment similarly is difficult to imagine in an organizational context driven by quarterly profits and shareholder returns unless the legal environment mandates it. And even then, we must pay attention to the politics of the process itself and examine how these politics shape it. That is the focus of this chapter.

Whether you formalize or not, when you design, you work out requirements. But reconciling the social and the technical in requirements activities should not simply mean performing trade-off analysis in a lone expert role to resolve any discrepancy. That would be an impoverished and unrealistic approach. Instead, good requirements work makes social concerns legible for the technical design sphere and makes technical features legible for the social design sphere. RE *as a discipline*, however,

Table 10.1 Systems design activities that reconcile technical and social issues (adapted from Becker et al. 2016)

Task	Standard current practice	Focus of a future sustainability design practice
Determine project objectives, scope, system purposes, and boundaries	Focus on the immediate business need and key system features. Don't question the project's or system's purpose.	Emphasize how the project affects sustainability in all dimensions. Strive to advance sustainability in multiple dimensions simultaneously. Experiment with different system boundaries to understand the changes' impacts.
Identify external constraints	Minimize constraints imposed by the direct environment of the system and its technical interfaces but include legal, safety, security, technical, and resources.	View constraints in each dimension as opportunities for design. Look for constraints from additional sources, starting with company corporate-social-responsibility policies, legislation, and sustainability standards.
Identify stakeholders	Minimize the number of stakeholders involved, and focus on those who have influence. Focus on internal stakeholders, and exclude unreachable stakeholders.	Maximize stakeholder involvement in an inclusive perspective integrating external stakeholders and involving those who are affected. Assign a dedicated role to be responsible for sustainability. Introduce surrogates to represent outside stakeholder interests.
Define success criteria	Focus on the financial bottom line at project completion. Measure the business outcome and financial return on investment.	Focus on advancing all dimensions of sustainability simultaneously, and take into account that most effects will occur after project completion.
Elicit requirements	Focus on the features and immediate outcomes the stakeholders want.	Help the stakeholders understand the system's enabling effects. Forecast potential structural impact (see figure 1.2).

(continued)

Table 10.1 (Continued)

Task	Standard current practice	Focus of a future sustainability design practice
Identify risks	Identify risks that threaten timely project completion within budget.	Include the effects on the system's wider environment. Include enabling and structural effects and risks that can develop over time.
Analyze tradeoffs	View tradeoff analysis as a question of prioritizing and selecting; let the influential stakeholders decide.	Strive to transform sustainability tradeoffs into mutually beneficial situations. Ensure that a wider range of stakeholders discuss sustainability tradeoffs. Beware marginalization.
Decide to go forward or not	Base the decision on feasibility, financial costs and benefits, and risk exposure to project participants—that is, internal stakeholders.	The decision is based on a consideration of positive and negative effects in all five dimensions on internal and external stakeholders. The evaluation is made auditable to show to external audiences how it considered sustainability indicators and enabling effects.
Validate requirements	Let key stakeholders verify that their interests are captured.	Ensure broad community involvement focused on understanding systemic effects.
Complete project	Verify whether success criteria are met on completion, then focus on maintenance and evolution.	Evaluate the effects in all five dimensions for a certain time after completion. Align with the expected timescale of effects.

has grown to take a certain approach to the task of reconciling the two spheres it connects, an approach heavily influenced by *engineering* and by the myths of computing. Understood *as a science* (Akkermans and Gordijn 2006), RE applies natural and social scientific methods to building requirements. While the field has never lost sight of the social context of technology (Yu 2011; Jarke et al. 2011), requirements in this worldview are the object of an applied scientific enterprise and the products of engineering activity. Its dominant frameworks are built on rationalist worldviews that assume differences in value positions can be reduced to common denominators and empirically adjudicated on the basis of universal reasoning frameworks. This entails the operationalism that comes with such an enterprise; it entails a focus on valuing precision, completeness, efficiency, quantification, and formality, reflected in its literature, textbooks, and standards; and it delineates the boundaries of what will be accepted into the enterprise as valid (scientific) knowledge.

Hence, the approach in RE to bridge the gap between the social and the technical has been to develop sophisticated conceptual models and practical methods of applying them in systems design situations. There are models, diagrams, and standards for every activity listed in table 10.1 and every imaginable relationship between the conceptual entities they relate to. I will briefly discuss some of the less formal approaches because they are more commonly encountered in practice and because their simplicity allows us to examine here how they are political.

In early scoping stages, it is common to present an *onion model* in which stakeholders are located in concentric circles around a center occupied by the product or service to be designed. The basic structure of these diagrams is shown in figure 10.1. Those stakeholders who use the system directly are explicitly centered while those who are affected in distant ways are placed

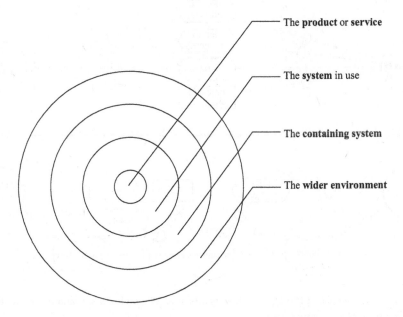

The **product** or **service**

The **system** in use

The **containing system**

The **wider environment**

10.1   The basic shape of product-centric stakeholder onion models.

at the margins. In some versions, the project team gets placed in an over-
lapping circle close to the center of the diagram. Drawing from information systems and management research and industry practice, the
field has developed sophisticated taxonomies of stakeholders (Alexander
2005), techniques for eliciting and representing their views (Zowghi and
Coulin 2005), and modeling approaches to connect this social sphere to
proposed technical elements. For example, the i* modeling language (Yu
2011) represents social actors and technical components as interdependent entities linked by the way they enable each other to perform tasks,
obtain resources, and achieve goals. This can be very useful to illustrate
and analyze how introducing technical components changes the relationships between people and software systems.

Figure 10.2 provides a crude model of an elderly living scenario. The
left depicts two stakeholders (shown as circles) and directional dependencies. For example, the elderly person is dependent on the caretaker to
monitor their well-being (a task), for their safety (a goal), and for their
time and attention (a resource). The caretaker is dependent on the elderly

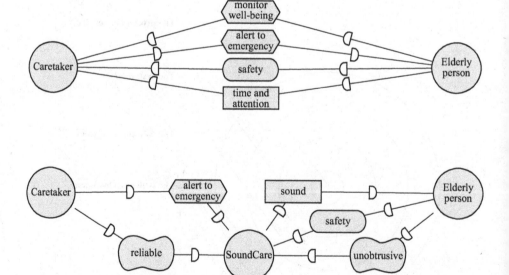

10.2 Two simplified i* diagrams show actors in an elderly living scenario before and
after technology is introduced.

person to alert them to an emergency (a task). These dependencies are altered significantly when a sensor system called *SoundCare* is introduced that detects accidents to alert the caretaker to a possible emergency. The SoundCare agent in the diagram on the bottom has an intermediary responsibility. People's dependency relationships change, and it appears that some of their interdependencies are rerouted through the product. The diagram also shows how people depend on technology to have certain qualities (reliability, unobtrusiveness). Social modeling provides a vocabulary to make these dependencies visible to facilitate reasoning processes and design decisions. But we should be mindful that knowing a vocabulary does not automatically bestow upon us an omniscient ability to represent reality:

> **goal modeling**: the illusion that everything that matters can be represented as instrumental achievement to be met; the delusion that anything that can not be represented as instrumental achievement to be met cannot possibly matter. Both are common in <u>requirements engineering</u>.

WHY REORIENT RE?

By deciding who to involve as stakeholders, how to elicit their views, how to consider these in design choices, and how to define success criteria, RE frames the scope of design and establishes the conditions for successful development. It is no surprise that the field is expected to address social and ethical concerns. The methods and techniques of RE are not programs, as we know, contrary to the myth of rational decisions. But they do structure social relationships and the meaning of work, and in doing so, they exert power. For example, formalized review meetings act as rituals that "maintain the appearance of rationality" (Rowlands and Kautz 2020, 18). We need to pay attention to the values and politics embedded in such models and methods. The professionals who practice RE engage stakeholders, facilitate and construct a shared understanding of what systems development should achieve, and represent that understanding in requirements specifications. They are thus often seen as translators, mediators, facilitators, or moderators between and among different discourses. Their position at these intersections affords them a unique opportunity to raise concerns of just sustainability, so they carry a moral responsibility.

RE today recognizes two paths to bridge the gap between the social and the technical. First, social modeling frameworks represent social actors and their motivations and intentions and document how they motivate and otherwise relate to designed artifacts and their qualities. Through these models, RE has addressed *domain complexity* exceedingly well. It is important to realize that approaches such as the i* social modeling language address social complexity only *as domain complexity*. They do not address the social complexity of a project's situation characterized by conflicting worldviews, unevenly distributed power, false consensus, or coercion (see chapter 8). Much less has been done to address such social complexity, and in particular, the uneven power relations between the different stakeholders involved in RE activities and affected by them (Milne and Maiden 2012).[2] Conventional RE simply has little to say about the politics of its models and methods and nothing about their formal biases. It considers itself a neutral technology and its process, a rational process.

Second, to address the well-known fact that many different problem formulations and system purposes are conceivable at the start of a project, experts often suggest the use of what they describe as "soft" approaches such as soft systems methodology (SSM). Techniques from these frameworks are suggested as tools to establish what the project purpose should be in ill-defined problem situations. However, a close look reveals that RE frameworks and methodologies themselves remain grounded in the scientific method, the view of engineering as its application, and a hard systems-epistemology.[3] As a result, they lack the vocabulary, concepts, and methods to address social questions of politics, morality, ethics, values, discourse, and belief systems. Instead, they treat problems and systems as given objects.

When practitioners move between the rich human and social worlds and the formalized technical models and methods used in RE, then, the expressive adequacy of technical models and the inadequacy of social theory are not balanced. Even if practitioners apply SSM more fully, as some do, SSM does not address the marginalization that inevitably arises out of power dynamics. As a result, those practitioners who recognize value tensions and conflicts can feel rather helpless. And it appears that their numbers are increasing as awareness of the role of IT in sustainability and justice rises. Ultimately, many requirements professionals are aware that they carry responsibility and have the opportunity to influence the outcomes of systems design, that this influence is not neutral, and that their will to change stands in conflict with existing power structures that define their jobs (Chitchyan et al. 2016). But RE practice offers little guidance in recognizing values, negotiating value tensions, and in particular, dealing with value conflicts under the surface (Thew and Sutcliffe 2017).

In other words, the myths of computing *depoliticize* RE: they make it look as if a neutral, objective method was used to discover and document preexisting requirements fairly. The outcome has the illusion of technical rationality and presents itself as a neutral object. Requirements become reified; the statement takes on the appearance of a fact. In this context, even if RE professionals approach their task with the best intentions,

how are they supposed to justify their work, their design decisions, and their actions if it is not considered feasible to estimate or predict possible effects over time and in many contexts; if they have no foundational education in social sciences, policy, or ethics; if they are embedded in industry projects with tight timelines, expectations of profit, and dispersed networks of potential stakeholders? (Duboc, McCord, et al. 2020, 18)

RE research and practice have tackled sustainability by introducing frameworks that allow the integrated consideration of direct, indirect, and systemic effects understood through environmental, social, and individual lenses in addition to the prevailing technical and economic views. Frameworks such as SusAF (see figure 1.2) continue the tradition of developing conceptual models to represent domain complexity and are now slowly taken up in industry projects.

These are important conceptual advances, but of the challenges of JSD, this framework addresses only domain complexity, not social complexity. It addresses an information problem: What is this system's sustainability debt? It does not provide an analytical framework to examine the question causally. It provides a framework to *represent* answers, but the designers need to find them elsewhere. This is why it is presented as an elicitation and awareness tool (Leticia Duboc, Penzenstadler, et al. 2020). That leaves open the substantive question of sustainability *analysis*. To begin to address it, we need to consider what happens when we interpret the outcomes of such analysis. The reorienting lens shows us: The answer to 'what is the system's sustainability debt' is predicated on how problems are framed and how system boundaries occlude their assumptions. These will always require critical examination (see chapter 8). The decisions made on the basis of the answers are not 'computed,' because they are human judgments (see chapter 7). Any consequences of these decisions carry value implications that are again in need of justification (see chapter 6). And to conduct this analysis ethically in a context of asymmetric vulnerability and moral corruption, we need the help of our critical friends (see chapter 5). In other words, we need to address the challenge of social complexity before we can hope to address the challenge of domain complexity.

THE ROOTS OF RE

Because of its conceptual orientation, conventional RE as characterized on the left side of table 10.1 cannot address the central challenges of JSD. Conceptually, it is simply too firmly grounded in a rationalist worldview. There is a historical explanation for this too: the roots of RE lie in the tradition of corporate engineering. The positioning of the requirements role within the organizational context of systems development projects is bound to serve those who are involved and, in particular, those who pay for the requirements activities:

Stakeholders include anyone with an interest in, or an effect on, the outcome of the product. The owner is the most obvious stakeholder, but there are others. For example, the intended users of the product are stakeholders. . . . Potentially dozens of stakeholders exist for any project. Remember that you are trying to establish the optimal value for the owner. (Robertson and Robertson 2012, 44)[4]

Even if placed in an ostensibly 'neutral' position, RE professionals[5] inevitably serve those who define their reference systems. This is not new: the subservience of engineers to industrial capital was at the heart of the job description when that job emerged over a century ago.

From the outset . . . the engineer was at the service of capital, and not surprisingly, its laws were to him as natural as the laws of science. If some . . . drew a line between technology and capitalism, that distinction collapsed in the person of the engineer and in his work, engineering. . . . "the dollar," [a leader] told Purdue engineering students, "is the final term in every engineering equation." . . . The economic inspiration inherent in technical work, of course, did not altogether rule out the possibility of conflict between the demands of technological supe-riority and of market expediency. When such conflict did arise, however, there was never any doubt about the outcome. [A leader's address to engineers] in 1896 had no trace of ambiguity: "The financial side of engineering is always the most important: . . . [the engineer] must always be subservient to those who represent the money invested in the enterprise." (Noble 1977, 34–35; see also 1984)

CRITICAL REQUIREMENTS ENGINEERING IN PRACTICE

If the myths of computing depoliticize RE, how can we restore an aware-ness that politics are always there? How can we invoke and apply the politics appropriate to the complexity of a project's situation and especially, the asymmetry of stakeholders? Agre proposed to develop leverage by working critically *within* the technical fields. He hoped to create a dialectic between constructive work and critical reflection. This would not be easy, and it takes time:

Successfully spanning these borderlands . . . will require a historical under-standing of the institutions and methods of the field, and . . . a praxis of daily work: forms of language, career strategies, and social networks that support the exploration of alternative work practices that will inevitably seem strange to insiders and outsiders alike. (Agre 1997b)

To develop a first version of a critical technical practice in RE, I brought RE into contact with critical systems heuristics (CSH). In discussing the project we conducted (first described in Duboc, McCord, et al. 2020), I focus on the first few activities of table 10.1, asking:

1. How can RE activities be augmented to uncover and negotiate implicit value positions?
2. How can RE critically interrogate the fact that RE professionals must always make selections and choices that are ultimately always political and partial, without becoming paralyzed?[6]

The SoundCare project described here focuses on a vulnerable popula-tion of elderly people living in their own homes. It addresses the long-established worry that fragile seniors are susceptible to accidents and may be unable to call for help. SoundCare brought networks of acoustic sen-sors to the rescue. A research group had developed algorithms capable of finely detecting acoustic events and was exploring a range of domains in which they may be useful. The project received funding for technology transfer research with an industry partner. You may notice a whiff of solutionism: technology was searching for a suitable problem to solve.

The product proposal describes a wireless acoustic sensor network used to capture and analyze sounds in the homes of elderly people to detect sounds that indicate accidents or changes that a caretaker might follow up on. The project focused on early-stage requirements engineering. We structured it using action research.[7] Our challenge was to perform the activities of table 10.1 using the approach in the right-hand column, albeit in a context focused on the approach on the left. The idea was quite simple: We would combine standard RE practice with CSH by performing iterative cycles of RE practice combined with reflections and critique. We wanted to plant one foot in the craft work of RE and the other in constructive critique, balancing our steps with the help of our critical friend CSH. During each cycle, the "requirements engineer," Leticia, would (1) interact with stakeholders to create new RE artifacts and (2) use the CSH questions to update what CSH calls the *ideal map*. This is not a visual map but a set of tentative responses to the twelve questions represented in table 10.2, posed in the *ideal mode*, for example, "Who *ought to be* the beneficiary of SoundCare?" Ideal maps in CSH are and always remain tentative and incomplete. They are not treated as a checklist but as probes for continued exploration.

After these two artifacts were created or updated, Leticia reviewed them jointly with one or two research partners in a *critical friend* role. We distinguished two such friends: one was closely involved in each iteration and thus became part of the team; the other remained more distant to represent a more critical, even polemic attitude. While the former helped Leticia to navigate the conceptual space of CSH, populate the map, and reflect on the questions and answers, the latter probed into the map's entries, using them as starting points for questions.

CSH can be used in a range of modes with differing intent.

1. *Critically heuristic self-reflection* is a form of reflective practice. "What are the boundary judgements presupposed in what I believe or claim to be true or right? What is the normative content of these boundary judgements, as measured not only by their underpinning value assumptions but also by their live practical implications, i.e., the ways they might affect other people? Should I consider alternative boundary judgements, and what would be their

Table 10.2 CSH questions for designing a system S (adapted from Ulrich 1987, 279)

	Specific concerns (stakes)	Social roles (stakeholders)	Key problems (issues)
Sources of motivation	What is/ought to be the **purpose** of S?	Who is/ought to be the intended **beneficiary** of S?	What is/ought to be S's **measure of success or improvement?**
Sources of control	What **resources** (i.e., conditions of success) are/ought to be under the control of S's decision maker?	Who is/ought to be the **decision maker** in control of the conditions of success for S?	What conditions of success are/ ought to be outside the control of S's decision maker, i.e., in the **decision environment?**
Sources of knowledge	What are/ought to be considered relevant **expertise**, knowledge, and skills for S?	What **experts** are/ought to be providing the relevant knowledge and skills for S?	What are/ought to be the assurances of validity for relevant knowledge for S: i.e., what is its **guarantor?**
Sources of legitimation	What are/ought to be the opportunities for **emancipation**, for the interests of those negatively affected to have expression and freedom in the worldview of S?	Who is/ought to be the **witness** representing the interests of those negatively affected but not involved with S?	What space is/ ought to be available for reconciling different **worldviews** regarding S among those affected and involved?

normative content? What ought to be my boundary judgements so that I can justify them vis-à-vis those concerned?" (Ulrich 1998, 7)

2. *Critically heuristic deliberation* is a form of dialogue: "Why do our opinions or validity claims differ? What different boundary judgements make us see different 'facts' and 'values'? How does my position look if I adopt my partner's boundary judgements, and vice versa? Can we agree on differing boundary judgements, and if we cannot agree, can we at least understand and respect why we disagree?" (Ulrich 1998, 8)

3. The *polemical employment* of boundary critique, finally, supports a more confrontational debate: "How can I make visible to others the ways in which my opponent's propositions depend on boundary judgements that have not been declared openly but which are debatable? How can I argue against an opponent's allegation that I do not know enough to challenge [them]? How can I make a cogent argument even though I am not an expert and indeed may not be as knowledgeable as the opponent with respect to the issue at hand?" (Ulrich 1998, 8)

Over the project iterations, we used CSH in all three modes. Leticia, who was new to CSH, used it for self-reflection. The *deliberation* stage in figure 10.3 employed the dialogical form of critically heuristic deliberation. We alternated this with a *critique* stage, which shifted between deliberation and polemic. We designed this configuration to ease the difficult conversations we can have with critical friends, and it often led to the discovery of underlying assumptions, making visible the *reference system* on which the ideal map was built (see figure 5.1).

The project underwent some significant shifts that I will illustrate with a focus on the ideal maps, highlighting CSH question categories in *italics* and key levers that re-integrate politics into RE in **bold**.[8]

Ideal map 1 was created by Leticia and one of the critical friends as the latter walked the former through the CSH process. It contained Leticia's observations from her previous discussions with the SoundCare development team, third-sector organization, and the caretakers of elderly people. Among other things, the map identified the elderly as the main beneficiary with the stated *purpose* of increasing their independence and enabling them to stay in their homes longer. It also set the elderly person's "independence and well-being levels" and the "number of years living alone at home" as *measures of improvement*.

Reflection revealed, among other things, that this first map represented the **privileged** views of those developing the system. The team consulted the notes of previous conversations with stakeholders to reflect their needs more accurately.

Ideal map 2 was created after the consultation of the notes from previous meetings and interviews with stakeholders. The new map added the families as a *primary beneficiary* and the health-care system as *secondary beneficiaries*, extending the **boundaries** of the system. The *purpose* now included peace of mind for families. Ideal map 2 also reviewed the concept of well-being and independence, among other things.

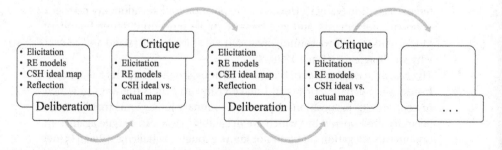

10.3 Iterating construction and critical reflection.

Reflection on the validity of our *measurements of success* made the team reflect on the legitimacy of our situated knowledges and the necessity to introduce different types of knowledge, most importantly, elderly people or at least professionals with expertise relevant to the situations of elderly people. For Ulrich, relying on incomplete or dogmatic perspectives is a major **source of deception** and can be a false guarantor that harms our understanding of the situation and our system's design. Reflection also raised a number of questions, including: Is "years at home" a suitable measure of well-being? Should the purpose be to increase independence or rather self-determination? What if the elderly *cannot be* the decision makers? This last question in particular raises important issues of fairness in representing the concerns of those at risk of marginalization.

What I personally found most interesting at this stage is how CSH helped make something obvious: In contrast to the stated project purpose, the real purpose of the product could not possibly be to exclusively benefit the elderly themselves. An honest articulation of its purpose had to prioritize the caretakers' benefits. That the system was more for them than for the elderly became clear when examining how the argument for benefits was constructed and operationalized via measures of improvement. Following the trace, from one CSH category to the next, made it very explicit that little would change for the elderly themselves. This was not a function of one particular question, but the relationships between them. CSH made it impossible to ignore the discrepancy between the ideal responses and the actual requirements statements. The critical friend's questions reshaped the scoping and purpose of the project. They introduced different values, different facts, and different boundaries (see figure 6.2). The logical consequence was to take a closer look at how the elderly may be negatively affected and seek out expertise on that matter.

Ideal map 3 integrated the views of a social worker with those of a psychologist, both specialized in issues of the elderly. The interviews highlighted several concerns, including that the system did not increase independence but rather security (as it cannot meet their physical and emotional needs), and that the number of distress calls from the elderly were a **false guarantor of success**. We also learned about common behaviors, coping mechanisms, and the importance of a trusted person. This map included new measures of success (e.g., increased social support, reduced anxiety of the caretaker, earlier detection of dementia) and professional scales used in psychology and social care to measure the well-being of elderly people and their caretakers. This shift was a direct outcome of the reflection on the types of knowledge that could be considered legitimate sources of evidence in the previous iteration. The map also included an extended list of decision makers and sources of knowledge.

Reflection raised doubts regarding the measurement of self-determination and early signs of dementia.

Ideal map 4 was developed after an interview with a practical philosopher who specialized in how technological projects affect ethics and privacy. We explored such questions as: How do you frame care and well-being? Can over-reliance of families on this technology lead to a loss of "human touch" and thus reduce well-being rather than support it? Can the technology reduce the autonomy of the elderly person, who should have the right to decide when to get help? Will third parties be interested in these data? Each of these questions brings **power imbalance** and the **politics** of stakeholders to the forefront of RE activities. Finally, we recognized the importance of public debate on such technologies, and we identified techniques from practical philosophy for uncovering stakeholder ethics, morals, and values. The new map included autonomy as a *primary aim*, offered a better definition for self-determination, identified the general public as a *desired expert*, recognized institutions that want data as a commodity as *undesired experts*, and incorporated possible worldviews about being old, supporting the elderly, living the good life, and surveillance technology.

Reflection led us to recognize we had been more concerned with people's perceptions than security. We identified a knowledge gap on security and decided to interview a security expert.

Several additional iterations followed (Duboc, McCord, et al. 2020). Each simultaneously shifted the boundaries of the system and made the underlying reference system visible. The project resulted in a requirements specification alongside reflections and template materials for the use of CSH in RE. But most importantly, the project enacted a *reoriented* requirements practice and showed that requirements work *can* be politically conscious. I will illustrate this in two effects of the critical friendship that stood out to me.

THE DILEMMA OF THE WITNESS

First, CSH put to the forefront the dilemma of the *witness category* illustrated in figure 10.4. Ultimately, only those living with SoundCare are in a position to legitimately approve the effects of the product's features and affordances on their lives. On that subject, they are the ultimate experts. Yet, it proved practically very difficult to involve them fully within the resource constraints of this project.

Experts such as social workers, psychologists, or neurologists speak for and on behalf those affected, but the *guarantor* for the legitimacy of their expertise ultimately rests on some form of authorization that lies outside their area of expertise. We may assume that the expert speaks as surrogate for the affected, but the surrogacy assumed by the experts is not innocent. Ultimately, there is no full substitute for participation. This points to the *sources of legitimation*. We were not initially in the position to empower at least some of those affected and move them from the right side to the center of figure 10.4. So what was the ethical thing to do: cancel our participation in this project or attempt to represent the interests of those affected but not involved? Similar to a classic study in critical systems thinking (CST) on boundary critique, we decided to cautiously proceed to the best of our abilities, engaging in continuous reflection on the boundaries of the views represented and the absent situated knowledges.[9] The team eventually succeeded in interviewing many more elderly people and caretakers. Despite the detailed attention to stakeholder models and the notion of *surrogacy*, conventional RE does not offer such systematic guidance to identify systemic marginalization. Nor does it offer the reflective capacity that CSH introduces.[10]

You might object that CSH did not discover anything new or surprising here. Other fields in computing have long demonstrated nuanced sensitivity to participation and marginalization in designing for *and with* vulnerable populations, drawing on other critical friends such as feminist STS and disability justice (Frauenberger, Good, and Keay-Bright 2011; Spiel

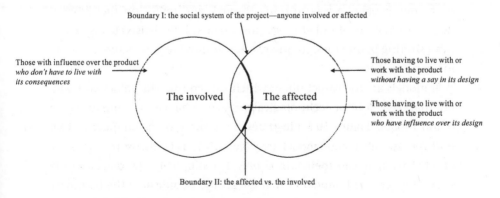

10.4 Stakeholder boundaries in systems design.

et al. 2019; Costanza-Chock 2020). From this perspective, it is obvious that those affected should be empowered to design. But that is precisely the point: CSH makes no claim to have the answers. Instead, it facilitated the emergence of the argument within RE and made it impossible to ignore it in a team with little prior expertise with participatory design. Less-than-ideal team composition is a frequent occurrence in RE practice. When we genuinely consider sustainability and justice in systems design, the implication of far-reaching, long-term concerns is that those affected will *never* be fully involved. CST made the argument for shifting participation, and it provides a framework to reflect on the inevitable limits to participation in relationship to the positionality of situated knowledge and the varying forms of participation.

THE SACRED AND THE PROFANE IN SYSTEMS DESIGN

The SoundCare project also illustrated how CSH turns the tables on our expectations of where legitimate knowledge comes from. Stakeholders who are nominally participating, for example as interviewees, are not automatically decision makers in CSH terms: that is, they do not automatically shape boundary judgments. Instead, they are often merely treated as a "data source" for design: their knowledge is used by those who design (Palacin et al. 2020). CST practice involves close attention to what STS calls positionality and situated knowledge via what Midgley (2000) calls *second order boundary critique*. It examines the boundaries of the social system that generates knowledge claims and how these boundaries evolve over time. It offers a way to think through and reflect on the different degrees of validity afforded to varying situated knowledge claims arising from shifting boundaries. In other words, it provides a structure to examine positionality.

If models are the carpet under which we can find the values and politics that have become technological artifacts, what can we find under an onion model? Figure 10.5 illustrates the same type of boundaries that underpin so-called onion models (see figure 10.1). Let's discuss their politics, or what we may call their formal bias. For simplicity the diagram only shows two crucial boundaries that are always established: the boundary of the system to be designed and the boundary that includes those parts of the environment that are considered relevant because they affect, or are affected by, design.

In the case of SoundCare, unless elderly people themselves *operate* the system, their claims and knowledges would be located not within the primary boundary but within the secondary, while technical and professional expertise are located in the primary boundary. Matters are further complicated by the presence of surrogate stakeholders such as elderly-appointed trusted decision makers and legally appointed guardians. When a conflict arises between two claims or value systems, it is often resolved by marginalizing one (Midgley, Munlo, and Brown 1998). And when claims about facts or values out of different spheres collide, they do not meet on equal footing. The views of this vulnerable population were therefore at risk of being dismissed by others who consider themselves to be properly trained, educated, and knowledgeable. In other words, their knowledge could easily be relegated to a profane, trivial status, while professional knowledge about RE, acoustic event detection, or elderly care may appear uncontestable.[11]

This happens in practice every day, whether we notice it or not. But when we probe the reference system of assumptions that justify and elevate some types of knowledges and marginalize others, the burden of proof shifts to the experts who need to justify not only that their claims

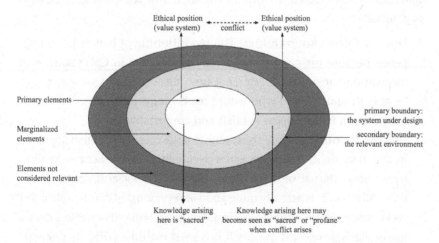

10.5 Marginalization and the sacred/profane (adapted from Midgley, Munlo, and Brown 1998).

are correct but, more importantly, that the expertise they use to justify their claims is legitimate. When it comes to evaluating whether their own knowledge is relevant and adequate, experts cannot rely solely on that knowledge itself to make the argument but need to rely also on boundary judgments. When it comes to justifying their reference system—the boundary conditions on which their application of knowledge is based—they are lay people like anyone else. In the face of the CSH questions, their knowledge is no longer sacred: "There are no experts in the systems approach" (Churchman 1979a, 232).[12] This decentering of codified knowledge is a crucial effect of CSH. It does not level the playing field altogether, of course, but constitutes a key step in reorienting RE for just sustainability.

CRITICALLY SYSTEMIC RE

The asymmetric vulnerability that characterizes the ethical question of just sustainability was mirrored, on a much smaller scale, within the microcosm of the SoundCare project. The project thus represents a central challenge to JSD: How can we justify the normative implications of systems design, as Werner Ulrich would frame it?

In the SoundCare project, CSH supported us in navigating the design space and counteracted the pull of the myths of computing. It made it easy to see:

1. **How problems are framings**: The repositioning of beneficiaries happened because the connected nature of categories in CSH's sources of motivation threw into sharp relief how success criteria were going to be operationalized and who would really stand to benefit. This made the problem frame appear explicit and contestable.
2. **How RE is political**: Once the framing of the original problem appeared in this new light, it also appeared decidedly political because it illustrated how alternative designs reconfigured the social relationships of those affected. I* is actually quite good at representing configurations of social relationships among stakeholders, but it will only enable a meaningful dialogue *about its politics* if it is used within a critically systemic framework such as the one presented here.

3. **How values become facts in RE**: The tensions that surfaced in the comparisons between ideal maps and actual maps, including what would appear in alternative i* diagrams, always connected statements of boundary, facts, and values. CSH made visible that a choice between different boundaries inevitably was a value-loaded choice that in turn created future facts.

By planting one foot in the craft work of RE and the other in the critically systemic work of CSH, we thus developed what I describe as a *critically systemic approach* to RE. The point here is not to argue that this approach was the best possible approach to conducting this kind of project. I would much rather defer to others better prepared to design for the elderly. Rather, the point is to illustrate how we can reorient RE to become critically appreciative and conscious of its political nature. In other words, CSH can help us thoughtfully reintroduce an awareness of politics into RE practice. CSH also offers concrete steps in design to what Sen (2009) describes as a "comparative approach" to justice: it is not aiming to define or design the "perfect" situation or system but aims to evaluate, compare, and improve the real situation in which design takes place.

In a retrospective interview about the role of CSH in SoundCare, Leticia brought up its role in supporting reflection, emancipation, and pluralism. She said: "What's nice about CSH is that it guides you into exploring these different aspects that I would probably not have asked about . . . CSH makes you aware of how your own views can shape the system." She thought in a standard RE process, "We would be the one taking all these inputs and then giving the privileged [those involved in the project] the task to reconcile them." In other words, the argument CSH makes for stakeholder emancipation shifted the modes of participation in a direction more consistent with participatory design and design justice. Leticia, however, considered the learning curve steep: reading the theoretical and philosophical work behind CSH, as she had attempted, proved difficult. Ulrich's and Reynold's case studies did not help to clarify how they worked with CSH. A crucial gap was simply that a critical technical practice, with one foot in technical craft and one in reflection, is still not part of mainstream computing education: "no one is being taught like this," she says, but "as you start developing the map, it becomes easier," and over time it can become a central component of a critical, reflective practice. She continues to work and teach with CSH, and others have begun to do the same.

If values beget facts, then pulling on the thread of facts begets values. To some degree, we can unravel the carpet of modeling to marvel at its composition and structure. And the use of CSH very effectively questions every framing and makes visible its associated sources of motivation, control, knowledge, and legitimacy. On that basis, it is difficult for any framing to become reified. Using these categories also sidelines correctness and efficiency in favor of legitimacy.

Such reorientation changes the activities listed in table 10.1. Instead of focusing on the immediate business need(s) and key system features without questioning the system's purpose, we experimented with different system boundaries to understand the impacts change would have. The team counteracted the tendency to minimize the number of stakeholders (meant to increase efficiency and focus on those who have influence). Instead, we widened the field of view and reflected on issues of surrogacy and legitimacy. Table 10.3 maps some of these early RE concepts and models to the issues that are often overlooked; the CSH questions that bring these issues to the surface; and the CST concepts that shine a light on them. Note this is not a checklist. Whenever we probe into one question, we are led to the others, so the example questions are an illustration, not a prescription for a method.

The practical effect of this type of practice will always be constrained by the political realities of the design context, and much remains to be explored even within this limitation. For example, CSH raises the importance of *reconciling* viewpoints but leaves it open to those involved how to interpret the meaning of reconciliation and how to address it. Decentering codified expertise opens a space for a pluralist view but does not populate it. How do we design such a process to facilitate RE for the "pluriverse"? And how will this process work in a situation where sustainability is to be evaluated, for example if SoundCare gets further developed as a smart sensing service?

There are pathways to pursue to build on this critical technical practice for RE. First, we need to gain more concrete experience with the approach and with other approaches to attain a reoriented, politically conscious,

Table 10.3 Mapping critically systemic concepts to stakeholder-focused RE concepts and activities

RE concepts and models	Issues	Example CSH questions	Critically systemic heuristic concepts
Stakeholder categories are used to identify and classify relevant stakeholders	Is the distribution of benefits and harms fair and just? On whose account?	Who is/ought to be the intended beneficiary?	Boundary I Boundary II Sources of motivation
Stakeholder viewpoints are used to represent divergent interests and perspectives	How are conflicting perspectives reconciled?	What space is/ought to be available for reconciling different **worldviews** among those affected and involved?	Worldview Sacred/profane
Stakeholder influence matrices are used to map stakeholders according to their power and interest	How do the designers conceptualize who has influence over the design?	What conditions of success are/ought to be *outside the control* of the decision makers?	Sources of control
Surrogate stakeholders represent interests of those affected but not involved	How do the designers make boundary judgments about participation and their own selectivity?	Who is/ought to be the **witness** representing the interests of those negatively affected but not involved?	Sources of knowledge Second-order boundary critique
Stakeholder **onion models** diagram stakeholder categories according to their distance to the product or project	Marginalization: How are inputs recognized as sacred or dismissed as profane? What does it mean to be a *distant/indirect* stakeholder?	What opportunities do those negatively affected have to express their view beyond the worldview of the designers, i.e., to emancipate themselves from the designers' worldview?	Marginalization Emancipation
Goal models and social models such as i*	Whose goals get represented? What are the value systems implied by boundary judgments?	What is ultimately regarded as valid *guarantor* of success?	Sources of knowledge Sources of motivation Reference system

critical RE. Second, we need to examine in more depth the formal bias of RE models, moving down the set of activities in table 10.1 and reworking each through the lens of critically reflective practice. And, finally, we should examine the hidden values and value tensions in RE practice and theory itself. How should RE as a role, a profession, and a scientific community deal with the political nature of what it does? For this question, we enrolled another critical friend, whose friendship I will briefly describe.

DECONSTRUCTING CRITICAL RE

Feminist STS scholar Doris Allhutter describes "sociotechnical design as situated, embodied practice configuring new alignments between the social and the material" (Allhutter 2012, 685). She draws on deconstructivist feminist theory to develop the method *mind scripting*. Its objective is to allow teams to uncover implicit values and implicit positions, reflect on how subjects construct their own identities through discursive and material acts, and better understand how the performance of categories such as gender, race, class or sexuality intersect with their work. The method "allows a collective to deconstruct unconscious sense making and enables negotiations on the adequateness of implicit assumptions guiding decision making" (Allhutter 2012, 686).

With colleagues in the RE field, she and I used mind scripting to examine RE as a field. In the first Critical RE Workshop in 2020, we brought together "the analytic and modelling strengths of requirements engineering with the critical and social theories that can help our community better reckon with the social forces that shape technology design through requirements" (Becker, Betz, et al. 2020).

How does the collective process of mind scripting work? There are several steps to this method. First, a series of interviews leads to the crafting of a prompt. Our prompt was: *When they pushed the status quo in requirements work to acknowledge value-laden concerns* . . . Then, each member of the collective writes down a brief text describing a memory in response to this prompt. In one memory text, a situation was laid out in which a researcher team worked with industry partners. They used the sustainability evaluation framework of figure 1.2 in conjunction with a negotiation process called *EasyWinWin*, in which "stakeholders move through a

step-by-step win-win negotiation where they collect, elaborate, and prioritize their requirements, and then surface and resolve issues" (Grünbacher
and Boehm 2001). Here is a part of this response:

First, the other participants were a bit unsure if the CEO really had understood the
different dimensions, but later it turned out that he actually had. For each of the
requirements the CEO wanted to discuss, the people in the workshop were thinking of potential issues regarding each dimension. However, for most of the time
the CEO was talking, which was a bit annoying, but his arguments were valid. He
managed to cover different viewpoints, but of course it would have been great to
actually have different stakeholders participating in the negotiation. For every
issue, they also tried to identify possible options—ways to mitigate or even fully
overcome the issue. Here, a lot of assumptions were made which made them
feel that the process is a bit fuzzy and more based on opinions than on actual
facts. Based on the list of issues and options for each issue, an agreement—
highlighting the "best" available option—could be achieved. (Memory text,
CREW2020)

During the workshop, these anonymous texts are collectively discussed,
one by one, guided by a series of questions. This process of *deconstruction*
"questions the normativity of discourses and practices by revealing the
constructedness of seemingly 'natural' sense making" (Allhutter 2012,
689). Guided by Dr. Allhutter, the group explored how power is reproduced in RE and what kind of identity the text presented. The conversation
was recorded, transcribed, and analyzed.

A central feature of mind scripting is that it "seeks absences and
silenced contradictions that obscure the mechanisms sustaining hegemo-
nies and power relations" (Allhutter 2012). As part of the method, silenced
contradictions are pursued to uncover how seemingly obvious positions
have been discursively constructed. At the second workshop, Dr. Allhutter (2021) showed us some results. The questions she surfaced included
issues with parallels to CSH categories. For example, the memories showed
that different kinds of knowledges were being negotiated. Facts about
causal relationships and tradeoffs between sustainability dimensions were
implicitly preferred over fuzzy opinions. During the deconstruction, the
discussion shifted to the assumptions and value positions that underpin
conflicting claims about tradeoffs. The method also surfaced power relationships between actors represented in memories. The author of the text
above was keenly aware of the tension between the real world and an ideal

world in which the workshop would have happened with different participation, in different forms. Deconstruction also revealed an uncomfortable
relationship between industry and academia: It felt to the researcher as
if their product was *examined* in a test of the team's ability to construct
something useful. Industry was calling the shots. In the common situation where short-term industry profits are in conflict with longer-term
sustainability debts, there may not be any win-win outcomes. With power
dynamics in play, the conventional method is unlikely to achieve its goals.

Even where mind scripting pointed to already familiar questions, it
offered a new view, and it revealed entirely new perspectives on what we
were trying to do that we could have never learned by following heuristics
alone. For example, our critical friend made us aware of identity struggles
and the drawing of boundaries between the "critically minded" and the
supposed "others." A recurring issue was how *being critical* was negotiated, and how critical subjects constructed their own position through
their texts. The discussion around my own memory text displayed this
powerfully: I described what I experienced at a sustainability-oriented
computer science workshop. A segment of it went as follows:

Some ideas had crystallized into a proposal: to design a universal experimentation and evaluation platform that would be almost like a crystal ball, a platform to do it all: gather information from anywhere, predict the future, help us
all transition to healthier lifestyles and policies. Some felt that the computing
disciplines present could provide enormous benefit to the planet. They could
put their forces together and help people make better decisions on the basis
of such predictions. Others thought that this proposal could only be a critical idea, a story to remind them of the inevitable failures to predict emergent
socio-ecological behavior, a reminder to explore the inevitable limits of success
of this idea. . . . The compromise that ensued, more implicitly than explicitly, was
that most of the proposal read like the system was to be designed, but a caveat
text was introduced (and fiercely defended on several iterations) to state it as a
"mechanism" or tool for examination. (Memory text, CREW2020)[13]

Deconstruction revolved around the difficult desire of intending to
be critical but not paralyzed and how the anonymous author (I) negotiated this question, following a common idea of a continuum between a
critical and a pragmatic attitude. By examining the deconstruction of my
memory by our collective mind, mind scripting showed us how our conversation simultaneously constructed the subjects and objects of critique.

An important insight I gained was the encouragement to shift the target of critique: Instead of focusing on *the other*, ask what are the structural conditions of knowledge production that create this opposition? As Dr. Allhutter framed it (2021), we "need to analyze our own political circumstance to make transparent the social antagonisms/ideologies that obscure the workings of power in our epistemic norms, values and practices."

CONCLUSIONS

These brief vignettes cannot fully capture the nuance of this critical friendship or its developing insights, but I hope they show why I consider it an important help in the ongoing effort to reorient RE. Mind scripting can uncover implicit norms that CSH does not reach. This is partly because it engages with affective interactions during deconstruction. More generally, the two friends have grown up on such different epistemological grounds that they offer contrasting viewpoints. But both of these critical friends offer uncomfortable insights that computing will not find elsewhere. "By making us uncomfortable, critique contains within itself a transformative orientation" (Bargetz and Sanos 2020, 511). These insights arise only if we seek out these friendships, care for them, and listen. Getting into a zone of discomfort is a key step to overcoming insolvency. And it takes two for a critical friendship to flourish.

In this chapter, I critically examined the state of RE in light of the challenges of JSD. I illustrated how the myths depoliticize RE methods and practice, and I presented an attempt to restore the acceptance that RE is already political by uncovering and negotiating implicit values and implicit assumptions. The resulting critical technical practice for RE that I have presented here is not a complete or comprehensive method. I present it as an orientation and starting point to allow us to engage with the craft of RE while simultaneously remaining with one foot in the craft of critique. The introduction of CSH into RE practice supports the CST goals of reflection, emancipation, and pluralism. In contrast to value-based engineering approaches, CSH succeeds in making transparent the values in the assumptions and implications of modeling choices. Mind scripting adds an entirely distinct voice to the reorientation proceedings, a voice with its own history, melody, ways of speaking, arguments, and attitude.

The overlap of concerns and the complementary views that arise from each is striking.

Do I think that this is enough, that we just need to add CSH to RE and all the challenges will be resolved? Categorically not, of course. But that's the thing about leverage points: they are small changes that lead to bigger changes. I believe we have barely started and a lot to do. We will continue to be reliant on our critical friends as we reorient RE and other systems design fields for just sustainability.

11

SEARCHING FOR JUST, SUSTAINABLE DESIGN DECISIONS

Many citizens are ready to sacrifice for the greater good. We just need institutions that help them do so.
—Hauser et al. (2014, 222)

What if you found out you had a choice between a 92 percent chance of getting a delicious cacao olive biscotto from Toronto's Biscotteria Forno Cultura when you have read this book or a 66 percent chance of getting *two* of these marvelous cookies one month later. Which would you choose?

Have you made up your mind? Congratulations: You have made one of the many "intertemporal" choices you make every day—"decisions that involve trade-offs between outcomes occurring at different points in the future" (Frederick, Loewenstein, and O'Donoghue 2002; Loewenstein, Read, and Baumeister 2003). You have made many such choices before. In my favorite intertemporal choice cartoon, little boy Calvin and his tiger Hobbes are outside looking at piles of fresh snow (figure 11.1). Calvin feels torn. He must contain his evil urges to receive Christmas gifts: "my immediate pleasure is pitted against my future greed!" Hobbes sighs. "Poor Susie." But Calvin insists: "It's not a foregone conclusion!"

Researchers investigating intertemporal choice have noted that "most—if not all—choices that individuals and organizations make in the real world are intertemporal" (Soman et al. 2005). It follows that many, if not most,

11.1 Calvin and Hobbes face an intertemporal choice. (CALVIN AND HOBBES © 1993 Watterson. Reprinted with permission of ANDREWS MCMEEL SYNDICATION. All rights reserved.)

systems design choices are intertemporal too. But their temporal nature is not always recognized, acknowledged, and considered.

Since sustainability is all about future outcomes, this should matter a great deal: short-term choices are definitely bad for sustainability. If design teams tend to discount the future in their decisions, then we should counteract that, as the Karlskrona Manifesto on Sustainability Design suggests (see table 1.1). We need to understand how this process actually takes place to have hopes of intervening in it, especially if the delayed systemic effects that need to be considered are so complex and ambiguous. These decisions are complicated by the fact that who is affected is no longer a straightforward question. *Psychological distance* indicates how far an event is removed from direct experience. The concept encompasses the dimensions of time, space, social distance, and how real an event appears to be (Liberman, Trope, and Stephan 2007). The consequences of psychological distance in consumers have been studied extensively, and some researchers have investigated the larger implications of the underlying insights, including "why climate change doesn't scare us (yet)" (Weber 2006; McDonald, Chai, and Newell 2015).

How do intertemporal choices occur in systems design practice, how does psychological distance manifest, and how does all this affect sustainability and justice?

This chapter's aims are twofold. First, I introduce intertemporal choice and psychological distance and describe what we know about *how* people make systems design choices that incur distant effects, as all design decisions related to sustainability and justice do. This view forms an important part of the diachronic perspective of just sustainability design. Then, I illustrate that in order to have a chance at genuinely shifting towards more just and more sustainable decisions, we need to reorient the theoretical frameworks and methods used in researching such decisions. By avoiding the normative fallacy discussed in chapter 7, we identify ways to ask better questions. I draw on my own research on intertemporal choices and psychological distance in software engineering (SE) decisions to make that argument, then connect it to studies of other researchers more directly focused on climate change.

Before that, another intertemporal choice. This time, you can be certain to get an entire box of delicious biscotti, and you have to pick: it's either delivered to you right when you finish reading, from a biscotteria that denies climate change, or about two months later, from a cooperatively owned shop that partners with women-owned organic farming cooperatives for all ingredients. Either shop will be paid for the purchase. Do these two biscotti boxes still seem equally delicious to you? How long would you be willing to wait for the second box? And how do you weigh these options?

JUST SUSTAINABILITY DECISIONS ARE INTERTEMPORAL CHOICES

Once we pay attention to the temporal nature of systems design decisions, it becomes apparent that intertemporal choices are ubiquitous and often ambiguous. Do I finish coding this new feature quickly so that it works for the deadline next week, or do I first design a robust architecture to make sure the feature can be easily tested and extended later? Do we spend another day testing the latest release or do we roll it out to our customers? Do I spend an extra hour writing documentation for my latest code while it's fresh on my mind, even though I don't seem to need it at the moment, or do I schedule two hours after the deadline to do it? All these are intertemporal choices in software development. The entire area of technical debt arises from design choices that are "expedient in the short term" but expensive later on (McConnell 2007).

Once we widen the horizon to consider the broader implications of design decisions, intertemporal choices are even more prominent, and the time horizon expands. Do we take the time up front talking to the community about which issues they identify, or do we create a prototype for the first problem *we* identify? Do we undergo the effort to equitably involve a marginalized stakeholder group? Should the success of this project be evaluated on the day of its release, or should the project remain unevaluated until one year later? Should we build a feature into the system that allows users to extract all their data out of the database when the system is retired at the end of its life cycle? Do we spend time educating ourselves about the proper way to represent gender, or whether we need to represent it at all, or do I reuse this existing code that does? On the other hand, by adding code to store and process gender, aren't we just wasting time? If my team lead wants us to reuse a module using binary gender code because "it already works," how hard do any of us try to educate them on the value of gender diversity and fluidity? How do we evaluate the latest proposal to redesign our transaction processing system on the basis of blockchain tech?

All these choices are intertemporal and will have material implications for sustainability, equity, and justice. But the intertemporal nature of design decisions that affect sustainability may not be apparent until we make it so. Some of the outcomes of these decisions are uncertain, ambiguous, and removed not just from our immediate experience but also from our primary planning scope. These characteristics can easily tilt the evaluation of what should be done such that distant outcomes are overlooked or not evaluated with the same attention or weight as closer ones. As a consequence, many design decisions that appear relatively harmless contribute to what I have described in chapter 2 as the debts of computing. Performing a sustainability evaluation, even just applying the sustainability awareness framework based on figure 1.2, is one way to make these outcomes more apparent and consider them more fully. But this in itself is an intertemporal choice: Should we spend a week on evaluating and comparing the sustainability effects of these two alternative system architectures or should we pick one and move ahead?

INTERTEMPORAL CHOICES IN JUDGMENT AND DECISION-MAKING RESEARCH

There is a plethora of research on intertemporal choices in the area of judgement and decision-making (JDM).[1] Recall from chapter 7 that a *decision* is a commitment to a course of action, a *choice* is a type of decision that involves the selection of an option, and *judgment* is a broader concept that involves reflective consideration of situational factors. When a decision or choice is intertemporal, it is often tricky to know what the best decision is. Individual choices will vary. *Temporal discounting* describes how a decision maker's valuation of an outcome changes when it shifts into the future. The classic model of intertemporal choice is the discounted utility model (Samuelson 1937). In this simple quantitative model, a person indifferent to the choice between receiving $100 in one year and $100 in two years is said to exhibit no discounting. Someone who requires an additional $100 in order to postpone the reward by a year is said to have a discount rate of 100 percent for that year.

A positive discount rate is common in studies, meaning that in general, people tend to favor positive outcomes when they are closer in time. But research shows that there is no permanent temporal preference built into our minds. A wide range of studies measuring discount rates in observed behavior across all sorts of decisions made by consumers have resulted in such a wild range of discount rates that a milestone review diagnosed "spectacular disagreement" (Frederick, Loewenstein, and O'Donoghue 2002). A natural explanation for this is that our brain is not a broken computer and that we consider many contextual factors when we make choices. Here we need to be cautious about avoiding the normative fallacy: the fact that we can describe behavior using a discount *rate* does not imply that the discount rate is the causal mechanism that causes this behavior. It is in fact extremely unlikely. The discount rate is an exponential function that compounds not unlike interest rates do. It does not take a lot of empirical research to understand we do not evaluate our biscotti preferences that way. If you picked the earlier biscotto, your inferred discount rate is 69.7 percent *per month*. Over a year, this amounts in removing 98 percent of the biscotto's value, leaving only a crumb. And how would you calculate the second version of the gamble using discount rates? Many would sooner give up on the biscotti than resort to this calculus.

Luckily, we do not need to model participants' intertemporal choice behavior as a simplistic function to ask interesting questions. There are other ways to characterize intertemporal choices, distinguish whether people exhibit temporal preferences, detect patterns, and explore what gives rise to these patterns. This requires a more systemic view. The *choice architecture* concept describes how choices are framed, organized, and presented to decision makers (Thaler, Sunstein, and Balz 2010). Changes in choice architecture can "nudge" decision makers toward preferable choices (Thaler and Sunstein 2008). It forms part of the broader context of decision-making, which includes such aspects as team dynamics, organizational incentives, and values. In JDM, the broader system of these elements is called the *macro-cognitive* system (G. Klein et al. 2003; Maarten et al. 2017). We will return to it later.

INTERTEMPORAL CHOICES IN SYSTEMS DESIGN

In software projects, where I have studied this subject myself, the most explicitly intertemporal decisions surface in technical debt management, architectural tradeoffs, test automation, feature prioritization, and project management decisions (Becker, Walker, and McCord 2017; Becker et al. 2018; Fagerholm et al. 2019). These kinds of decisions specifically deal with options that have outcomes at different points in the future. However, intertemporal choices also surface in less obvious ways.

When I began to study these decisions myself, I opted for a conservative approach. To demonstrate to the SE research community that there was *something worth studying*, we had to conform to the expectations that would be enforced in peer review. So initially, we designed a behavioral experiment in the spirit of the rationalist tradition, aiming to evaluate to what extent software developers exhibit temporal discounting at all. In this recent study (Becker, Fagerholm, et al. 2019), replicated in several countries (Fagerholm et al. 2019), we examined whether software developers discount future outcomes. We chose an inconspicuous project management decision relatively unrelated to just sustainability to minimize ambiguity and complexity in the study design. We simply asked participants to indicate what time savings they would require to consider an uncertain positive outcome at different times in the future (potential effort savings) as equally valuable as a comparable closer outcome (feature development). In step one, we established an initial baseline for a fixed time horizon. In step two, we asked them again but for a set of time horizons, as illustrated in figure 11.2. This is a standard design for eliciting temporal preferences adopted from behavioral economics

You are managing an N-years project. You are ahead of schedule in the current iteration. You have to decide between two options on how to spend your upcoming week. Fill in the blank to indicate the least amount of time that would make you prefer Option 2 over Option 1.

Option 1: Implement a feature that is in the project backlog, scheduled for the next iteration. (five person days of effort).

Option 2: Integrate a new library (five person days effort) that adds no new functionality but has a 60% chance of saving you _____ person days of effort over the duration of the project (with a 40% chance that the library will not result in those savings).

(The only difference here is the timeframe.)

For a project time frame of 1 year, what is the smallest number of days that would make you prefer Option 2? _____

For a project time frame of 2 years, what is the smallest number of days that would make you prefer Option 2? _____

For a project time frame of 3 years, what is the smallest number of days that would make you prefer Option 2? _____

For a project time frame of 4 years, what is the smallest number of days that would make you prefer Option 2? _____

For a project time frame of 5 years, what is the smallest number of days that would make you prefer Option 2? _____

For a project time frame of 10 years, what is the smallest number of days that would make you prefer Option 2? _____

11.2 Intertemporal choice task used in our study, step two.

research (Frederick, Loewenstein, and O'Donoghue 2002; Hardisty et al. 2011; Becker, Fagerholm, et al. 2019; Fagerholm et al. 2019). I will use this example to illustrate how we should study decisions involving psychological distance before speaking more directly about just sustainability.

Figure 11.3[2] shows the aggregate responses for different project time horizons, with outliers above one hundred days omitted. The responses vary significantly but trend upward with increasing timeframe. Does this mean that our participants exhibit temporal discounting?

To answer this question, we sidestep the normative fallacy inherent in the temptation to calculate discount rates. Instead, we use a descriptive approach that does not rely on a normative model (Myerson, Green, and

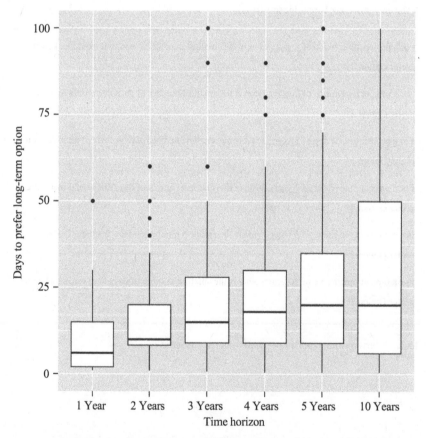

11.3 Responses from a replication study on intertemporal choice in software projects.

Warusawitharana 2001; Fagerholm et al. 2019). We use the first response as the present value. For years two to ten, we calculate the ratio between the respective responses and the present value. The result expresses how temporal distance affects the responses and allows us to distinguish between three response patterns: some participants prefer temporally nearer outcomes, some are indifferent to changes in time, and some in fact prioritize distant outcomes over nearer outcomes (Fagerholm et al. 2019).

BEWARE THE NORMATIVE FALLACY

We still have to be careful interpreting these results. What do they actually show? All we know is that some people behave *as if* they would perform temporal discounting in this particular situation, and some (fewer) exhibit future preference. We have not identified how or why this effect takes place. We have most certainly not identified a processing module in their brain that computes a discount function. Performing an MRI scan of our participants would not change that. Nor do we have a "gold standard" of optimal decision-making. There is no "correct" decision to be made in the presented scenario. Many reasonable factors influence the evaluation of uncertain future outcomes. Many professional situations may be structured in ways that makes temporal discounting perfectly reasonable, whether because of job rotation and turnover, incentive structures, divisions of labor, business models, project cycles, system life cycles, or other factors.

We could still pursue the rationalist interpretation in the tradition of the heuristics and biases program, as many in the field do. We could treat any discounting as a mistake, focus on cognitive biases and supposed heuristic shortcuts, and aim to "fix" the discounting in our participants. The rationalist approach treats the situation presented in figure 11.2 as a case of choice under probabilistic uncertainty. The response is to handle it as a quantitative tradeoff problem and model its probabilities and benefits to recommend the optimal choice. But that is a problemist orientation to sustainable decision-making. Many practitioners will take a pass (Dittrich 2016; Becker et al. 2018). Because it treats the cognitive process as machinery, the rationalist model simply does not address the lived experience of practitioners and the difficulties they face when they try to exercise careful judgement.

I prefer to approach this topic as a human and social phenomenon and investigate why some participants took a long-term view without being asked for it; why many participants in their feedback inquired about a wide range of contextual factors that they felt were omitted; and to ask *how did you make your decision?* Only by understanding how they perceive the situation, how they reason about the factors they consider relevant, and how they evaluate their options, can we hope to identify paths to more sustainable and just design decisions in general.

WHAT LIES BEYOND THE MYTH OF RATIONAL DECISIONS?

How do we get a better understanding of intertemporal choices in systems design then? In another study with fifty participants from Colombia, Greece, and Sweden, my team and I began to use the insights represented in chapter 7 more fully. Our participants performed the task discussed above as a think-aloud protocol study, followed by a semistructured interview session in which they explained how they had reasoned and reflected on intertemporal choices in their professional life. The research design is a configuration of cognitive task analysis (CTA), the central toolbox of naturalistic decision-making researchers (Crandall, Klein, and Hoffman 2006).[3] What did we learn?

Half of our participants behave as if they discounted, half do not. For many professionals, shifting time frames do not merit discounting at all. Instead, they explicitly emphasize that the timeframe should not make a difference and therefore show an entirely flat discounting curve. One said: "Personally I'd place the same days of effort, it doesn't matter to me how long it is and how much is left . . . I felt that it's not relevant how long the project takes."

Intertemporal choices are ubiquitous in systems design practice. Almost all participants were quick to recall similar examples of intertemporal decisions they had personally faced. While they thought the scenario was artificial, they recognized the pattern of intertemporal tradeoffs. One mentioned that "this happens many times" and added that "in software development, this is constantly an issue . . . not that specific but in some form."

Intertemporal choice situations are often as ambiguous as they are uncertain.
Participants often emphasized a lack of information, particularly with
regard to precise numeric data on effort and probabilities of success, as
in this case: "really this is . . . pretty much the way it usually looks . . .
maybe there is even less information. . . . It's a bit harder in reality to
[make the choice] at least from our team's point of view. We have not
dealt so much with numbers."

Numeric methods are considered irrelevant. Many participants mistrusted
the probability estimate provided by the scenario. When asked, "What
information would you seek?," they requested (1) information to evaluate the chances of success, even though the probability was fixed at 60
percent; and (2) information to evaluate the trustworthiness of that
estimate: Where did it come from? Who created it and how? What
factors were considered? Many participants distinguished instinctively
between uncertainty and ambiguity, in line with Camerer and Webers
(1992, 325) finding that ambiguity, or the "uncertainty about probability," matters significantly in people's preferences. As a result, many of our
participants seemed to ignore the numeric 60 percent estimate. Some
explicitly mistrusted it, saying for example "Saving effort is difficult to
quantify and also to understand." In addition to the raw number, then,
our participants considered aspects such as the risk of secondary ripple
effects of both options, even though those aspects were nominally captured in the estimate. In this and other matters, our participants demonstrated high reflective awareness. One put his unease with the limited
perspective that methods often entail very succinctly: "When it comes
to making decisions about time, using a method courts laziness."

Professionals appear to be more reasonable than methods. Many participants
were aware of numeric methods but regarded them as inapplicable
because they are unable to integrate nuanced qualitative aspects of reasoning that are more important, more reasonable, and more accurate
than an artificially reduced numeric value. Even though the framing of
the study encourages rationalistic reasoning (it suggests that numeric
analysis is the appropriate angle for "solving the problem"), there is
more evidence for cognitive processes of the naturalistic kind than
for numeric reasoning. Participants relied on mental simulation and

heuristics to establish initial boundaries. For example, "at a minimum, the effort saved must equal the effort expended." Ultimately they considered a much broader range of factors than what was presented and contained in the probability estimate, and they drew on their experience to identify how to interpret the presented situation and evidence. Some thought it more important than the "mechanical" consideration of methods to bring multiple perspectives to bear on the question by involving other team members: "One perspective is not enough, cannot be enough, I need to hear other perspectives." When normative models insist on reducing all these aspects of uncertainty and ambiguity into numeric variables, they disregard the importance of judgment and thereby render themselves unreasonable and irrelevant.

Time perception is proportional. Amounts of time are not evaluated as numbers but in relationships to other amounts, such as the total duration of the project and the amount of time left. Most participants expressed that they perceived a specific length of time as long or short, or a future moment as "near" or "far." Participants tended to represent the meaning of numeric amounts of time as relative valuations, e.g., "a long project," "almost no time needed to finish the task," "very soon," or "in a long time." Relative valuation over time affects *both* options, but it affects them differently. That implies that it can shift the matching number because that number is relative to the perceived value of both options. This lends credibility to the relevance of *mixed outcome* discounting (Soman et al. 2005). In technical debt management, for example, paying down a technical debt involves a loss (effort investment) and a gain (architectural quality). Just as redeeming a coupon in Soman's study, the relative value of both loss and gain shifts differentially over time, but the actual choice involves both. That can imply that the attractiveness of paying down debt appears great as long as both loss and gain are distant, but it loses its appeal once loss and gain get closer.

Perspective matters. What distinguishes individual approaches to intertemporal choice? Different professional roles bring distinct perspectives. For example, a product owner communicating to a team might choose information that has a shorter time scale than what team members themselves see. This is also reflected in their decision-making process in this task: participants in a specific role assume a certain frame with a certain set of information and have assumptions about other parts of the organization and the information they provide. In particular, participants asked for the source of the estimations given in the scenario and indicated that they would interpret them differently depending on where they came from.

Professionals demonstrate situational awareness. Several participants considered timeframes beyond those suggested. Some incorporated realistic longer-term effects into their evaluation. For example, even if the project would end in six months, if the project is successful, the system will evolve and then the decision to choose option 2 will have additional value. That value can hardly be quantified and occurs beyond the given time horizon, but it is quite reasonable to consider it. Similarly, several participants highlighted that they would need to understand the project environment to make a reasonable recommendation: They pointed out that company values and cultures, customer relationships and priorities, and established methodologies should all be considered. Decisions do not fall neatly into any particular method but are rather socially situated in the organization. As one participant noted, "each company is a different universe, it has a different culture, different methodologies. Even though many companies say they are agile, each [company] does [agile] differently, has different competencies, and different talent." This situational awareness is not a defect, it's a strength.

Judgment matters. Meaningful boundaries for supposedly technical decisions span social and temporal distances. Seemingly technical decisions always involve a range of social concerns. For most participants, decision-making begins with talking to a range of stakeholders covering the aspects entailed by the decision to be made. They recognize that the outcomes of intertemporal choices often extend beyond the current project and beyond operational measures. "We have a project right now, in fact, an internal tool for disseminating skills. I am probably a little more hesitant towards it than the team, but it is a huge morale boost for them, and it can work. . . . What tips the scales for me is above all the fact that they want to build it."

JUST SUSTAINABILITY DECISIONS INVOLVE
PSYCHOLOGICAL DISTANCE

What do these initial findings about intertemporal choices tell us about the pressing challenges of decisions affecting just sustainability, where temporal distance seems to conspire with social distance and ambiguity to lure us into unjust, unsustainable choices? Yes, some participants sometimes act as if distant future outcomes mattered little to them. But others showed remarkable foresight. The outcomes are not simply a function of temporal distance. The participants in our study provide a much richer texture of how they reason about intertemporal choices than what a survey instrument or rationalist model can capture.[4] Their responses show us starting points we can use to develop a roadmap for the more complex, difficult decisions we need to work on. How do those participants reason who emphasize long-term perspectives? In our continuing analysis, we explore the factors our participants considered and the cognitive moves employed in their decision-making. We aim to identify patterns of short-termism and long-term thinking and develop interventions that help designers take a long-term perspective.

When we widen our horizon from internal technical perspectives to the larger, longer-term implications of design decisions, the intertemporal nature of decisions is complicated by the fact that the question *who* is affected is no longer straightforward. The decision-makers of the present may not be the ones who bear the consequences of their designs in the future. Instead, others with differing social proximity to the decision-makers will be influenced. Once again, it is the gap between the involved and the affected. Figure 11.4 loosely visualizes this "field" of decision-making, with its constantly evolving present horizon of commitments.

Psychological distance describes how the removal of an event from direct experience affects "the perception of when an event occurs, where it occurs, to whom it occurs, and whether it occurs" (Trope and Liberman 2010, 4). The four dimensions commonly thought to constitute it are thus temporal, spatial, social, and hypothetical. There is ample evidence that the dimensions are related but not in a trivial way. For example, we tend to use spatial metaphors to reason about temporal distance (Boroditsky 2000), and in many studies, the effects of social and temporal distance are

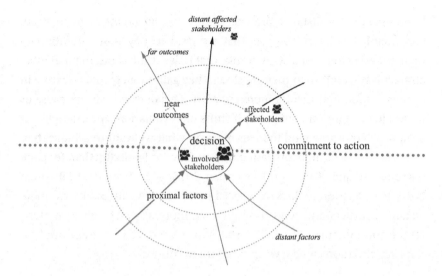

11.4 A sketch of the decision field.

marked when participants believe that the effects are real but much less pronounced when they are considered hypothetical (Pronin, Olivola, and Kennedy 2008, 233).

When an outcome is perceived as more distant on any dimension, it requires the mind to travel. Construal-level theory (CLT) suggests that our ability to do this is in fact a remarkable and rather unique ability of the human mind, and it involves some form of abstraction. We construe distant events in a different form than close events:

The key to traveling across psychological distance, CLT suggests, is cognitive abstraction. Abstraction is a reductive process that entails extracting the essential, core (i.e., gist) aspects of objects or events while ignoring surface-level, or verbatim, features . . . abstraction allows one to extract those features that are likely to be invariant across different manifestations of objects and events . . . *high-level construal* . . . is central to traveling mentally to distant times, places, perspectives of other people, and possibilities. . . . High-level construal, however, does not necessarily produce more impoverished representations. Instead, it connects us to remote content, those things that may not be apparent in the here and now. . . . One can thus have rich and elaborated, yet abstract, representations of distant entities. As events become more proximal, we can incorporate the idiosyncratic and unique information that becomes increasingly reliable and available into our representations via a process of concretization. (Fujita, Trope, and Liberman 2015, 406)

So effects at a distance are not simply *less important* to us. Instead, construal-level theory suggests that we tend to pay more attention to higher-level features of distant events, that is, we think about them in more abstract ways, and when they are closer, they gain more granular texture in our perception and mental representation. This does not always cause us to devalue distant outcomes, but it shifts our evaluation. For example, for distant outcomes we tend to focus more on what we want to achieve than how, and we focus more on desirability and less on feasibility than for close outcomes (Fujita, Trope, and Liberman 2015, 415; Trope and Liberman 2010, 19). Decision-makers with social power perceive themselves as more distant to others and tend to emphasize more abstract representations (Trope and Liberman 2010, 27). This shift may have important implications for situations where we give advice to others:

Any decision made about an issue generally affects decision makers to a greater degree than advisors. As such, advisors are more socially distant from decisions relative to decision makers. This change in social distance may impact the advice preferences of the two parties. Whereas advisors may prefer to provide information that preferentially weights desirability over feasibility, decision makers seeking advice may prefer to receive information that weights feasibility over desirability. (Fujita, Trope, and Liberman 2015, 414)

When we designed our first studies on intertemporal choice, we were intrigued by a prior study in which software developers exhibited a curious effect of psychological distance. They were asked to evaluate code for technical debt and then suggest whether the identified quality defects should be fixed or not. Remarkably, people were much keener on deciding for other people's defects to be fixed rather than their own (Amanatidis et al. 2018). In our own studies, however, participants made no difference between deciding for themselves or advising someone else.[5]

In all this, human decision-makers demonstrate cognitive flexibility, or the ability to adapt their reasoning to the situation at hand. The fact that some of our participants discount and others don't suggests that they perceive and evaluate different aspects of the situation. The diversity in demonstrated preferences should be reason for hope, not despair, as long as we are not looking for the "intertemporal defect" but rather what Klein calls the "sources of power": the remarkable cognitive abilities that human decision-makers demonstrate when they make decisions under uncertainty.

These abilities are especially pronounced in decision-makers with rich and varied experience (G. A. Klein 1998). In our study, we looked for any correlation between participant background and discounting. We thought that perhaps the amount of education or experience or the degree of responsibility would make a difference. Curiously enough, there was only one correlation, and it was none of those. Instead, only the breadth of experience, measured as the number of *distinct* responsibilities held in past jobs, made a difference—broader experience was correlated with reduced discounting (Fagerholm et al. 2019). With what we know now, it is not that surprising that those with more diverse experience would find it easier to traverse psychological distance, mentally simulate what would happen, and take it into account. This can carry significant practical implications for hiring, stakeholder participation, team composition, and the value of rotating responsibilities.

The abstraction implied in construal-level theory is consistent with the observation that "people pay less attention to subjective experience when that experience belongs to psychologically distant selves, that is, future selves and others, rather than when it belongs to psychologically immediate (present) selves" (Pronin, Olivola, and Kennedy 2008, 225). As a result, some suggest that "the salience, vividness, and emotional impact of choices decreases with psychological distance" (Pronin, Olivola, and Kennedy 2008, 234). But others emphasize that it is not necessarily the degree of emotional affect that changes but the type: "Research suggests, for example, that whereas low-level construal promotes the experience of lust, high-level construal promotes the experience of love" (Fujita, Trope, and Liberman 2015, 421).

These insights can help explain how the removal of the climate crisis and many aspects of social justice from the direct lived experience of many privileged people in the Global North may affect their willingness to act on it. In 2006, Elke Weber suggested that "the absence of (visceral) concern about global warming on part of the general public" is in large part because what she describes as the analytical processing system takes second rank to the affective system when we make risky decisions under uncertainty: "people's visceral reactions to risky situations often have little correspondence to more objective measures of risk that quantify either the statistical unpredictability of outcomes or the magnitude or likelihood of adverse consequences. Instead, visceral judgments of risk (which fuel self-protective action) are determined by other situational characteristics that elicit affective reactions as part of our evolutionary heritage" (Weber 2006, 104). She highlights that empirical research on decision-making has been limited by the fact that researchers "have almost exclusively employed choice situations where the outcomes of risky choice options are (statistically) described . . . rather than personally experienced over time" (Weber 2006, 109).

TRAVERSING PSYCHOLOGICAL DISTANCE

Like others, I refuse to succumb to the fatalistic belief that we humans will boil each other on this planet like the proverbial frog in the water glass. But we have evidently no time to lose to activate our ability to "traverse psychological distance" (Liberman and Trope 2014) and make wise decisions. In *The Good Ancestor*, a plaidoyer and guidebook for long-term thinking, Tomas Krznaric (2020) argues that it matters a great deal how we think of ourselves. Our view of ourselves will shape who we become: "Changing the story about who we are makes a difference. If we keep telling ourselves that we are primarily driven by short-termism and instant gratification, it is likely we will exacerbate such traits" (40).

These stories about ourselves are not as innocent as they appear. In Calvin and Hobbes, Hobbes's name is key. For decades, the heritage of Hobbesian views, banishing any whiff of teleology from scientific accounts, has served to suppress ideas of purposive action in psychological and social science research. But more recently, the recognition has mounted that we do have some teleological capacity for purposeful action in our minds after all. *Prospection* enables our minds to mentally simulate navigation, the minds of other people, possible arguments for or against ideas, and counterfactual accounts of the past (Seligman et al. 2013). Come to think of it, it is quite remarkable.

And in contrast to the Hobbesian view that humans are wolves to each other, decades of research have recognized what Indigenous knowledges around the world have always emphasized: that humans are cooperative beings who live in and through relationships with the rest of the world.

SEARCHING FOR JUST, SUSTAINABLE DESIGN DECISIONS   273

When disaster strikes, the most natural and immediate reaction is not theft and destruction but mutual aid (Solnit 2010). And when humans are left to govern public goods in a way that resembles *the commons*—for example, partially renewable resources like fish—the outcome is not at all an inevitable destruction of all resources (Kimmerer 2013; Linebaugh 2014; Ostrom 2015). In a fascinating experiment, a group of researchers designed an *intergenerational goods game*, a version of a popular systems thinking game in which a group of fishers has to decide, through a number of rounds, how much fish to remove from an ocean. In the original game, the fish regenerate, up to a point, but it is common for the fish population to collapse due to overfishing (Sweeney and Meadows 2010). When simple communication mechanisms are established, however, it is common for groups to self-organize effectively and govern the common pool resource successfully. (That is in fact what happened when I played this game the first time with my students at the University of Toronto after they read Ostrom [2016] instead of Hardin. Mindset matters.) An important feature of this game is that everyone is in the room and can communicate with each other and that collapse may be at a distance at the time of fishing, but it happens very fast.

In the intergenerational goods game, asymmetric vulnerability is encoded into the rules of the game. Decision-making is fragmented into discontinuous groups, and future generations (groups) have no way of influencing past generations. The findings show that even under unregulated conditions, most participants cooperate with the future, at significant cost to their own success. But a small minority does not, and its behavior inevitably exhausts the common resource. When a very basic regulation by a democratic institution is introduced, however, two things happen. First, a basic robust voting mechanism is sufficient to contain the divergent greedy minority, if there is a mechanism to enforce it. In this mechanism, everyone votes, and everyone receives the amount of the majority vote. Second, cooperative behavior in this scheme is more common than in the unregulated condition. The authors' conclusions: "Many citizens are ready to sacrifice for the greater good. We just need institutions that help them do so" (Hauser et al. 2014, 222).

HOW TO STUDY JUST SUSTAINABILITY DESIGN DECISIONS

With these ideas at hand, I suggest that decisions in systems design should in general be characterized in terms of their commitment to action, uncertainty, psychological distance, situated cognitive processes, and context.[6]

1. The *context* in which the decision occurs is understood in the widest sense as anything that influences the decision.

2. *Commitment* describes which actions are available to commit to, and which is committed to. Decision-making is not always a selection out of explicitly enumerated options. There may sometimes appear explicit, well-defined "options" to choose between. But often, there are myriad ways in which to proceed, and some or all of the actions are generated by the decision-makers in the course of decision-making.

3. *Uncertainty* covers uncertain properties of the options and possible outcomes as well as their ambiguity. Uncertainty, or risk, refers to the objective probability of potential outcomes. *Ambiguity*, on the other hand, means that only vague information about the probabilities is available (Ellsberg 1961). It may be uncertain whether something will happen or not; to whom it will happen; and what it will mean at the time if it happens. The distinction between the two matters because they are and must be handled differently. Uncertainty about probability complicates how people think about possible outcomes when they decide.

4. *Psychological distance* comprises temporal, spatial, social, and hypothetical distance. The temporal dimension separates possible outcomes related to sustainability across time and can involve multiple timescales that need to be considered simultaneously. Time always introduces uncertainty about the outcomes and often also ambiguity regarding both the options and the outcomes. The social dimension often manifests as a distance between those involved in systems design and the justice-focused outcomes of their decisions.[7]

5. The *situated cognitive process* involves individual decision-makers possibly acting as a group. Psychological distance raises difficult questions about cognition that are not adequately understood yet. For example, people differ in their attitudes towards ambiguity: some are drawn to ambiguous options while others avoid them. Several studies indicate that attitudes towards ambiguity depend on the likelihood of the

uncertain events, the domain of the outcome, and the source of the uncertainty (Trautmann and van de Kuilen, 2015). This means that decisions cannot be understood only through the temporal separation of the outcomes. It is crucial to understand how the outcome uncertainty is perceived by decision-makers.

Like other decisions, those with special import on questions of just sustainability should be characterized in terms of commitment, uncertainty, psychological distance, situated cognitive processes, and context. But in just sustainability, *psychological distance must take center stage*. The central sustainability choice could not be more clearly intertemporal: Do we save the planet now that we still can, or do we salvage the scraps later? But the truth is, we are not facing one big choice to save the world. We are facing a myriad of small and large choices. In systems design, commitments to future actions are constantly made and remade. For many of these commitments, their distant implications may not be salient and are significantly removed from direct experience.

To understand and improve the degree to which these decisions take distant outcomes more fully into account, we will need the whole toolbox of cognitive theories and the whole range of scientific methods. The legitimate role of the rationalist model here is to support a quantitative behavioral description of "as if" to guide our attention to the differences in behaviors, so that we ask the right people some good questions, and later, to measure how any interventions we may design affect behaviors. To understand how people actually reason, however, we need to rely on macro-cognitive methods, carefully reflecting on their scope and assumptions (Klein and Wright 2016). To support such research, we developed a framework for studies on decision-making including figure 11.5, a guiding map of key aspects that orients researchers toward central elements and relationships they should consider when studying decision-making.

CONCLUSIONS

In this chapter, a group of critical friends from JDM has helped us reorient our view of decision-making in systems design. This group too has already received some attention in computing before, but it has much more to offer us when it comes to understanding how psychological distance, and

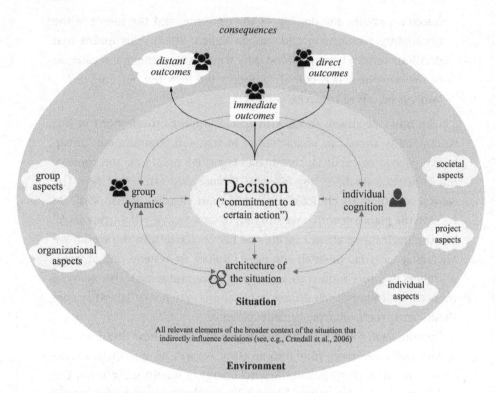

11.5 A macro view of decision-making in systems design.

asymmetric vulnerability, shape the dynamic of just sustainable design decisions. When the effects of decisions are removed from our direct experience into a distance, we tend to shift our perception, evaluation, and preferences about them. The reasons for that are manifold. Psychology's most robust theory on this topic, construal-level theory, explains that our mental representations of distant outcomes focus more on high-level features, whereas closer outcomes are represented with concrete incidental detail. We also seem to perceive and represent temporal distance in proportional ways similar to spatial distance: future events appear smaller to us. As a result, decision makers in some circumstances behave *as if* they discounted future or socially distant outcomes. But we should be cautious about conclusions, predictions, and assumptions.

In describing, explaining, and prescribing how to design, we need to be mindful of human capacities and take a macro-cognitive view of design situations. This attention can allow us to relate cognitive strategies, strengths, and weaknesses to the features in the environment that enable or constrain them. By recognizing which constellations are more conducive to traversing psychological distance, we have the best chances of making just and sustainable design decisions. In understanding how to change practice, we need to talk about redesigning decision-making situations as much as changing individual behavior. If the incentive structures we are all in have such a strong influence in generating unsustainable, unjust outcomes, then we must face the larger social context and structures surrounding those situations.

The human ability to envision outcomes at a distance is cause for hope. We are remarkably capable of moving across psychological distance. We possess what we might call temporal flexibility, which allows us to adapt our timescales and behaviors, switching from rapid chess under time control to strategic foresight to swift moves on a dance floor. We already possess the prospective capacity to mentally simulate the future and other distant places. And in the right circumstances, we are very capable of making decisions that are good for those affected at a distance, even if those affected are so completely removed that they have no way of thanking us.

But it must be said that we rarely are in the right circumstances these days. Systems designers too often get caught up in incentive structures that reward temporal and social discounting, set short-term objectives, ignore social and environmental debts, and structure employee behavior to fit that mold. The growth-obsessed mindset that favors short-term thinking dominates the capitalist IT industry. It is not conducive to cooperating with the future. There is no doubt that we need to redesign that environment. Some have already begun doing that.

12

A SILICONE RING
SOCIAL RESPONSIBILITY AND COLLECTIVE ACTION

From the outset, therefore, the engineer was at the service of capital, and, not surprisingly, its laws were to him as natural as the laws of science. If some political economists drew a distinction between technology and capitalism, that distinction collapsed in the person of the engineer and in his work, engineering.
—D. Noble (1977, 34)

If democracy is going to mean anything, it is the ability to all agree to arrange things in a different way.
—Graeber (2011, 390)

At 5:31 p.m. on August 29, 1907, in the span of fifteen seconds, the world's largest cantilever bridge collapsed. Thousands of tons of steel, twisted and torn like melted plastic, sank seventy meters into the Saint Lawrence stream near Quebec City. The bridge was not completed yet, and eighty-six workers fell with it. They were about to leave the site for the day when the iron under their feet gave way. Seventy-six workers died, half of them buried at the bottom of the river, never to be recovered. In the days before, several had raised concerns about the obvious bending of the iron chords holding up the structure, but their fears had been dismissed. Subsequent inquiries into the causes of the tragedy absolved all involved parties from negligence. But a reexamination of evidence suggested that the leading engineers could have averted the disaster had they paid attention. "There was no trace of humility in any of the senior engineers involved with the design of the Quebec Bridge. Their arrogance and absolute confidence in their work prevented them from realizing what the workers building the bridge already knew: that the bridge was failing under its own weight" (Levert 2020).

The collapse left a mark on the psyche of Canadian engineering. Not long after the tragic disaster, a group of engineers and engineering professors introduced a ritual into the profession that persists to this day. Upon graduation, young engineers complete a private ceremony, away from public sight and celebration, in which they take an Obligation and a ring. Both remain unchanged since 1925. The Obligation reminds them of their "assured failures and derelictions" (see Raymond Francis 2006). It is not quite a Hippocratic Oath, but close. The Iron Ring bestowed on the engineer is worn on the little finger of the working hand so that it rubs on working materials as a constant reminder of humility and dedication.

We will never know if the Obligation oath and the wearing of the ring have materially improved the safety of Canadian infrastructure. There is no doubt, however, that the Obligation and its material symbol impress the minds of engineers and serve as a powerful reminder of the responsibilities they carry (Levert 2020) as they use specialized technical expertise to shape the world around us.

HOW (NOT) TO AVOID HARM

In 2018, I still had hope. Hope as in confidence in the ability of institutional action to establish effective guides for ethical conduct. I felt confident that the organization I thought best represented me as a professional and researcher—the Association for Computing Machinery (ACM)—would take meaningful action to address the social responsibility of computing in the twenty-first century. ACM was finally revamping its outdated code of ethics (ACM Committee on Professional Ethics 1992), which was in dire need of overhaul and improvement. It spoke of "avoiding harm," of course—that is considered the absolute minimum for a code of ethics, after all. But its examples for harm were tellingly narrow: "'Harm' means injury or negative consequences, such as undesirable loss of information, loss of property, property damage, or unwanted environmental impacts. . . .

Harmful actions include intentional destruction or modification of files and programs leading to serious loss of resources or unnecessary expenditure of human resources such as the time and effort required to purge systems of 'computer viruses'" (ACM Committee on Professional Ethics 1992). In its focus on property and material damage, the ethics code was, as always, a product of the times. As ACM president Cherri Pancake wrote in 2018, "in 1992, many of us saw computing work as purely technical" (ACM Committee on Professional Ethics 2018, 1).

The process for updating the code looked promising. Significant consultation seemed built in. Under the leadership of an interdisciplinary taskforce, multiple rounds of public feedback were to shape the new edition of the code, informed by critical insights into the conflicted role of computing in our societies. Multiple drafts were released for comments. Many excellent changes were introduced to update and strengthen the code and consider feedback of the community. When I read the final draft, however, alarm bells went off. The code contained crucial changes to a central principle that weakened it substantially and fundamentally. The last public draft of *Principle 1.2 Avoid Harm* looks harmless enough at first glance, but note the highlighted passages.

In this document, "harm" means negative consequences to any stakeholder, especially when those consequences are significant and unjust. Examples of harm include *unjustified* physical or mental injury, *unjustified* destruction or disclosure of information, and *unjustified* damage to property, reputation, and the environment.

Why is each example of harm an *unjustified* damage? Is it only harm when it isn't justified? By whom?

Well-intended actions, including those that accomplish assigned duties, may lead to harm. *When that harm is unintended*, those responsible are obligated to undo or mitigate the harm as much as possible. Avoiding harm begins with careful consideration of potential impacts on all those affected by decisions.

How actionable is the distinction based on intentions in today's computing landscape? How are those responsible supposed to mitigate harm that they have not intended in the first place? Why are they not instructed to avoid harm?

When harm is an intentional part of the system, those responsible are obligated to ensure that the harm is ethically justified and to minimize *unintended* harm. To minimize the possibility of indirectly harming others, computing professionals should *follow generally accepted best practices*. Additionally, the consequences of emergent systems and data aggregation should be carefully analyzed. (Gotterbarn et al. 2017, 124)

The closer we look, the more dubious it sounds. When harm is intentional, it does not have to be minimized? How are computing professionals supposed to ethically justify the harm caused by their work, and to whom? And since when are "generally accepted best practices" sufficient guidance when they lead us to a state of insolvency?

In ethics, the doctrine of *double effect* allows negative harm for a positive outcome under certain conditions. But "applications of double effect always presuppose that some kind of proportionality condition has been satisfied. Traditional formulations of the proportionality condition require that the value of promoting the good end outweigh the disvalue of the harmful side effect" (McIntyre 2014). For example, a surgeon may cause significant harm to a patient's skin to save their heart. In academic research, whenever I collect personally identifiable information, I need to justify the value of doing so in relationship to the risk posed to participants. Crucially, I am not allowed to make that judgment myself: an institutional review board makes it for me.

The new code introduces the double effect without addressing proportionality at all. The text above instructs those responsible to minimize unintentional harm without requiring them to minimize intentional harm. The word "unjustified" hints at a proportionality condition but wraps it up within the examples, without opening a conversation about it. Its use to articulate all examples suggests that justified negative consequences would not count as harm at all. At no point does the text address the crucial question of accountability: the act of justifying is left to those who are involved. The result is a principle that spells "avoid harm" while expressly condoning intentional harm, suggesting that justified harm is no harm at all, *and* leaving it up to the designers to decide which is which.

Consider the effect on two examples, one benign, one murderous. In the process of coordinating a collective response, my colleagues and I discussed the action of alerting all members of a dormant community email list to the danger posed by the proposed principle. This involved minor harm (the annoyance of yet another unsolicited email) for a greater public good (the wider discussion of an ethics code). It may have also involved unintentional harms of some minor kind, but we are unable to anticipate those with any certainty. Maybe these could be some people's possible irritation about a standpoint they do not share or the carbon footprint of emails. The draft Principle 1.2 required us to minimize these unintentional harms—which we could not feasibly anticipate—but it did not require us to minimize the intentional harm caused by yet another unsolicited email. It seems obvious and common sense, though, that we should aim to minimize that, for example by taking care to explain the rationale behind the message.

The principles fail more strikingly in an example more typical of ethics literature. Consider a team that designs a bomb to end a war. They face a choice between developing a bomb that would neutralize electronic equipment and developing a hydrogen bomb. According to Principle 1.2, the responsible agents would have to minimize unintentional harm, but not intentional harm. This would make the hydrogen bomb the more appealing solution *according to the ethics code*. Yes, it is absurd.

To understand how this text came to be, I examined the publicly available documentation. The distinction between intentional and unintentional harm had been introduced in the final draft to expand the code's applicability to arms manufacturers (pardon, "the defense industry") (Gotterbarn et al. 2017). To voice my concerns, I published a Twitter thread and reached out to colleagues, who helped me draft an analysis (Becker 2018b). The ethics committee swiftly responded on social media, publicly asserting they would meet with me to discuss the concerns.[1] They did not. Instead, I was informed in writing that the text had already been approved by ACM Council. The committee's response did not address my concerns, but eventually, the final text *was* changed to address one of the issues my colleagues and I had objected to: The final principle states "In either case, ensure that all harm is minimized" (ACM Committee on Professional Ethics 2018), just as I had proposed. It was a bittersweet moment to see that bare minimum change implemented. No other issue was addressed.

ETHICS WON'T SAVE US

I tell this story as a reminder that vested interests capture ethical concepts under the cover of the objectivist illusion. Engineers tend to look to codes of ethics as a seal of approval to absolve them from responsibility for misconduct. From the orthodox standpoint of insolvent computing, the "ethical problem" posed by just sustainability is solved at one clearly identified point: the code of ethics. The codes of ethics of professional associations like ACM and IEEE provide the principles of actions that guide ethical conduct. The development of ethics codes is the central point at which governance and value judgments are admitted into a clearly circumscribed location of this picture. Within the expressly political process of developing a new code, ethical and moral philosophy are translated into prescriptions by formalizing supposedly democratic decisions into ethical principles. Outside the process, the resulting codes formalize the criteria on which the ethicality of design can be adjudicated. Methods further translate these principles into actionable guidelines. If those methods are followed properly, so they say, then the outcomes will be ethical. By implication, failure can be blamed on the individual (Metcalf, Moss, and boyd 2019). This is why it is so important that the process carry the democratic appearance of legitimacy. When I raised concerns at a late stage, the institution ACM was more concerned with that appearance than with the substantive questions I raised. The episode made it clear to me personally that these codes have very dubious legitimacy.

Artificial intelligence (AI) ethics is a booming business today, and still, some of the public discussion evolves around the content of ethics codes. With much less attention paid to the question of *how* ethical principles and directives come to be, and how they work in practice, it is common for industry to shop around for the set of ethics codes that best fits their existing business practices, to use these ethics codes to make exaggerated claims about the business's social responsibilities, to lobby for self-imposed ethics regulation over externally imposed legislation, and to remove unethical components of computational systems, such as the invisible labor behind so-called AI, from highly regulated countries, only to re-import the resulting technologies (Floridi 2019).

This is why ethics is often critically seen as an instrument of control and complacency. Western moral philosophy has long tried to pretend that its reasoning takes place in a vacuum. While there is still value to moral philosophy in AI ethics (Bietti 2020), the critical friends highlight that counter to the "ideal" tradition that dominates it, ethical reasoning always already happens on a ground of inequity and injustice

(A. Hoffmann 2020). On this ground, industry sometimes hires critical researchers with public visibility to conduct critical work with the intent of using them as "legitimation workers" (Whittaker 2020). The central purpose of their employment is not to conduct substantive critical research, but to perform the appearance of critical research. Many noticed what happened when Timnit Gebru and Margaret Mitchell, two such legitimation workers, dared to interpret their job at Google differently: they got fired, "revealing just how much companies will prioritize profit over self-policing" (Hao 2021b, 51).

Some pointed out that I had been naïve to expect anything but moral bankruptcy from an institution like the ACM, and they may be right. ACM has had a long history of entanglement with the military industrial complex, after all. It is hard to draw the line between the principles discussed earlier and a satirical comment such as this one: "*impact assessment* (ph)—A review that you do yourself of your company or AI system to show your willingness to consider its downsides without changing anything" (Hao 2021a).

The theory of economic regulation distinguishes between two forms of *regulatory capture*. Financial capture refers to material incentives, often of dubious legality, which shape the behavior of a regulating entity. Cultural or *cognitive capture* refers to the shaping of regulatory thought by industry. Both types are at play in the capture of AI regulation by Big Tech. Tech companies do not just operate the revolving door connecting their offices to the halls of government, they are also "startlingly well positioned to shape what we do—and do not—know about AI and the business behind it, at the same time that their AI products are working to shape our lives and institutions . . . [they] control the tooling, development environments, languages, and software that define the AI research process—they make the water in which AI research swims" (Whittaker 2021). In this context, institutional declarations of goodwill mean little if the actions of the institution do not correspond. And we know that to identify an organization's purpose, we must look at its behavior, not at what it declares its purpose to be (D. H. Meadows 2008, 14).

Amid regulatory capture, ethics washing, and the inadequacy of ethical codes to guide responsible design, it is clear that even good ethics codes alone won't save us. So what might?

BIG TECH WON'T SAVE US

Not to worry about sustainability and the fate of the global poor, say the founders of Big Tech. We don't need ethics codes to do good. In fact, they just hold us back. Step aside and let us save the world. We know what we are doing. After all, we have already proven to you how capable we are, and we have nothing else to prove. We'll save you. Bill Gates has no geoengineering credentials but has published an entire book on it, in which he proudly declares: "I think more like an engineer than a political scientist" (Milman and Rushe 2021). Elon Musk markets autonomous electric cars to rescue us from fossil fuel dependency while his company exploits lithium reserves in the Global South. Anyway, he plans to retire on Mars. And Jeff Bezos claims that Amazon will be net zero by 2040—which would be too little, too late—but for now, Amazon continues to market its cloud services to the fossil fuel industry (Palmer 2020). It does not take a lot of scrutiny to become suspicious of these claims. But a common perception is that they must have some point. After all, they are highly successful entrepreneurs, and that means they have judiciously taken high, calculated risks and demonstrated visionary leadership in developing innovative technological breakthroughs with real value to the world. Or have they?

We often hear that origin myth of Big Tech. It tells us that innovation begins at places such as Silicon Valley, where technological entrepreneurs devise ingenious and entirely novel ideas—ground-breaking innovations that lead to breakthrough technologies. Shouldn't computing then deserve the profits it earns and the soaring stock market evaluation of its shares? While no one disputes that innovative entrepreneurs should make money, economic studies show that it is the state, not private entrepreneurs, who takes the biggest risk, funds the real breakthroughs, and takes the long-term view necessary to make these breakthroughs happen. Economist Mariana Mazzucato (2013) has compiled overwhelming evidence showing that private investors, focused on short-term profits, are more risk-averse than the state and rarely fund the kind of technological breakthrough that enables the businesses they invest in. Instead, they come in afterwards to turn these technological innovations into profits. This is true especially for the iconic successes of Silicon Valley, such as the iPhone. From the touchscreen to the internet, from GPS to voice recognition and from battery technology to

HTML, each component that makes it work was made possible by risky, strategic, long-term state funding: "far from getting out of the way of private innovation the State paves the way" (Barendregt et al. 2021). There is a long history in which government funding enables true breakthrough innovation in IT and private business harvests the profits.[2]

Instead of placing our bets on Big Tech getting it right, "we should be striving for a much more radical agenda that envisages the wholesale transformation of the computing profession, putting an end to the technological solutionism of Silicon Valley, turning it into a humble enterprise that places human dignity first," as my colleague Steve Easterbrook puts it (2021). As I put it with a group of coauthors in a new platform we launched after a series of workshops on collective action:

Now is the time to radically redirect the future of tech, by reclaiming the purposes of technology development, and redistributing the associated responsibilities and benefits, in the service of our collective and sustainable well being . . . to redirect Big Tech's excessive revenue flow, we must transform the conditions and funding structures that enable it. The aim is to free up resources to support a wide range of socially beneficial ends, not least community-based and community-oriented initiatives to develop digital infrastructures that better serve the public interest. While we are not calling for the demise of Big Tech, we are calling for radical reform. (TechOtherwise Collective 2021)

In fact, defunding Big Tech is not as radical a position as it may appear. It simply means that societies ought to collectively reap the rewards of their collective investments. To achieve that, every participant in our societies must exercise our agency to reduce the outsized influence of corporate tech giants over how our lives on this planet are organized and for whose benefit. As Graeber's epigraph puts it so well: democracy should mean that we can all agree to change course. The proposal offers steps for tech companies, policymakers, tech professionals in industry and academia, civil society organizations, and individuals. One of these groups gives me the most hope.

TECH WORKERS FOR SOCIAL RESPONSIBILITY

While the executives at Amazon & Co are busy brokering deals with the oil and gas industry, employees at the same companies are organizing for sustainability and justice. In 2019, more than 8,000 Amazon employees signed an open letter and supported a shareholder proposal to accelerate the company's commitment to carbon neutrality. Throughout the pandemic, however, while Amazon's carbon intensity sank, its overall environmental footprint grew. Google workers may have made the biggest headlines. In spring 2018, thousands joined a walkout in protest over a Pentagon weapons contract (Tarnoff 2020). Later that year, walkouts from offices around the globe protested the company's mishandling of sexual misconduct. With success: the company stopped the Pentagon project that year and updated its processes for handling misconduct.

Protest organizers faced retaliations, and prominent organizers like Meredith Whittaker left the company, explaining: "The reasons I'm leaving aren't a mystery. I'm committed . . . to organizing for an accountable tech industry—and it's clear Google isn't a place where I can continue this work." (Vincent 2019). Ben Tarnoff, who has studied the "tech worker movement" for years, highlights that tech specialists such as software engineers occupy a middle ground. They are located between executives and the outsourced service workers who keep the posh corporate campuses running and who also should be understood as tech workers. In fact, the latter were the first to organize, including on issues of environmental justice (Pellow and Park 2002). The experience of their organizing was profoundly influential on how IT specialists understood their own role.

By virtue of their position between labor and capital, they inhabit what Erik Olin Wright famously called "contradictory class locations" . . . they are pulled in two directions. . . . The tech worker movement offers a fascinating illustration of the latter phenomenon. It involves many members of the middle layers coming to see themselves as workers. This new identity has in turn enabled those individuals to act collectively as workers: to use their leverage over the spaces where profit is made in order to demand more control of their work and their workplaces. They are wielding the weapons of working-class struggle while speaking its language. Most significantly, they are building relationships of solidarity with their working-class colleagues in the tech industry and coordinating their efforts in order to advance their campaigns together. (Tarnoff 2020)

During the pandemic, tech workers around the world have organized, lobbied, protested, and unionized. The wave of collective organizing in tech companies that we are witnessing today demonstrates that many tech workers are keenly aware of the tension between corporate values and their own sense of professional social responsibility, and that they are ready to act collectively on their commitments, even if this puts their financial security at risk. This tension is not all new. As briefly seen in chapter 10 and recalled in the chapter epigraph, the room for action available to specialists in technology fields has always been shaped and constrained by their employers' interests—directly, through incentives, responsibilities, and rules, and indirectly, through the production and reproduction of ideas through education and the cultural imagination. These tensions emerged together with the engineering disciplines and professions themselves.[3]

As the chapter's epigraph illustrates, Noble provides a sharp analysis of this history. Strikingly, the leaders of previous centuries did not find it necessary to hide their intentions. One told engineering students that "the dollar is the final term in every engineering equation." Another spoke to engineers about the role of the engineer, saying "he must always be subservient to those who represent the money invested in the enterprise" (D. Noble 1977, 34–35).

For over a century, some tech workers have disagreed very strongly. Already in the 1910s, some progressive engineers in the United States rejected the idea that the dollar should always have the last word. Most were committed to corporate liberal reform, however, and others remained peripheral (D. Noble 1977, 50–63; Wisnioski 2012, 5). By the heady 1960s, the growing power of computing and fields such as operations research fueled the idea that engineering could be applied to society at large. The scaling up of engineering education had also shifted the labor dynamic: by the 1960s, the number of engineers in the US had risen from a very small cadre of highly respected engineers to a workforce exceeding 800,000. With that came a shift into the dynamics of a nameless workforce and the values of team play and compliance (Wisnioski 2012, 31). The growth in number also entailed ever-increasing degrees of specializations. And for each subdivision of specialized expert labor, politics are supposedly just outside. As Feenberg highlights, "the differentiation of specializations gives specialists the illusion of pure, rational autonomy. This illusion masks a more complex reality. In reality, they represent the interests which presided over the underdetermined technical choices that lie in the past of their profession" (Feenberg 1999, 139). But at the same time, activism grew in many professions, including engineering fields (Wisnioski 2012; Hoffman 1989).

As traditionally defined, professional work is the work of "experts"—those who apply a given body of abstract knowledge to specific problems . . . the traditional model asserts that the dependence of the client, the knowledge of the expert, and the importance of the task, make it necessary to maintain an impersonal orientation to the task at hand, a hierarchically structured relation between client and professional, and professional control of most if not all aspects of professional work. For critics, on the other hand, professionalism is not a "best solution" to an inherent set of dilemmas, but a form of practice that maximizes professional control to maximize professional self-interest. (Hoffman 1989, 6)

Hoffmann's study of political activism focuses on medicine and urban planning, but her characterization of professionals' struggles and strategies apply to other fields too. Some professionals changed how services were delivered to address deficiencies in bureaucratic institutions. Some aimed to create movements for social change by workplace organizing and intellectual leadership. And some worked on empowering people:

Professions were seen as bureaucratic actors in organizations which served dominant interests. Professionalism, defined as a system of dominance and dependence, was the culprit, and deprofessionalization, the solution. The problem for minorities and the poor was not lack of experts and services per se, but awe of professionals and service bureaucracies. The activist roles were to mobilize client communities to help themselves and to democratize service delivery by transferring expertise. (Hoffman 1989, 8)

At the same time, "a small and vocal minority attempted to redefine engineering by rethinking the nature of technology through collective action" (Wisnioski 2012, 6). In the 1980s, computing professionals and academics followed suit and organized as *Computer Professionals for Social Responsibility* (CPSR 2007), galvanized by concerns about the role of computing in nuclear war (Ornstein, Smith, and Suchman 1984). Despite early connections to immigrant labor organizations concerned with working conditions, health hazards, and environmental justice, the CPSR focused on technical arguments about the limits to reliability inherent in computing and initially argued for limitations on computing's reach. This technical focus lent credibility to the moral argument its members made (Finn and DuPont 2020). As the threat of nuclear war faded in the 1990s and the interests of its leadership shifted, the organization struggled to reorient itself. It tried to integrate issues such as privacy, elections, and community networks into its portfolio. For a number of reasons, including financial insecurity, lack of focus, and diminished enthusiasm on the part of the leadership, its membership dwindled, and the organization eventually folded in 2013.

An important shift takes place among all these dissenters: from professional responsibility, defined and shaped by professional obligations and corporate institutions, to a social responsibility informed by social critique and understood as standing in tension to profit motives. In a small study of requirements professionals, many were similarly aware of the tension between what they considered their broader social responsibilities for sustainable IT and the corporate incentive structures they work in today (Chitchyan et al. 2016). Alternative approaches to design have often emerged from outside this corporate context. For example, the practices that crystallized into design justice principles developed in community contexts (Costanza-Chock 2020). Their approach aligns, in key aspects, with Hoffmann's characterization of professional activists. One effect is that it is proving very difficult to transfer them into professionalized corporate contexts (Spitzberg et al. 2020). But we need to introduce different approaches inside the mainstream of systems design practice because that is still where the majority of design happens.

That this shift is happening despite all structural obstacles should remind us: Hobbes was wrong. Mutual destruction is not a foregone conclusion. Since the 2010s, concerns over computing's many disasters are shared by a wider audience than ever, spanning all groups: tech workers, users, executives, researchers, NGOs, lawmakers, and everyone else. These concerns are taken up by new groups springing up all over the globe. There are too many to list, and there is no doubt that these voices are stronger, better coordinated, and sharper in their critique than ever before. Take the successful international campaign against the use of facial recognition technology, which combined the technical expertise of computer scientists with legal scholarship and critical social theory, organizing effectively to intervene in publishing, in policy, and in product development. In a workshop held to debate whether computing needs a new CPSR (Becker, Light, et al. 2020), the view prevailed that computing already has many voices speaking up in critical conscience on substantive matters.

What will we see in the 2020s? In North America, the tech worker movement is growing strong. Summer schools now support tech workers in social responsibilities and organizing (Logic School 2021). Ifeoma Ozoma, who blew the whistle on Pinterest's racism, created a handbook of resources to guide tech workers in speaking up about public issues and helped introduce legislation to protect whistleblowers in California (2021). And in April 2022 as I was completing this manuscript, workers at Amazon achieved a major milestone as they formed their first union. "We want to thank Jeff Bezos for going to space because while he was up there, we were signing people up," lead organizer Chris Smalls said with a triumphant smile. Meanwhile, the #TechWontBuildIt hashtag unites thousands of tech workers rejecting job offers on the ground of moral objections to the potential employers' practices. Will we soon see toolkits allowing socially conscious system designers to vet employers for ethical practices?

One way to think of this shift in how tech workers understand their profile is by contrasting it to conventional ideals of the perfect tech worker touted by business. The most frequently repeated of these ideals is the T-shaped professional, who has deep subject expertise in one narrowly defined technical area and broad shallow understanding of a range of fields (Hansen and Oetinger 2001; Neeley and Steffensen 2018); and the Π-shaped professionals, which combine deep technical expertise with deep domain expertise for some application domain of interest. I think of responsible tech workers instead as W-shaped designers (figure 12.1), who combine technical expertise with social theory, bridged by a critically systemic view, which enables them to work together with community partners and simultaneously form critical friendships to make sure they learn those lessons that may be hard to accept with humility.

A SILICONE RING

Rumor had it that the original iron rings for engineers in Canada were manufactured from the wreckage of the collapsed Quebec Bridge. There is no evidence that this is true, but it's a powerful image. The persistent rumor shows how much we care about the idea that we can be humble and learn from our mistakes.

Computing has already collected too much wreckage. Books on safety, reliability, and security have no shortage of examples on offer. The Cambridge Analytica scandal and others stand for privacy. Weapons of Math Destruction can be found across many areas of algorithmic

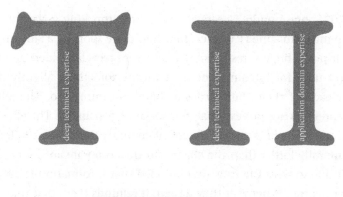

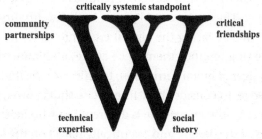

12.1 The W-shaped systems designer.

decision-making, from college admissions and recidivism assessment to credit scores. Facial recognition is so obviously toxic that it's called the "plutonium of AI" (Stark 2019). But because the nature of IT often removes the visible disaster from the time or place where and when a system is designed, constructed, and operated, it is harder to point to a particular smoking gun that spells out "just sustainability." Our collective computing debts are mounting, and we cannot wait, counting it, until all that impact has materialized.

I propose that we draw inspiration from the Iron Ring of Engineers, that we consider what kind of ritual would be appropriate for the twenty-first century. That we begin to develop such a ritual as a symbol of responsibility and a reminder of humility. After all, "it is the prevailing culture of ideas that shapes the direction of a society, that determines what is thinkable and unthinkable, what is possible and impossible. Yes, factors like economic structures, political systems and technology all play vital roles, but never underestimate the power of ideas" (Krznaric 2020, 19).

In 2025, one hundred years after the first Iron Ring graced an engineer's hand, I imagine IT workers here and there sporting a Silicone Ring on their pinky fingers, responsibly made from recycled wafers and CPUs formerly used for bitcoin mining. It means something slightly different for each of them, but carries a shared meaning, too. The Silicone Ring represents not professional, but social responsibility. The allegiance it reminds them of is to our planet. It reminds them that its flourishing is morally loftier than the client and the corporate authority. Those who decide to wear the ring are reminded that information technology is never neutral. Whenever they design, it reminds them that their work is not neutral either. They carry an obligation in the twenty-first century, after all, in which we are facing the biggest crises humanity has ever had to overcome. The ring reminds them to be aware of asymmetric vulnerability and the danger of moral corruption; to reflect on the limits of their technical knowledge; to consider how its exercise shifts power; to design with, not for, those whose knowledge is different but who must live with the consequences of the design; and to critically reflect on the legitimacy of their expertise. Some see it as a reminder of the simple fact that there is no Planet B and that all tech comes from somewhere and goes somewhere. Others simply see it as a symbol for collective action. As Roman Krznaric writes, "the key question is not 'how can I make a difference?' but 'how can *we* make a difference?' A mere shift of pronouns has the power to change the world. The urgency of our current crises demands strategies of change based on collective action directed at those in power more than isolated personal actions" (2020, 205).

CONCLUSIONS

Computing's dominant responses to the climate crises, injustice, and ethical challenges have been to double down on rationalist reasoning and the myths of computing. But codes of ethics will not materially shift or even reorient the practices and impacts of computing. Instead, tech workers are called on to find their collective and individual leverage points and organize collective actions.

We are beginning to witness the emergence of a greater sense of collective responsibility and collective action in tech workers of all kinds, driven by their growing awareness of the urgency and importance of changing social course and by a renewed sense of solidarity and "the collective, the communal, the commons, the civil, and the civic" (N. Klein 2014, 460).

You too can reorient your work: your design, your coding, your modeling, your development, your engineering, your research, your teaching, your organizing, and your volunteering. Everywhere you can, find the levers to apply force to amplify the small changes you have control over. And, in all of this, find allies and make critical friends and work together.

CONCLUSION
THIS CHANGES COMPUTING

Climate change—if treated as a true planetary emergency akin to those rising flood waters—could become a galvanizing force for humanity, leaving us all . . . with societies that are safer and fairer in all kinds of other ways as well. . . . It is a vision in which we collectively use the crisis to leap somewhere that seems, frankly, better than where we are right now.
—N. Klein (2014, 14)

Sustainability and justice urge us to change the direction of where computing is headed. This book has argued that the hegemonic orthodox form of computing, expressed in computational thought and rationalist design approaches, is incapable of doing that. It is insolvent: unable to pay the debts owed to the planet and its societies. It is stuck in ill-conceived assumptions about the nature of problems, the workings of the human mind, and the politics of technology. Its mythology evacuates history and politics from design and creates an illusion of neutral technical rationality: a calm cockpit, it appears, from which technology is steered to a better future, one solved problem at a time. But as coherent as it appears, the problemist illusion hides the collateral damage and suffering it produces. In problemism, unreflective solutionism in the unquestioned service of dominant interests reinforces the inequitable status quo, one reified problem at a time.

Understanding where we come from and how we got here helps us decide where to head next. This book spends considerable energy on understanding the myths and their history to demonstrate, step by step, how they shape important tendencies and values in computing research, education, and practice. It describes how these dominant beliefs are misguided and misleading, and it illustrates how they shape and distort what we can talk about when we talk about the role of computing in sustainability and justice. Thus, we progress on Phil Agre's path of critical inquiry, excavating the ground beneath our feet, not to dig us deeper into a hole but to develop an alternative practice of computing.

RESTRUCTURE AND REORIENT: COMPUTING FOR JUST SUSTAINABILITY

The diagnosis of insolvency motivated the *restructuring* performed in part II. By replacing some orthodox assumptions of computing, we gained new perspectives. Aided by computing's critical friends, we are "concerned not with destruction but with reinvention" (Agre 1997a, 24).

Setting the myth of value-neutral technology aside and recognizing the value-loaded nature of systems design makes room to reason about the values that shape it—explicitly and implicitly. Once we stop assuming that technology is neutral and accept that it bears values, we can shift the question from how to avoid misuse to the more meaningful questions. How does this system shift power to the powerless? How does this system help to end oppression? How does this system bring more justice to the world? On what timescale? If we accept that design and development turn values into facts, we are free to ask: Which values do we want to become facts? We become a lot freer to articulate our values and let them shape the designs we are working on. The critical friends of computing offer historical lessons, orienting principles, and concrete design techniques to do that. It should be liberating to see this return of a political understanding of design happen on a broader scale.

Setting the myth of rational decisions aside and recognizing the importance of situated, embodied decision-making in design opens up a space for better understanding the facets and effects of psychological distance in systems design. Once we stop thinking of people as flawed computers, our thinking about design is liberated. Instead of focusing on nudging users through manipulative framings or minimizing their ability to commit errors, we may begin by considering how our system can expand their horizon of experience; how it can amplify their wisdom and judgment; how it can empower their reasoning. This profoundly changes how we understand and counteract short-sighted or narrow-minded design decisions and support just and sustainable decisions.

Setting the myth of objective problems aside and recognizing the politics of problem framing opens a space for pluriversal design in which contrasting views can meet. Once we refuse to accept the first problem framing thrown our way, we are free to consider the rich perspectives inherent in any problematic situation, to facilitate systemic conversations, and to think more broadly about social responsibility and collective action. The critical friends bring principles, methods, and design tools to help facilitate these conversations.

Each shift changes how we can approach questions of sustainability and justice in computing. Together, these shifts help us overcome problemism. The result can be described as a paradigm shift in research, practice, and education towards what I describe as just sustainability design.

To *reorient* systems design practice and research on this newly leveled playing field, I focused on facilitating and instigating systemic change in engineering-oriented fields to shift their perspectives from within. After all, we have good reasons to believe that ideas can indeed shape our future. As you will have noticed, these chapters describe early paths of just sustainability design rather than a comprehensive roadmap. I have focused on attempts to apply leverage within engineering-focused fields and perspectives, despite the obvious fact that these attempts have not concluded yet and that I have little proof of their global impact on the direction of IT. I focused on requirements practice and decision-making research for two reasons: First, requirements hold a key leverage in systems design for shifting tangible outcomes. In the space of reconciling the social and the technical lies an excellent vantage point for making this shift happen. Second, I identified a gap in our understanding of how those involved in design, in the broadest sense, make their way across the psychological distance involved in just sustainability, and how and why they often fail to go far enough. If we want designers to take more responsible, sustainable, and just decisions, we certainly can't treat them as operating systems on which we have to install a new design program. Instead, we have to understand how they make those decisions in the first place, how contextual factors influence them, and how this influence shapes the outcomes of design.

COMPUTING AFTER INSOLVENCY

In part III, I thus illustrated how critical friendships between technical computing disciplines and critical social theory can help us to *reinject politics* into design theory and practice: by recognizing the role of values in shaping technological artifacts, the range of reasoning capabilities exhibited in systems design practice, the politics of problem framing, and the power of collective action. The tangible impacts of this reorientation may not be too impressive yet. Undoubtedly, I have failed to mention many actions and ideas of others that are perhaps more inspiring than the story I have told. But I hope that the principles of just sustainability design and the shifts in worldviews articulated here, as incomplete as they are, can serve as a stepping stone for others, perhaps in a form akin to what Paolo Freire (2000) called *conscientização* (conscientiousness)—"learning to perceive social, political, and economic contradictions, and to take action against the oppressive elements of reality" (35). You need to find your own ways to amplify the systemic impact of these shifts toward a sustainable and just role of IT. As Bonnie Nardi (2019) writes,

We cannot return to any particular past, but nothing is stopping us from finding new ways to apply the wisdom of organizing human activity around community, simplicity, equality, and care, and directing design efforts toward these desiderata. It is time to stop designing stuff and high time to use design wisdom and techniques to address the massive problems before us. Each of us has a different path to follow according to our talents, interests, and skills, but there is a path if we choose it. (14)

The critical friends help us locate ourselves at a historical juncture and choose a path. They critique because they care, and so do I: "critique as care means subjecting our ideas, assumptions, and commitments about the world to constant scrutiny . . . it means making us more open to others and less certain of ourselves" (Fernando 2019, 15). It may make us uncomfortable. This is important: "By making us uncomfortable, critique contains within itself a transformative orientation" (Bargetz and Sanos 2020, 511). It is important to make those involved in systems design "uncomfortable" by opening up the discussion at every stage of design and product development to ideas they are not used to, paths they have not taken, and thoughts that have not occurred to them yet. Many more critical friendships exist beyond those discussed here, and others are just waiting to be formed. Healthy discomfort is often a central intent of art and the humanities. By "defamiliarizing" us from what we thought we knew, good art reveals to us new meanings around us, beyond those which we already know and take for granted. We rely on such critical friends to experience the "otherness" of situations we can't otherwise ever expect to inhabit.

The grim diagnosis of insolvency is then softened by the realization that many tech workers are already at work restructuring the conceptual foundations and everyday practices of systems design to reorient computing. Working together with computing's critical friends, they are not just avoiding the traps set by the old myths of computing but reinventing and building a different computing. They know that IT is never neutral, that the human mind is so much more than an information processor, that any problem framing is not only debatable but *must be* debated. They organize to propose and implement collective actions to reorient the trajectory of IT so that it follows a more sustainable and just path. They teach differently, they design differently, and they research differently from the rationalist mainstream. And their work has the potential to reorient computing so that it becomes a genuine force for just sustainability. But they cannot reorient computing without your help. The task is too urgent, the stakes too high, the gravity wells too deep, and the playing fields too tilted.

SYSTEM GOALS AND PARADIGMS: COMPUTING AFTER GROWTH

If just sustainability design and related efforts change computing from inside, other changes are afoot on the outside. As we reorient computing for just sustainability through collective action, we must align with and learn from these broader shifts.

Most centrally, the harrowing influence of the growth-addicted economic development paradigms of the twentieth century resurfaced throughout this book. As introduced in chapter 1 and discussed in chapter 8, sustainable development is still considered a central compass for sustainability-oriented work in computing (e.g., Hansson, Cerratto Pargman, and Pargman 2021). But there is a genuine conflict between the hegemonic "sustainable development" approach and the need of our planet and our societies. "Sustainable development and its more recent reincarnation 'green growth' depoliticize genuine political antagonisms between alternative visions for the future. They render environmental problems technical, promising win-win solutions and the impossible goal of perpetuating economic growth without harming the environment" (Kallis 2015). That the latest SDGs still postulate economic growth targets renders them oxymoronic. And in fact, not a single "developed" country can claim to have developed sustainably (Fanning et al. 2022). If "what is to be sustained with sustainable development, more than the environment or nature, is a particular capitalistic model of the economy" (Escobar 2018, 43), then sustainable development cannot be the guiding frame of reference for just sustainability design. Finding a new compass has become an urgent question:

The dilemma, once recognised, looms so dangerously over our future that we are desperate to believe in miracles. Technology will save us. Capitalism is good at technology. So let's just keep the show on the road and hope for the best. This delusional strategy has reached its limits. Simplistic assumptions that capitalism's propensity for efficiency will stabilise the climate and solve the problem of resource scarcity are almost literally bankrupt. We now stand in urgent need of a clearer vision, braver policy making, something more robust in the way of a strategy with which to confront the dilemma of growth. (T. Jackson 2009, 188)

The alternatives to this bankrupt paradigm of cancerous growth are already here. They are not speaking with one voice but many. That may appear as a weakness, but it is also a strength. There is significant alignment between the globally diverse alternatives to "development" and the arguments of the primarily European degrowth movement (Demaria, Kallis, and D'Alisa 2015). These alternatives constitute a kind of discordant pluralism that, significantly, includes knowledges of the Global South:

Our project of deconstructing development opens into a matrix of alternatives, from universe to pluriverse. Some visions and practices are already well-known in activist and academic circles. For instance, *buen vivir*, "a culture of life" with various names throughout South America; *ubuntu*, emphasizing the southern African value of human mutuality; *swaraj* from India, centered on self-reliance and self-governance . . . there are thousands of such transformative initiatives around the world . . . [w]hile many terms have a long history, they reappear in the narrative of movements for well-being, and . . . co-exist comfortably with contemporary concepts such as degrowth and ecofeminism. (Kothari et al. 2019, xxviii)

Together, all these shifts paint "a vision in which we collectively use the crisis to leap somewhere that seems, frankly, better than where we are right now" (N. Klein 2014, 14). And it is important to understand that a shift away from the growth paradigm does not end innovation in computing. Innovation does not need endless growth. Innovation means to do things differently, not to do more every year. Growth and innovation are rather different then, and it is entirely possible to innovate without expanding economic activity. But doing so widely requires forms of organization different from the dominant institutions that today control *how* computing innovates. "Post-growth-oriented organization" for innovation involves a shift in nine dimensions, including shifts from profit to social justice, from competition to cooperation, from commodification of common resources to democratic control over commons, from intellectual property as artificially scarce resource to open-sourced knowledge sharing, and from massively scalable to widely replicated technology (Pansera and Fressoli 2021, 392). These shifts are already taking place in computing work too. As one example of many, the *eReuse* platform for digital device commons (Franquesa and Navarro 2018) embodies each of these organizational innovation shifts. It is part of a trend of *post-growth computing* emerging above all in the LIMITS community. Work in this space already defines its goals not in terms of sustainable development but contributing to human flourishing within ecological boundaries (Nardi et al. 2018; Knowles, Bates, and Håkansson 2018; Nardi 2019; Kaczmarek et al. 2020; Mann, Bates, and Maher 2018). While most are growth-agnostic rather than tackling growth actively, these initiatives and designs demonstrate that there is plenty of space for imagination, creativity, and innovation.

Collapse informatics (Tomlinson et al. 2013) asks us to design in a present of abundance for a future of involuntary scarcity following collapse and rapid contraction of economic activity. But "a degrowth transition is not a sustained trajectory of descent, but a transition to convivial societies living simply and in common" (DeMaria and Latouche 2019, 148). The authors write, degrowth . . . calls for a democratically led redistributive downscaling of production and consumption in industrialized countries as a means to achieve environmental sustainability, social justice, and well-being . . . the emphasis should not only be on less, but also on different. In a degrowth society everything will be different: activities, forms and uses of energy, relations, gender roles, allocations of time between paid and non-paid work, relations with the non-human world. (148)

So a *degrowth informatics* would need to ask: what is the role of informatics in a voluntary degrowth transition? How can we design to make degrowth a reality? What informatics would a degrown society need? Existing work guided by concepts such as care, commons, cooperatives, community, and conviviality can point the way. We can expect that everything may be organized in subtly different ways: design activities, energy use, system lifecycles, system use and sharing, divisions of labor in design, the priorities of different aspects of system quality, the form of tradeoff analysis between competing features, the business models and reporting requirements, and how we understand our relationships to other teams working on similar issues.

One of the branches of degrowth informatics may be *doughnut informatics*: the use of informatics to help societies flourish within what economist Kate Raworth (2017) calls the doughnut: a "safe and just space for humanity" visualized like the baked good, in which humanity meets basic social life support needs (the inner circle, measured by indicators such as housing, water, food, education, and gender equality) while respecting ecological boundaries (the outer circle of nine planetary boundaries discussed in chapter 1). One of the challenges of doughnut informatics will be to reconcile in its models the involved scales of activity. Doughnut economics scales the macro-level framework of planetary boundaries down from a global level (Persson et al. 2022) to a national (Hickel et al. 2022) and regional level (Boffey 2020). We will need to situate smaller-scale interventions via systems design in such models, drawing on the frameworks discussed early in this book and the lessons of systemic sustainability analysis discussed in chapter 8. The multi-scalar evaluation frameworks that are necessary to evaluate systemic effects of technology design will require the reintegration of hard systems approaches within critically systemic participatory frameworks highlighted there.

In other words, we will need to innovate. And innovative, reinventive change in computing is already underfoot. The concepts above—from buen vivir to swaraj, from conviviality to ubuntu, from cooperatives to autonomy—already motivate and invigorate some work in computing. The challenge is to amplify and replicate these shifts. But this is not a challenge that computing should address on its own. As Naomi Klein (2014) writes in *This Changes Everything*, "any attempt to rise to the climate challenge will be fruitless unless it is understood as part of a much broader battle of worldviews, a process of rebuilding and reinventing the very idea of the collective, the communal, the commons, the civil, and the civic" (460).

COMPUTING OTHERWISE

Those who work in the emerging paradigm of "computing after growth" often speak in a very different tone than the problemists. For example, Nardi (2019) makes clear that technology is not the savior in this paradigm but one role in many: "Technology can play a critical role in better futures, but not as an exogenous factor. Well-designed technology must enter as an element of a reasoned program of economic transformation with non-negotiable goals of promoting human and non-human well-being" (7). In other words, even if it is a computing team who gets the funding, it may be best to hand the lead to a critical friend.

Where to start? I am wary of telling others who I don't know well what to do. Deciding what to do begins with understanding who you are, where you are, and when. I hope this book offers many starting points for action, from sustainability evaluation frameworks to pathways for critical reflective practice and from collective action in the workplace to each person's role in defunding big tech.

The stakes could not be higher. One million species are on track to being extinguished by one: us (Díaz et al. 2019). One of many devastating findings in the 2022 Intergovernmental Panel on Climate Change (IPCC) report on climate change *Impacts, Adaptation and Vulnerability* is that up to 700 million people in Africa—half the continent's population—are on track to being displaced by water stress *by 2030* (IPCC 2022). At the same time, microplastics, petrochemical products heavily promoted by the fossil fuel industry (Brigham 2022), are now found in human blood (Carrington 2022). There is a good chance they are in yours too. Their health implications are the stuff of nightmares. And yet we are supposed to believe that the most important thing is to grow the economy?

The rude awakening comes when we finally realize that no one else is going to come and fix this for us. This is on us all. And that raises uncomfortable questions. We cannot avoid them a day longer. We need to be prepared to speak up, to ask them, and to realign what we do with the consequences of the responses. The more privilege we have, the more responsibility we have to take risks and act.

Like many, I want a computing that supports me in living my life, connecting with others, and staying alive and well *without* sacrificing the planet or my bodily autonomy for it. Like many, I want a computing that works for all, not just for me, and is driven not by excessive growth imperatives and extractivist logics but by the goals of helping all human and nonhuman nature to flourish. This is not too much to ask. It is not even a radical ask, is it? And the ask is not for a stagnant computing, simply that innovation be organized as ecologically sound and socially beneficial. To think about computing *otherwise* is "to move away from binaries of tech or not (though we may sometimes need to make that judgement), in favour of how we could make tech differently, in the service of our collective and sustainable well being" (TechOtherwise Collective 2021). There is so much to do for computing research, practice, and reality to achieve these lofty goals. What just sustainability design asks us to do is not *less* computing, even if some computing has to stop; and it is certainly not a *lesser* computing. It is a richer, ecologized computing. "Ecologizing society . . . is not about implementing an alternative, better, or greener development. It is about imagining and enacting alternative visions to modern growth-based development" (Kallis 2015). Imagining and enacting these alternative visions certainly requires some subverting of conventions, and "the next act of subversion is . . . to embrace a journey toward regeneration and greater equity" (Light 2022, 37).

How will all this change computing? It is up to you.

NOTES

INTRODUCTION

1. Economists call effects such as pollution externalities because they are external to the established economic system. That these are real and enormous has been known for a long time (e.g., Ayres and Kneese 1969).

2. By computational system, I understand any sociotechnical system whose behavior is partly determined by algorithms implemented by computing devices. I will use IT as the overall term for information, communication, and computing technology, and IT systems or computational systems for identifiable sociotechnical systems that incorporate IT.

3. Ken Fleischmann, who was not at the workshop, has written about this (e.g., Fleischmann and Wallace 2005; Fleischmann, Hui, and Wallace 2017).

4. The original *Devil's Dictionary*, published by Ambrose Bierce in the early twentieth century, contains such gems as "PLAN, v.t. To bother about the best method of accomplishing an accidental result" (1906).

5. Systems thinker Russell Ackoff pointed out that observing the US system of "healthcare" shows that its purpose is not to produce health: "It is not a health-care system, but a sickness- and disability-care system. . . . Whatever the intentions of the individual servers, the system produces and preserves sickness and disability" (Ackoff and Pourdehnad 2001, 200).

6. Disability scholars use the term "crip time" to describe a more flexible approach to time that takes into account diverse abilities and needs (Price 2011, 62–63). "Rather than bend disabled bodies and minds to meet the clock, crip time bends the clock to meet disabled bodies and minds" (Kafer 2013, 27).

7. Feenberg defines "margin of maneuver" more narrowly as the "reactive autonomy" of those marginalized by technological rationalization (2002, 84–85).

Nevertheless, the concept has informed my thinking about the room for maneuver available to all stakeholders involved in systems design.

8. For a gentle introduction to systemic thought, see D. H. Meadows (2008) but be mindful it calls system dynamics "systems thinking." For an excellent overview of approaches, their underpinning epistemologies and methods, and how each approach can and cannot be brought to bear on sociotechnical questions of design, see Michael C. Jackson (2003; 2019). For a sense of the vast diversity of perspectives to be found in systems thinking, see Ramage and Shipp (2009). And for the critical turn, see Flood and Jackson (1991c) and Flood and Romm (1996).

9. Patricia Hill Collins also labels her work "black feminist thought," not theory, for related but distinct reasons, but she characterizes black feminist thought "as a critical social theory" (Hill Collins 1990, 22).

CHAPTER 1

1. To name just two examples, the $100,000,000 Schwartz Reisman Institute for Technology and Society at the University of Toronto quickly adopted the #LetsSolveIt hashtag on social media, and the $1 billion MIT College of Computing will "bring the power of computing and AI to all fields of study at MIT" so that students can "responsibly use and develop AI and computing technologies to help make a better world," emphasizing that "computing reshapes our world" and will "fundamentally transform society" (MIT News Office 2018). Note the absence of the reverse. What would a $1 billion initiative look like that brings the power of all fields of study to computing?

2. See, for example, Bijker et al. (2012); MacKenzie and Wajcman (1999); Gitelman (2013); and Iliadis and Russo (2016).

3. The NATO Software Engineering Conference in 1968 is commonly taken as the founding moment of the discipline (Naur and Randell 1969).

4. The resistance to acknowledging this in software engineering runs deep, as existing dogmatic beliefs remain insistent on an exclusive focus on software artifacts (Ralph and Oates 2018). See also Kaczmarek et al. (2020).

5. This figure is redrawn based on figure 1 in (Becker et al. 2015). The figure on the top left is adapted from ("Sustainability: Can Our Society Endure?" n.d.), the figure on the right from (*Engineering for Sustainable Development: Guiding Principles*. 2005).

6. See Agyeman et al. (2003, 5). In a widely cited book, Neumayer (2013) explores and contrasts strong and weak sustainability from an economic perspective. In his characterization, weak sustainability distinguishes the utility derived from natural resources from the resources themselves, based on the assumption that the same utility can often be derived from a different resource. Strong sustainability rejects the substitutability assumption and requires that the resources themselves must be maintained. Neumayer shows that on an aggregate scale, the general principle of substitutability cannot be defended.

7. Neumayer (2013) concludes his discussion of quantitative cost-benefit analysis (CBA): "Whether and how to act on climate change cannot be decided on the basis of

'hard numbers' because there are no 'hard numbers' when it comes to climate change. To outsiders, the CBA studies of economists may suggest otherwise. But those who understand what the studies do, also know two things. First, many effects of climate change simply cannot be adequately monetarily valued. Second, what can be valued needs to be transformed from values in the far distant future to present-values and any CBA recommendation is therefore crucially dependent on the discount rate used, which is in turn inextricably linked to normative value judgments" (44). Neumayer shows that the debate between strong and weak sustainability cannot be resolved by theoretical argument nor empirical evidence, since both approaches rely on nonfalsifiable assumptions. "The contest between WS and SS cannot be settled by theoretical inquiry. Neither can it be settled by empirical inquiry since such an inquiry would be dependent on information that is only 'forthcoming in the always receding future,' where 'predictions are clouded by uncertainty regarding preferences, human ingenuity and existing resource availability' (Castle 1997, p. 305)" (Neumayer 2013, 95).

8. What appears to be missing or not fully developed in their approach is a full appreciation of the challenges of intergenerational justice and the concerns of critical systems thinking about handling the limits of equitable participation. Because soft systems thinking assumes a basic and free goodwill among participants, the proposed approach relies to a large degree on their willingness to communicate in the negotiations fairly and sincerely without being unduly restricted. This is rarely the case in situations where sustainability is a major concern, and it is impossible to ensure full participation given that many affected stakeholders live in the future. It will thus be necessary to introduce critical systems thinking concepts into this type of framework (cf. Martin Reynolds 2007).

9. An overview of sustainability and IT (Hilty and Aebischer 2015a) cites a cybernetics conference in the 1970s as a precursor and points to the emergence of *environmental informatics* in the 1990s (Avouris 1995), as well as the surfacing of sustainability in the proceedings of the IFIP TC9 Human Choice and Computers conference series in the late 1990s.

10. This overview makes no claim to be comprehensive. Other areas could certainly be included here. For example, *environmental informatics* has a well-established scope and direction, leveraging IT for environmental research (Avouris 1995; Huang and Chang 2003).

11. See Kelly (2019); Jessica Shankleman et al. (2017); and Crawford and Joler (2018).

12. Multiple tech companies have been sued over human rights violations in cobalt mining (Kelly 2019; Frankel, Chavez, and Ribas 2016). The Whanganui River and other natural entities in New Zealand have been declared legal persons (Hickel 2020; Lurgio 2019).

13. (Crawford and Joler 2018). It received widespread praise and attention and was ultimately acquired by the Museum of Modern Art.

14. A systematic review in *The Lancet* (Grant et al. 2013) errs on the side of extreme caution, but still reports "extensive evidence of a causal link between exposure to individual chemical compounds common in e-waste and negative health outcomes."

The authors emphasize that there is a dearth of studies aiming to collect empirical evidence and highlight that there is of course "strong biological plausibility of an association between e-waste exposure and health outcomes," urging the international community that "the health effects of exposure to e-waste must become a priority for the international community" (358).

15. The Ethereum "merge," which switched from proof of work to proof of stake in fall 2022, is a different, interesting case. The efficiency gain is enormous with a factor of about 1000. That bodes well unless it triggers an avalanche of new uses.

16. This is shown by Tsao et al. (2010), later clarified in Saunders and Tsao (2012).

17. Easterbrook made an excellent argument for the importance of this shift (2014a). We will return to this argument later.

18. An early compilation of such work is represented in Calero and Piattini (2015). A series of workshops at ICSE has focused on "Green and Sustainable Software" as well (http://greens.cs.vu.nl/).

19. As discussed in Hilty and Aebischer (2015a), the distinction between orders of effects goes back to Berkhout and Hertin (2001). Hilty and Aebischer also present a revised model called LES, for life-cycle impact, enabling impact, and structural impact. The LES model drops the distinction of problem and solution in favor of distinctions that vary between levels: production, consumption and disposal (L); production, consumption and technological change (E), and economic structures and institutions (S). Because the LES model has not received as much attention as the previous matrix, and because its economically focused structure makes it difficult to identify positive and negative effects, I summarize the two models more broadly in this section.

20. While Airbnb brands itself as a "sharing economy" platform, I follow the compelling arguments in Martin (2016) that this rhetorical branding move has to be resisted. Nothing is shared on platforms like Airbnb or Uber. The Airbnb example was first used to explain structural impact in Becker et al. (2016).

21. For example, Airbnb backed the development of a condo tower project (Sharf 2019) and the specialized operator Zeus Living (Josh Constine 2019).

22. A workshop series on Computing within Limits started 2015 (https://computingwithinlimits.org/), and an article on the subject appeared in the *Communications of the ACM* in 2018 (Nardi et al. 2018).

23. Collapse informatics refers to the "study, design and development of sociotechnical systems in the abundant present for use in a future of scarcity" (Tomlinson et al. 2013).

24. The Karlskrona Manifesto emerged from the Requirements Engineering for Sustainable Systems workshop (RE4SuSy) following my proposal in a paper presented at the workshop (Becker 2014).

25. Among others, a workshop at the 2015 iConference brought SE and ICT4S researchers together with HCI researchers and information scholars (Penzenstadler et al. 2016). Other workshops are listed at https://www.sustainabilitydesign.org/publications/.

26. Redrawn from C. Becker et al. 2016.

27. Notable exceptions include the "Denver Manifesto on Values in Computing," which took explicit inspiration from the Karlskrona Manifesto ("The Denver Manifesto—Values in Computing" 2017).

28. That ground however is still shaky, as misinterpretations abound. At least some work cites the Karlskrona Manifesto yet frames sustainability as a quality of the software system (e.g., Lago et al. 2015).

29. On an individual level, some of the beliefs highlighted as misperceptions in the Karlskrona Manifesto were identified. A lack of knowledge and experience can be addressed by concerted education efforts (Penzenstadler et al. 2018), while a lack of methodology and tool support motivates ongoing research in requirements engineering approaches for sustainability such as Mahaux, Heymans, and Saval (2011) and Chitchyan et al. (2015). On the level of professional norms, the lack of accepted responsibility for long-term consequences of software systems development is deeply embedded into the narrow framing of existing codes of ethics. On the level of the concrete professional environment, a more conflicted picture emerges. Some businesses are keen on opportunities to distinguish themselves, and some are taking very active steps to that end, but significant obstacles make it difficult for employees at lower and middle levels to find and enact their room for maneuver within their own practice. A lack of higher management support, coupled with a reliance on external stakeholders to advocate for sustainability, can exclude the concern from systems design practice. When sustainability is framed as a tradeoff with other goals, it is an uphill battle to get it addressed. This is amplified by a culture of risk aversion and a fear of lost income, exacerbated by short-term incentive structures.

CHAPTER 2

1. Many agree that sustainability always implies ethical judgments. For example, Neumayer (2013) highlights that even within a purely economic paradigm (optimizing the allocation of resources to maximize utility over a time period), the choice of discount rates cannot be theoretically or empirically resolved: "there is no 'right' discount rate," and the choices made in economic analyses "necessarily derive from ethical value judgments that, because they are normative judgments, can and will always be contested" (44).

2. It is worth highlighting that the common trope of the "tragedy of the commons" harks back to a highly problematic piece by a known white nationalist (Hardin 1968) that can only be described as baseless cold war propaganda (Mildenberger 2019; Southern Poverty Law Center n.d.). The tragedy of the commons has been their enclosure (Linebaugh 2014; Hickel 2020).

3. For comprehensive reviews on further developments in environmental justice, see also Mohai, Pellow, and Roberts (2009), and Agyeman et al. (2016). On the evolution of the environmental justice paradigm, see D. E. Taylor (2000).

4. Hickel dissects the available evidence at length in chapters 3 and 4 (Hickel 2020). See also Wiedmann et al. (2020); Demaria, Kallis, and D'Alisa (2015); and Kallis et al. (2020).

5. Hickel (2020) aptly explains that "a recession is what happens when a growth-dependent economy stops growing: it's a disaster. Degrowth is . . . about shifting to a different kind of economy altogether—an economy that doesn't *need* growth in the first place . . . organised around human flourishing and ecological stability, rather than around the constant accumulation of capital" (149–150).

6. The term "unmarked" refers to users whose position in the matrix of domination is left unspecified, which often activates implicit assumptions that they occupy privileged positions (see Costanza-Chock 2020, 69–102).

7. An earlier well-known confrontation is given in Wajcman (1991).

8. See https://facctconference.org/.

9. It is not clear how aware technical debt researchers are of the linguistic theories that explain the underlying conceptual mapping structures of the metaphor. Even direct reflections that emphasize the term "metaphor" make no reference to the linguistic literature on the subject (e.g., Kruchten 2012).

CHAPTER 3

1. The polarized view of language will sound familiar to anyone concerned with the "capture" and formalization of stakeholder input to technical development processes in systems design and resonates with Goguen's (1994) writing about the "wet" and the "dry" in requirements engineering.

2. To be clear, the problematic dualism that shines through this statement by Lukács, which artificially separates "nature" from "society," is a distinct conceptual issue that CST has also grappled with in depth (Midgley 2000; Stephens 2013; Stephens, Taket, and Gagliano 2019). But the central point is that scientific reason has no tools or concepts to address questions around purpose that ask *what is worth striving for?*

3. Marcuse (1964) puts it very succinctly: "The terms 'transcend' and 'transcendence' are used throughout in the empirical, critical sense: they designate tendencies in theory and practice which, in a given society, 'overshoot' the established universe of discourse and action toward its historical alternatives (real possibilities)" (xliii).

4. Others have made this case too (Winograd and Flores 1986; Escobar 2018; Ulrich 1983).

5. See https://twitter.com/ylecun/status/1274782757907030016.

6. In my own searching for compelling examples within the domain of software systems through the literature of values in technology, I have also found that examples tend to be drawn from domains that appear far-fetched for those in computing—whether it is the electric power grid (Hughes 1993), the bicycle (Bijker et al. 2012), bridges in New York (Winner 1980), or the architecture of Nazi concentration camps (Katz 2005), the examples are too easy to dismiss for an audience in computer science. Recent studies in machine learning have begun to change that situation (Birhane et al. 2021).

7. This last point is pressing as I write these lines: Recent mobilizations of activists and scholars against the use of facial recognition technology (FRT) on precisely such

grounds have led some corporate vendors to cease offering concrete products to US law enforcement, and some municipalities and states across the United States have enacted legal bans. While this is a success for those fighting FRT, it illustrates what they are up against: the a priori assumption is that the technology is neutral, and the burden of proof lies on them to show otherwise.

8. The marginalization of disadvantaged perspectives will be a central theme of this book. How that burden has played out is illustrated perhaps best by books such as Eubanks (2018) and S. Noble (2018) that demonstrate the undue burden carried by marginalized communities and the practical, political difficulties in proving the unwelcome news that they are in fact being oppressed by technological developments.

9. This will be explored in more detail in chapter 6.

10. Neither can "mistakes" be easily rectified after the fact—despite the possibility to alter software code in principle, it is well established that changing systems is just as hard as building them. The *path dependence* of technological change means that adverse consequences cannot be so easily rectified (Liebowitz and Margolis 1995; Arthur 1994).

11. This crude characterization can of course not do justice to the nuanced debates about what objectivity *should* mean. See Reiss and Sprenger (2017) for an overview of objectivity in science and B. Smith (1998) on what it means for something to even be considered an *object*.

12. This is illustrated by the contrasting definitions of what "a system" is. Even for Ackoff, "A system is a set of interrelated elements" with certain relationships (Ackoff 1971). For Checkland, the concept "embodies *the idea of* a set of elements" (Checkland 1981, emphasis added). Checkland thereby rejected the realist notion that the social world *is* systemic. In his view, models of systems do not have to correspond to anything; they only have to be useful in structuring a conversation.

13. In his history of automation, industrialization, and the emergence of engineering and management within the capitalist context, David Noble has shown that this is no accidental development: The role of scientific method in engineering, and the way it shapes the agency of specific roles in the division of labor, was significantly shaped by the particular interests of capitalist actors (D. Noble 1977; 1984).

14. I am no stranger to MCDM: my doctoral thesis centered on an MCDM approach to software component selection. I developed a decision-making process based on utility theory (Becker et al. 2009), a decision support system that was widely used and evaluated in digital preservation, and a suite of tools to automate evaluation based on a taxonomy of criteria and measures (Becker and Rauber 2010).

15. The idea of the "geek gene" itself is a dangerous myth in computing education (Patitsas et al. 2019).

16. This is not surprising because Checkland's writing itself shifted over the years (Checkland 2000) and because adopting his real intentions requires a significant rethinking of fundamental assumptions and metaphors.

CHAPTER 4

1. See https://twitter.com/ConsentTracker/status/1120623599868624897.

2. See https://twitter.com/ConsentTracker.

CHAPTER 5

1. I believe Jon Whittle made me first aware of the term and has written about it in an agile design team context, where one person was assigned the "critical friend" role "to raise concerns about design decisions that may interfere with agreed values" (Whittle et al. 2019, 110).

2. I make no claim to have compiled a comprehensive list of critical friends. Critical race studies may be the most important example not currently covered at depth in this chapter, and it offers significant insights to computing particularly when it comes to social justice and environmental justice, as discussed in chapter 2.

3. Advocates of participatory design may disagree with the stark distinction made here between the roles of supposed experts and lay people and instead propose that true collaboration in design will involve everyone in the translation effort. Otherwise, the danger is that their participation turns into "pseudo-participation," where participants are treated not as sources of insights but as sources of data to be consumed by experts (Palacin et al. 2020). I suspect that Feenberg would not disagree, but the quote seems to reinforce the boundaries between professional designers and others.

4. Reading this strange book, I was puzzled and found myself in a situation comparable to Agre. I expected a dialectic resolution, a final synthesis to resolve the dilemmas. Somehow, I expected the rational planner to find a way to deal with the enemies, because I had been brought up intellectually in a group of fields in applied computing where each contribution is expected to pose a problem, propose a solution, and demonstrate the solution's efficacy. But that synthesis never comes. Churchman never resolved his fundamental questions (Ulrich 1985).

5. See M. Jackson (1982, 1985a); Ackoff, Checkland, and Churchman (1982); and M. Jackson (1983).

6. See M. Jackson (1991, 2019); Flood and Jackson (1991c); Flood and Romm (1996); and Ulrich (1983).

7. In Ulrich's language, "the social system S to be bounded" contains "social actors defining the normative content of S," with the disjunct subsets "Those involved in the planning of S" and "Those affected by S but not involved in its planning" (Ulrich 1983, 248).

8. Ulrich references Marcuse in his use of that term.

9. This becomes visible when we consider the metaphors that underlie their paradigmatic image of social reality. Jackson drew on the use of metaphor in organizational theory (Morgan 2011; 2006) to distinguish a wide range of systems approaches according to the images that underpin their understanding of the structure of organizations as machines, as organisms, as brains, as cultures, etc. (M. Jackson 2003; 2019).

Focusing just on the main paradigm of cybernetics, Ulrich distinguishes the original (physical) cybernetic paradigm and its reliance on the "machine" metaphor of servomechanisms from the later cybernetic paradigm, which relies on the organism as metaphor. In Stafford Beer's work on managerial cybernetics, this becomes explicit (Beer 1972). In contrast, the organizing metaphor for Ulrich's view of social reality is the purposeful *polis*—the social group formed around an issue (Ulrich 1983, 333).

10. This is crucial for CST practice because it routinely incorporates rationalistic methods within a carefully designed framework *if* their application is appropriate to the context and legitimate (Flood and Jackson 1991a; M. Jackson 2019; Midgley 2000).

11. The validity of Habermas's theory of knowledge-constitutive interests and its role and relevance for critical systems thinking has been discussed at length (Reynolds 2002; Midgley 2002; M. Jackson 2019). It is important to recognize that the role of this theory is not foundational for the development of the critical categories of heuristics in CSH, and that later critique of Habermas and this particular part of his work, as well the evolution of his own position away from the emphasis on technical control in the original argument, do not invalidate the argument and conceptual framework introduced here (see also Reynolds 2002; Ulrich 1993; M. Jackson 2019; Habermas 1985).

12. Purposeful systems are also a central element of soft systems methodology. Its central modeling techniques represent hierarchically contained networks of purposeful activities and their dependencies on each other. Purposeful activity models capture systems relevant to the problem situation. They are used to structure a discussion among everyone involved in the situation (Checkland 1981; 2000).

13. For critiques and further commentary on CSH, see also M. Jackson (1985b, 2003, 2019); Ivanov (1991); Flood and Jackson (1991a, 1991b); M. Brown (1996); Willmott (1989); Mingers (1992); Romm (1995, 1994); and Midgley (1997).

14. See, for instance, The Refusal Conference 2020 (https://afog.berkeley.edu/programs/the-refusal-conference) and the Resistance AI workshop at NeurIPS 2020 (https://sites.google.com/view/resistance-ai-neurips-20/home).

CHAPTER 6

1. The idea of VNT is not innocent. The examples in (Pitt 2014) are explicitly about guns, and a counterexample brings up the technology of the holocaust to demonstrate the absurdity of VNT (Katz 2005). But it is perhaps precisely because those technologies are so viscerally murderous that it is too easy to focus on what is different in computing—even if automated weapons targeting systems are not all that different in their attempt to scaling up the extermination of human life.

2. This distinction was a crucial part of establishing the social sciences *as sciences* and deeply influenced the debate around the nature of objectivity in science (Reiss and Sprenger 2017; Douglas 2011).

3. In their recent case study of three maps, D'Ignazio and Klein (2021) mobilize the commitments of data feminism to similarly compare whose values are turned into maps.

4. The uncritical lens of VBE shines through in some misinterpretations of prior work too. For example, Spiekermann cites Ulrich but misrepresents the purpose of CSH as a tool to select stakeholders (Spiekermann and Winkler 2020, 7). CSH is referred to as a "checklist" and Ulrich as "the originator of CST" (Spiekermann 2016, 202), which is inaccurate. In contrast, Ulrich emphasizes that the entire point of CST lies not in determining correct answers but in continued questioning of categories and answers. The point is not to optimally select stakeholders considering *their* motivation, but to critical *reflect on* the sources of motivation, power, knowledge, and legitimacy underpinning *the claims made when selecting* stakeholders. In a sense, Spiekermann mistakes the surface boundary judgment of including or excluding a stakeholder with the boundary of the *reference system* (see chapter 5).

CHAPTER 7

1. For examples, see behavioral studies of design practice (Cross 2006; Dorst 1995; Dorst and Cross 2001) and this review of behavioral studies in software engineering (Lenberg, Feldt, and Wallgren 2015).

2. The principal focus on description, explanation, and analysis on the one hand, and prescription or action on the other hand, distinguishes two paradigms: Empirical research aims to describe and explain behavior, while normative research establishes standards to evaluate behavior (Ralph 2018). Normative models such as SE methods, process models, and quality models establish standards for evaluation. But there is a significant grey area because the object of empirical SE research is prescribed by normative models. Because research in SE often combines normative and empirical elements, even research that understands itself as empirical often relies on theories that are normative.

3. For their brief overview of the rationalist tradition in design, see chapter 3. I use the term *rationalist* here consistent with their critique and the critique of decision-making researchers to refer to streams of research that take rational ideals too far, at the detriment of situated knowledge and other, broader forms of reasoning.

4. On the relationship between Simon's work and that of contemporaries, and the various streams of work that gave rise to the metaphor of the mind as a computer, see Crowther-Heyck (1999, 2005); Gigerenzer and Goldstein (1996).

5. This is a complex argument but very worth retracing (Maturana 1980; Maturana and Varela 1992, 141–76; Varela, Thompson, and Rosch 1991, 133–214). The falsification of information-processing theories of mind is perhaps most compellingly demonstrated, and most easily traced, by studies on color perception (Varela, Thompson, and Rosch 1991, 147–184).

6. Crowther-Heyck (2005) reports that Simon reacted strongly to his critics but failed to see their point. Instead, he seemed so beholden to his rational worldview that his critics appeared to him as mystics, and he felt he could not reach them.

7. I will here treat rationalist theories broadly, acknowledging that there are of course significant differences too. Simon's work pursued a path distinct from Tversky and Kahneman, for example, and placed more emphasis on the metaphor of the brain as computer than their work. This does not affect my observations.

8. This is still leaving aside that in reality, ambiguity will prevent such clear numeric probabilities from being precise estimates. Once we appreciate ambiguity more fully, we realize the normative fallacy is often even more pronounced than in this example, because it strongly suggests that people use these probabilities even when they don't.

9. The empirical validity of that calculation is dubious. Recall that even statistically, the expected value is not a reasonable approximation of a single gamble, only of a long series of identical gambles.

10. Actual research will of course often use a combination of theoretical frameworks. This can make it difficult to detect instances when normative assumptions slip into descriptive research.

CHAPTER 8

1. They have been further discussed, grouped, and refined over the years (Termeer, Dewulf, and Biesbroek 2019; Alford and Head 2017; Brown, Harris, and Russell 2010; Head and Alford 2015).

2. Ulrich is decidedly democratic in his ideals here. "The democratic consensus of affected citizens is a mandatory source of legitimation for 'rational' planning *whenever the involved and the affected are not identical*"—and admits that "Of course, there ultimately remains a rationally unverifiable (decisionistic) value judgment in this ideal" (1983, 297). Ulrich's use of the term "citizen" stems from his primary interest in policy and social planning and the language common at the time. While Graeff (2020) and others advocate for the term "citizen professional" or "citizen engineer," I will not continue this use of the term "citizen" because it suggests to many that important stakeholders who are affected by tech, including noncitizen residents or undocumented immigrants, are excluded.

3. The lineage of this diverse body of work is best appreciated by following in chronological order the key works in CST (Flood and Jackson 1991c; Flood 1990; Flood and Romm 1996; Ulrich 1983; Midgley 2000; M. Jackson 2019; Stephens 2013).

4. Escobar (2018) highlights the arrogance of the development discourse and summarizes critics such as (Redclift 2005): "critics have pointed out that such a definition is oxymoronic in that the interests of development and the needs of nature cannot be harmonized within any conventional model of the economy" (Escobar 2018, 43). Instead of the "impoverished" sustainable development agenda (225), Escobar calls for the elimination of "the structures of unsustainability that maintain the dominant ontology of devastation" (7). See also Chambers (1997) and Kothari et al. (2019).

CHAPTER 9

1. This is not to say that these are the only myths and counternarratives. For example, if we consider the tendency to disregard the historicity of IT as a myth, we can understand the work of historians of technology as producing the counternarrative (e.g., Hicks 2017).

2. For example, data feminism, design justice, and autonomous design all present a critically appreciative, value-sensitive view of systems design for a more just and

sustainable world. From that vantage point, the shift away from VNT is overdue, and the view of problems as contingent framings is accepted. I do not suggest that JSD should replace these approaches but that its framework can help reorient other approaches, such as requirements engineering (RE). In addition, JSD may still contribute two concrete insights: first, it strengthens the capacity to speak to and with the rationalist scientific programs that are also needed to address the domain complexity of just sustainability, and second, it introduces a contemporary perspective of judgment and decision-making.

CHAPTER 10

1. This understanding is also supported by many in the *requirements engineering* field. See for example widely used textbooks such as Robertson and Robertson (2012), Alexander and Beus-Dukic (2009), and Pohl (2010). Within the context of software development, Mohanani and Ralph have criticized the framing of "requirements" as an "illusion" and highlighted that a fixation on perceived requirements can reduce the creativity and quality of software development (Ralph 2013; Mohanani, Ralph, and Shreeve 2014). But their study refers to the narrow traditional concept of requirements statements in the second sense, as documented in requirements specification standards. For example, IEEE Std. 610.12 and later ISO 24765 define a *requirement* as "(1) A condition or capability needed by a user to solve a problem or achieve an objective; (2) A condition or capability that must be met or possessed by a system or system component to satisfy a contract, standard, specification, or other formally imposed documents; (3) A documented representation of a condition or capability as in (1) or (2)" (ISO 2010).

2. In fact, well known work in RE on stakeholders and onion models (Alexander 2005; Alexander and Beus-Dukic 2009) draws from literature that considers only those *with influence* as stakeholders and severely misrepresents Midgley's argument (e.g., Coakes and Elliman 1999).

3. Most writers' interpretation of soft systems approaches is rooted in hard systems thinking, reflecting the assumptions corresponding to the left column in table 8.1. For example, one classic textbook states that "[in] the early stages of analysis . . . the application domain, the problem, and the organisational requirements must be understood . . . 'hard' models . . . are inflexible" (Kotonya and Sommerville 1998). While the authors recommend "soft" systems approaches instead, they imply that domain, problem, and requirements exist prior to requirements work. A more recent book explains that "'Soft' here means that the work is ill-defined, involving many partially known concerns, not susceptible to deterministic approaches" (Alexander and Beus-Dukic 2009, 13), neglecting to embrace the epistemological shift from hard to soft systems thinking represented in table 8.1. This misunderstanding underpins a misinterpretation of techniques, methods, and the role of models: For example, Checkland's "Rich Pictures" are described as models of reality ["a diagram of what is happening in a business . . . an informal but very useful view of a system's context and scope" (Alexander and Beus-Dukic 2009, 77–78)]; Checkland would disagree (Checkland 2000; Checkland and Poulter 2007). Another textbook suggests that one of the

structuring techniques used in SSM called CATWOE means that "you consider the Customers, Actors, Transformation Processes, World view, Owners, and Environment *for the project*" (Robertson and Robertson 2012, 64; emphasis added). In SSM, CATWOE is a set of concepts used to shape a thought model of one of multiple relevant systems (which are *ideas*), with the central aim being that the participants develop multiple such ideas from different world views as an aid to structure a conversation about the situation; it is not a way to specify "the project." Another textbook, finally, makes no mention of SSM at all (Pohl 2010) and instead is firmly grounded in a hard systems view. The adoption of SSM techniques into other types of methods is typical, and one might argue there is nothing inherently wrong with it. (The Robertsons' textbook provides a sophisticated method for carefully scoping out a project and developing useful requirements.) But it changes the nature of the techniques, and these examples illustrate that the writers have not completed the epistemological shift involved in SSM. The myth of objective problems shapes their texts.

4. Note that this definition of stakeholders excludes those who are subsequently affected but unaware. In the world of this textbook, they are not even recognized as stakeholders. Fortunately, others are more sensible (Alexander 2005).

5. The role is rarely labeled "requirements engineer" but more typically "business analyst" or "systems analyst." Very often requirements work is spread across multiple roles and/or done jointly with project management work, software architecture work, etc. (Robertson and Robertson 2012; Rozanski and Woods 2012). This does not affect the argument since all these roles are positioned in that same context. I use the terms "requirements work," "RE professionals," and sometimes "requirements engineer" to refer to this spectrum of work, roles, and people.

6. These questions are adapted from Becker, Betz, et al. (2020).

7. The research team was composed of Leticia Duboc, a researcher at La Salle University in Barcelona, my PhD student Curtis McCord, and myself. Leticia had been involved in conversations with the third-sector company and other stakeholders and had close contact with the technical and business developers of SoundCare but no previous knowledge of CSH. Curtis and I were knowledgeable in CSH. We took a mentoring role and helped to critically reflect on the models created by Leticia.

8. The description of these iterations is adapted from Duboc, McCord, et al. (2020).

9. Midgley reports on a structural parallel in a community project developing housing services for elderly people. The participation of those affected in a key meeting could not be arranged, despite the researchers' attempts, because the project sponsor insisted this was for their managers only (those involved, but not affected). The researchers faced the ethical dilemma of canceling the project or representing, as surrogates, the interests of those affected. Based on a reflection of their own positionality, they configured the meeting such that some of them became dedicated spokespeople for those affected (Midgley, Munlo, and Brown 1998, 472–473).

10. The presence of surrogacy is mostly implicitly accepted. Some discuss it at length (Alexander 2005), but it is worth pointing out that in his reliance on Coakes's flawed

interpretation of Midgley, Alexander, despite his sensitivity to stakeholder surrogacy, misses Midgley's substantive contributions on marginalization and boundary critique.

11. Of course, professional fields such as medicine and social work have a history in contesting and reflecting on the nature of professional expertise and the ethical and political nature of such knowledge (Hoffman 1989).

12. In a different project, we used CSH in polemic form to contest the process of marginalization. Sidewalk Toronto was a flagship project for Alphabet Sidewalk Labs to design and build (and operate) a prime waterfront district in Toronto as a "sustainable Smart City." The consultation processes sounded participatory enough, the shiny brochure had wooed many, but civic advocates rightfully criticized the corporate urban land grab as a project that would benefit Alphabet, but not Torontonians (Wylie 2020; Editorial 2018). We used CSH to "explore how the professed values guiding the project were contentiously enacted, and we showed how key stakeholders and beneficiaries in the planning process significantly constrained the emancipatory and transformative potential of the project by marginalizing the role of residents in determining project purposes" (McCord and Becker 2019).

13. You can read the result of this struggle in the *Communications of the ACM* (Kienzle et al. 2020).

CHAPTER 11

1. For overview texts, see (Frederick, Loewenstein, and O'Donoghue 2002; Loewenstein, Read, and Baumeister 2003; Loewenstein, Rick, and Cohen 2008; Soman et al. 2005). My coauthors and I review these and others and explain their relevance for decision-making in SE (Becker, Walker, and McCord 2017; Becker et al. 2018).

2. Figure © 2019 IEEE. Reprinted, with permission, from Fagerholm et al. 2019, "Temporal Discounting in Software Engineering: A Replication Study," ACM/IEEE International Symposium on Empirical Software Engineering and Measurement (ESEM), 1-12, 10.1109/ESEM.2019.8870161.

3. More detail about the method is reported in Fagerholm et al. (2022).

4. And this is before we have spent time embedded with teams, ethnographically studying decision-making in practice. How much could we learn doing this? I hope to find out.

5. This may have been due to many factors. Maybe it was because in our study, the situation was hypothetical, or because the design already construed a high-level view of the question, while in the mentioned study, the situation was real and full of incidental detail.

6. The definitions of these five elements are adapted and broadened from (Fagerholm et al. 2022).

7. Hypotheticality in practice is a direct function of uncertainty and ambiguity, but in research designs, matters are complicated by the question of how real the situation appears to the participants.

CHAPTER 12

1. See https://twitter.com/ChriBecker/status/1014938612746551296 and https://twitter.com/ACM_Ethics/status/1015280727137705984.

2. Recent decades have added tax evasion to that history (Fair Tax Mark 2019; Schaake 2020). The pandemic of 2020 has raised broader awareness of the erosion of public goods that comes with this history (S. Noble 2020).

3. Software engineering is often not recognized as a "true" engineering discipline. The graduates of SE programs are not chartered engineers—a source of much soul searching in the discipline's community and scholarship. Nevertheless, the term "engineering" is often used. I use it here to emphasize commonalities across a range of professionalized work in which specialized technical expertise is applied in evolving collaborations in a spectrum between independent contracting and employed labor to shape the material environment, including IT. In Noble's (1977) account of the early emergence of professional engineers, he focuses on mechanical, chemical, mining, and civil engineers (33–49). Accounts of later developments naturally include newer engineering disciplines such as electrical engineering (Wisnioski 2012). The structural similarities of this work across "technical" fields far outweigh the differences in certification status, and many of the struggles of IT professionals today with technical rationality and its limitations, as well as grappling with issues of equity, mirror earlier struggles (Wisnioski 2012; Tarnoff 2020).

REFERENCES

Abson, David J., Joern Fischer, Julia Leventon, Jens Newig, Thomas Schomerus, Ulli Vilsmaier, Henrik von Wehrden, et al. 2017. "Leverage Points for Sustainability Transformation." *Ambio* 46 (1): 30–39. https://doi.org/10.1007/s13280-016-0800-y.

Ackoff, Russell, Peter Checkland, and West Churchman. 1982. "Three Replies to Jackson." *Journal of Applied Systems Analysis* 9.

Ackoff, Russell L. 1958. "Towards a Behavioral Theory of Communication." *Management Science* 4 (3): 218–234. https://doi.org/10.1287/mnsc.4.3.218.

Ackoff, Russell L. 1971. "Towards a System of Systems Concepts." *Management Science* 17 (11).

Ackoff, Russell L. 1977. "Optimization + Objectivity = Optout." *European Journal of Operational Research* 1 (1): 1–7. https://doi.org/10.1016/S0377-2217(77)81003-5.

Ackoff, Russell L. 1999. "Disciplines, the Two Cultures, and the Scianities." *Systems Research and Behavioral Science; Chichester* 16 (6): 533.

Ackoff, Russell L., and John Pourdehnad. 2001. "On Misdirected Systems." *Systems Research and Behavioral Science* 18 (3): 199–205. https://doi.org/10.1002/sres.388.

Ackoff, Russell Lincoln. 1999a. *Ackoff's Best: His Classic Writings on Management*. New York: Wiley.

Ackoff, Russell Lincoln. 1999b. *Re-Creating the Corporation: A Design of Organizations for the 21st Century*. New York: Oxford University Press.

ACM Committee on Professional Ethics. 1992. "1992 ACM Code." *ACM Ethics* (blog). https://ethics.acm.org/code-of-ethics/previous-versions/1992-acm-code/.

ACM Committee on Professional Ethics. 2018. *ACM Code of Ethics and Professional Conduct*. ACM.

Agre, Philip. 1997a. *Computation and Human Experience (Learning in Doing)*. New York: Cambridge University Press.

Agre, Philip. 1997b. "Toward a Critical Technical Practice: Lessons Learned in Trying to Reform AI." In *Social Science, Technical Systems, and Cooperative Work: Beyond the Great Divide*, 131–157. Mahwah, NJ: Lawrence Erlbaum Associates.

Agyeman, Julian. 2013. *Introducing Just Sustainabilities: Policy, Planning, and Practice*. London: Zed Books.

Agyeman, Julian, Robert Doyle Bullard, and Bob Evans. 2003. *Just Sustainabilities: Development in an Unequal World*. Cambridge, MA: MIT press.

Agyeman, Julian, David Schlosberg, Luke Craven, and Caitlin Matthews. 2016. "Trends and Directions in Environmental Justice: From Inequity to Everyday Life, Community, and Just Sustainabilities." *Annual Review of Environment and Resources* 41 (1): 321–340. https://doi.org/10.1146/annurev-environ-110615-090052.

Aho, Alfred V. 2012. "Computation and Computational Thinking." *The Computer Journal* 55 (7): 832–835. https://doi.org/10.1093/comjnl/bxs074.

Akkermans, Hans, and Jaap Gordijn. 2006. "What Is This Science Called Requirements Engineering?" In *Requirements Engineering, 14th IEEE International Conference*, 273–278. IEEE. http://ieeexplore.ieee.org/xpls/abs_all.jsp?arnumber=1704077.

Albers, Susanne. 2010. "Energy-Efficient Algorithms." *Communications of the ACM* 53 (5): 86–96.

Alcott, Blake. 2005. "Jevons' Paradox." *Ecological Economics* 54 (1): 9–21. https://doi.org/10.1016/j.ecolecon.2005.03.020.

Alexander, Ian F. 2005. "A Taxonomy of Stakeholders: Human Roles in System Development." *International Journal of Technology and Human Interaction (IJTHI)* 1 (1): 23–59. https://doi.org/10.4018/jthi.2005010102.

Alexander, Ian F., and Ljerka Beus-Dukic. 2009. *Discovering Requirements: How to Specify Products and Services*. 1st ed. Wiley.

Alford, John, and Brian W. Head. 2017. "Wicked and Less Wicked Problems: A Typology and a Contingency Framework." *Policy and Society* 36 (3): 397–413. https://doi.org/10.1080/14494035.2017.1361634.

Allhutter, Doris. 2012. "Mind Scripting: A Method for Deconstructive Design." *Science, Technology, & Human Values* 37 (6): 684–707. https://doi.org/10.1177/0162243911401633.

Allhutter, Doris. 2021. "The Politics of Critical Requirements Engineering." In *The Critical Requirements Engineering Workshop (CREW)* 2021. https://criticalrequirementsengineering.wordpress.com/.

Allhutter, Doris, and Bettina Berendt. 2020. "Deconstructing FAT: Using Memories to Collectively Explore Implicit Assumptions, Values and Context in Practices of Debiasing and Discrimination-Awareness." In *Proceedings of the 2020 Conference on Fairness, Accountability, and Transparency*, 687. FAT* '20. New York: Association for Computing Machinery. https://doi.org/10.1145/3351095.3375688.

Allhutter, Doris, and Roswitha Hofmann. 2012. "Deconstructive Design as an Approach for Opening Trading Zones." In *Computer Engineering: Concepts, Methodologies, Tools and Applications*, 394–411. Hershey, PA: IGI Global.

Amanatidis, Theodoros, Nikolaos Mittas, Alexander Chatzigeorgiou, Apostolos Ampatzoglou, and Lefteris Angelis. 2018. "The Developer's Dilemma: Factors Affecting the Decision to Repay Code Debt." In *Proceedings of the 2018 International Conference on Technical Debt*, 62–66. TechDebt '18. New York: Association for Computing Machinery. https://doi.org/10.1145/3194164.3194174.

Ames, Morgan G. 2019. *The Charisma Machine: The Life, Death, and Legacy of One Laptop Per Child*. Cambridge, MA: MIT Press.

Angwin, Julia, Jeff Larson, Surya Mattu, and Lauren Kirchner. 2016. "Machine Bias." Pro Publica. https://www.propublica.org/article/machine-bias-risk-assessments-in-criminal-sentencing.

APA. 2020a. "Decision Making." *APA Dictionary of Psychology*. 2020. https://dictionary.apa.org/decision-making.

APA. 2020b. "Judgment." *APA Dictionary of Psychology*. 2020. https://dictionary.apa.org/judgment.

Appiah, Kwame Anthony. 2018. "The Myth of Meritocracy: Who Really Gets What They Deserve?' *The Guardian*, October 2018. http://www.theguardian.com/news/2018/oct/19/the-myth-of-meritocracy-who-really-gets-what-they-deserve.

Aragon, Cecilia, Clayton Hutto, Andy Echenique, Brittany Fiore-Gartland, Yun Huang, Jinyoung Kim, Gina Neff, Wanli Xing, and Joseph Bayer. 2016. "Developing a Research Agenda for Human-Centered Data Science." In *Proceedings of the 19th ACM Conference on Computer Supported Cooperative Work and Social Computing Companion*, 529–535. CSCW '16 Companion. New York: Association for Computing Machinery. https://doi.org/10.1145/2818052.2855518.

Arthur, W. Brian. 1994. *Increasing Returns and Path Dependence in the Economy*. Ann Arbor: University of Michigan Press.

Augier, Mie, and James G. March, eds. 2004. *Models of a Man: Essays in Memory of Herbert A. Simon*. Cambridge, MA: MIT Press.

Aurum, Aybüke, and Claes Wohlin. 2003. "The Fundamental Nature of Requirements Engineering Activities as a Decision-Making Process." *Information and Software Technology* 45 (14): 945–954. https://doi.org/10.1016/S0950-5849(03)00096-X.

Avgeriou, Paris, Philippe Kruchten, Ipek Ozkaya, and Carolyn Seaman. 2016. "Managing Technical Debt in Software Engineering (Dagstuhl Seminar 16162)." Edited by Paris Avgeriou, Philippe Kruchten, Ipek Ozkaya, and Carolyn Seaman. *Dagstuhl Reports* 6 (4): 110–138. https://doi.org/10.4230/DagRep.6.4.110.

Avouris, Nicholas M. 1995. *Environmental Informatics: Methodology and Applications of Environmental Information Processing.* Dordrecht, Holland: Kluwer.

Ayres, Robert U., and Allen V. Kneese. 1969. "Production, Consumption, and Externalities." *The American Economic Review* 59 (3): 282–297.

Balde, Cornelis P., Vanessa Forti, Vanessa Gray, Ruediger Kuehr, and Paul Stegmann. 2017. *The Global E-Waste Monitor 2017: Quantities, Flows and Resources.* United Nations University, International Telecommunication Union, and International Solid Waste Association. https://collections.unu.edu/view/UNU:6341.

Bardzell, Shaowen. 2010. "Feminist HCI: Taking Stock and Outlining an Agenda for Design." In *Proceedings of the SIGCHI Conference on Human Factors in Computing Systems*, 1301–1310. CHI '10. Atlanta, GA: Association for Computing Machinery. https://doi.org/10.1145/1753326.1753521.

Barendregt, Wolmet, Christoph Becker, EunJeong Cheon, Andrew Clement, Pedro Reynolds-Cuéllar, Douglas Schuler, and Lucy Suchman. 2021. "Defund Big Tech, Refund Community." *Tech Otherwise*, February. https://doi.org/10.21428/93b2c832.e0100a3f.

Bargetz, Brigitte, and Sandrine Sanos. 2020. "Feminist Matters, Critique and the Future of the Political." *Feminist Theory* 21 (4): 501–516. https://doi.org/10.1177/1464700120967311.

Baron, Jonathan. 2012. "The Point of Normative Models in Judgment and Decision Making." *Frontiers in Psychology* 3:1–3. https://doi.org/10.3389/fpsyg.2012.00577.

Barthes, Roland. 1972. *Mythologies.* London: J. Cape.

Baskerville, Delia, and Helen Goldblatt. 2009. "Learning to Be a Critical Friend: From Professional Indifference through Challenge to Unguarded Conversations." *Cambridge Journal of Education* 39 (2): 205–221. https://doi.org/10.1080/03057640902902260.

Bass, Len. 2013. *Software Architecture in Practice.* 3rd ed. SEI Series in Software Engineering. Upper Saddle River, NJ: Addison-Wesley.

Bauer, Monika A., James E. B. Wilkie, Jung K. Kim, and Galen V. Bodenhausen. 2012. "Cuing Consumerism: Situational Materialism Undermines Personal and Social Well-Being." *Psychological Science* 23 (5): 517–523. https://doi.org/10.1177/0956797611429579.

Baumer, Eric P.S., and M. Six Silberman. 2011. "When the Implication Is Not to Design (Technology)." In *Proceedings of the SIGCHI Conference on Human Factors in Computing Systems*, 2271–2274. CHI '11. New York: Association for Computing Machinery. https://doi.org/10.1145/1978942.1979275.

Beach, Lee Roy, and Raanan Lipshitz. 1993. "Why Classical Decision Theory Is an Inappropriate Standard for Evaluating and Aiding Most Human Decision Making."

In *Decision Making in Action: Models and Methods*, by Gary A. Klein, Judith Orasanu, Roberta Calderwood, and Caroline Zsambok. Norwood, NJ: Alex Publishing. http://psycnet.apa.org/psycinfo/1993-97634-002.

Beath, Cynthia Mathis, and Wanda J. Orlikowski. 1994. "The Contradictory Structure of Systems Development Methodologies: Deconstructing the IS-User Relationship in Information Engineering." *Information Systems Research* 5 (4): 350–377.

Becker, Christoph. 2014. "Sustainability and Longevity: Two Sides of the Same Quality?" In *Proceedings of the Third International Workshop on Requirements Engineering for Sustainable Systems Co-Located with 22nd International Conference on Requirements Engineering (RE 2014)*, 1216:1–6. CEUR Workshop Proceedings. http://ceur-ws.org/Vol-1216/paper1.pdf.

Becker, Christoph. 2018a. "Metaphors We Work By: Reframing Digital Objects, Significant Properties, and the Design of Digital Preservation Systems." *Archivaria* 85 (0): 6–36. http://hdl.handle.net/1807/87826.

Becker, Christoph. 2018b. "An Analysis of Principle 1.2 in the New ACM Code of Ethics." *ArXiv*:1810.07290 [Cs], October. http://arxiv.org/abs/1810.07290.

Becker, Christoph, S. Betz, R. Chitchyan, L. Duboc, S. M. Easterbrook, B. Penzenstadler, N. Seyff, and C. C. Venters. 2016. "Requirements: The Key to Sustainability." *IEEE Software* 33 (1): 56–65. https://doi.org/10.1109/MS.2015.158. https://hdl.handle.net/1807/74682.

Becker, Christoph, Stefanie Betz, Fabian Fagerholm, and Doris Allhutter. 2020. "1st Critical Requirements Engineering Workshop (CREW2020)." 1st Critical Requirements Engineering Workshop (CREW2020). September. https://criticalrequirementsengineering.wordpress.com/.

Becker, Christoph, Ruzanna Chitchyan, Stefanie Betz, and Curtis McCord. 2018. "Trade-off Decisions across Time in Technical Debt Management: A Systematic Literature Review." In *Proceedings of TechDebt '18: International Conference on Technical Debt*. IEEE Press. https://doi.org/10.1145/3194164.3194171.

Becker, Christoph, Ruzanna Chitchyan, Leticia Duboc, Steve Easterbrook, Martin Mahaux, Birgit Penzenstadler, Guillermo Rodriguez-Navas, et al. 2014. "The Karlskrona Manifesto for Sustainability Design." *ArXiv Preprint ArXiv:1410.6968*. http://arxiv.org/abs/1410.6968. https://hdl.handle.net/1807/74698.

Becker, Christoph, Ruzanna Chitchyan, Leticia Duboc, Steve Easterbrook, Birgit Penzenstadler, Norbert Seyff, and Colin C. Venters. 2015. "Sustainability Design and Software: The Karlskrona Manifesto." In *Proceedings of the 37th International Conference on Software Engineering—Volume 2*, 467–476. ICSE '15. Piscataway, NJ: IEEE Press. http://dl.acm.org/citation.cfm?id=2819009.2819082.

Becker, Christoph, Gregor Engels, Andrew Feenberg, Maria Angela Ferrario, and Geraldine Fitzpatrick. 2019. "Values in Computing (Dagstuhl Seminar 19291)." *Dagstuhl Reports* 9 (7): 40–77. https://doi.org/10.4230/DagRep.9.7.40.

Becker, Christoph, Fabian Fagerholm, Rahul Mohanani, and Alexander Chatzigeor-giou. 2019. "Temporal Discounting in Technical Debt: How Do Software Practitio-ners Discount the Future?" In *2019 IEEE/ACM International Conference on Technical Debt (TechDebt)*, 23–32. https://doi.org/10.1109/TechDebt.2019.00011.

Becker, Christoph, Hannes Kulovits, Mark Guttenbrunner, Stephan Strodl, Andreas Rauber, and Hans Hofman. 2009. "Systematic Planning for Digital Preservation: Evaluating Potential Strategies and Building Preservation Plans." *International Journal on Digital Libraries* 10 (4): 133–157. https://doi.org/10.1007/s00799-009-0057-1.

Becker, Christoph, Ann Light, Chris Frauenberger, Dawn Walker, Victoria Palacin, Syed Ishtiaque Ahmed, Rachel Charlotte Smith, Pedro Reynolds Cuéllar, and David Nemer. 2020. "Computing Professionals for Social Responsibility: The Past, Present and Future Values of Participatory Design." In *Proceedings of the 16th Participatory Design Conference 2020—Participation(s) Otherwise—Volume 2*, 181–184. PDC '20. Manizales, Colombia: Association for Computing Machinery. https://doi.org/10.1145/3384772.3385163.

Becker, Christoph, and Andreas Rauber. 2010. "Improving Component Selection and Monitoring with Controlled Experimentation and Automated Measurements." *Information and Software Technology* 52 (6): 641–655.

Becker, Christoph, Dawn Walker, and Curtis McCord. 2017. "Intertemporal Choice: Decision Making and Time in Software Engineering." In *Proceedings of the 10th International Workshop on Cooperative and Human Aspects of Software Engineering*, 23–29. CHASE '17. Piscataway, NJ: IEEE Press. https://doi.org/10.1109/CHASE.2017.6.

Beer, Stafford. 1972. *Brain of the Firm: The Managerial Cybernetics of Organization.* London: Allen Lane; Penguin Press.

Belkhir, Lotfi, and Ahmed Elmeligi. 2018. "Assessing ICT Global Emissions Foot-print: Trends to 2040 & Recommendations." *Journal of Cleaner Production* 177 (March): 448–463. https://doi.org/10.1016/j.jclepro.2017.12.239.

Bell, David E., Howard Raiffa, and Amos Tversky, eds. 1989. *Decision Making: Descriptive, Normative, and Prescriptive Interactions.* Cambridge: Cambridge University Press.

Bell, Simon, and Stephen Morse. 2008. *Sustainability Indicators: Measuring the Immeasurable.* 2nd ed. London: Earthscan.

Bender, Emily M., Timnit Gebru, Angelina McMillan-Major, and Shmargaret Shmitchell. 2021. "On the Dangers of Stochastic Parrots: Can Language Models Be Too Big? 🦜." In *Proceedings of the 2021 ACM Conference on Fairness, Accountability, and Transparency*, 610–623. FAccT '21. New York: Association for Computing Machinery. https://doi.org/10.1145/3442188.3445922.

Benjamin, Ruha. 2019. *Race after Technology: Abolitionist Tools for the New Jim Code.* Polity.

Berkhout, Frans, and Julia Hertin. 2001. *Impacts of Information and Communication Technologies on Environmental Sustainability: Speculations and Evidence.* Report to the OECD.

Bernoulli, Daniel. 1954. "Exposition of a New Theory on the Measurement of Risk." *Econometrica* 22 (1): 23–36. https://doi.org/10.2307/1909829.

Betz, Stefanie, Christoph Becker, Ruzanna Chitchyan, Leticia Duboc, S. Easterbrook, B. Penzenstadler, Norbert Seyff, and Colin Venters. 2015. "Sustainability Debt: A Metaphor to Support Sustainability Design Decisions." In *Proceedings of the Fourth International Workshop on Requirements Engineering for Sustainable Systems Co-Located with the 23rd IEEE International Requirements Engineering Conference (RE 2015)*, 1416:55–63. Ottawa, ON: CEUR Workshop Proceedings. http://ceur-ws.org/Vol-1416/.

Bierce, Ambrose. 1906. *The Devil's Dictionary*. Wordsworth Editions.

Bietti, Elettra. 2020. "From Ethics Washing to Ethics Bashing: A View on Tech Ethics from within Moral Philosophy." In *Proceedings of the 2020 Conference on Fairness, Accountability, and Transparency*, 210–219. FAT* '20. New York: Association for Computing Machinery. https://dl.acm.org/doi/abs/10.1145/3351095.3372860.

Bijker, Wiebe E., Thomas P. Hughes, Trevor Pinch, and Deborah G. Douglas. 2012. *The Social Construction of Technological Systems: New Directions in the Sociology and History of Technology*. Cambridge, MA: MIT press.

Bijker, Wiebe E., Thomas Parke Hughes, T. J. Pinch, and American Council of Learned Societies., eds. 1987. *The Social Construction of Technological Systems: New Directions in the Sociology and History of Technology*. Cambridge, MA: MIT Press.

Birhane, Abeba. 2020. "Tweet on GPT3." @Abebab. September 24. https://twitter.com/Abebab/status/1309137018404958215.

Birhane, Abeba, Pratyusha Kalluri, Dallas Card, William Agnew, Ravit Dotan, and Michelle Bao. 2021. "The Values Encoded in Machine Learning Research." *ArXiv:2106.15590 [Cs]*, June. http://arxiv.org/abs/2106.15590.

Blackwell, Alan F. 2006. "The Reification of Metaphor as a Design Tool." *ACM Transactions on Computer-Human Interaction (TOCHI)* 13 (4): 490–530.

Blevis, Eli. 2007. "Sustainable Interaction Design: Invention & Disposal, Renewal & Reuse." In *Proceedings of the SIGCHI Conference on Human Factors in Computing Systems*, 503–512. CHI '07. San Jose, CA: Association for Computing Machinery. https://doi.org/10.1145/1240624.1240705.

Bødker, Susanne. 2006. "When Second Wave HCI Meets Third Wave Challenges." In *Proceedings of the 4th Nordic Conference on Human-Computer Interaction Changing Roles—NordiCHI '06*, 1–8. Oslo, Norway: Association for Computing Machinery Press. https://doi.org/10.1145/1182475.1182476.

Boffey, Daniel. 2020. "Amsterdam to Embrace 'Doughnut' Model to Mend Post-Coronavirus Economy." *The Guardian*, April 8, sec. World News. https://www.theguardian.com/world/2020/apr/08/amsterdam-doughnut-model-mend-post-coronavirus-economy.

Booch, Grady. 2014. "The Human and Ethical Aspects of Big Data." *IEEE Software* 31 (1): 20–22. https://doi.org/10.1109/MS.2014.16.

Borges, Jorge Luis, and Andrew Hurley. 1999. "On Exactitude in Science." In *Collected Fictions*. Penguin Classics Deluxe Edition. New York: Penguin Books.

Boroditsky, Lera. 2000. "Metaphoric Structuring: Understanding Time through Spatial Metaphors." *Cognition* 75 (1): 1–28. https://doi.org/10.1016/S0010-0277(99)00073-6.

Bowker, Geoffrey C., and Susan Leigh Star. 1999. *Sorting Things out: Classification and Its Consequences*. Cambridge, MA: MIT Press.

Breslin, Samantha. 2018. "The Making of Computer Scientists: Rendering Technical Knowledge, Gender, and Entrepreneurialism in Singapore." Doctoral, Memorial University of Newfoundland. https://research.library.mun.ca/13499/.

Brey, Philip. 2000. "Disclosive Computer Ethics." *ACM Sigcas Computers and Society* 30 (4): 10–16.

Brey, Philip. 2010. "Values in Technology and *Disclosive* Computer Ethics." In *The Cambridge Handbook of Information and Computer Ethics*, 4:41–58. Cambridge: Cambridge University Press.

Brigham, Katie. 2022. "How the Fossil Fuel Industry Is Pushing Plastics on the World." *CNBC*, January 29. https://www.cnbc.com/2022/01/29/how-the-fossil-fuel-industry-is-pushing-plastics-on-the-world-.html.

Britton, Ren, Claude Draude, Juliane Jarke, and Goda Klumbyte. 2020. "Crafting Critical Methodologies in Computing: Theories, Practices and Future Directions." In *Prague: EASST*. https://www.easst4s2020prague.org/accepted-open-panels-knowledge-theory-and-method/.

Brown, Mandy. 1996. "A Framework for Assessing Participation." In *Critical Systems Thinking: Current Research and Practice*, edited by Robert L. Flood and Norma R. A. Romm, 195–213. Boston: Springer US. https://doi.org/10.1007/978-0-585-34651-9_11.

Brown, Tom, Benjamin Mann, Nick Ryder, Melanie Subbiah, Jared D. Kaplan, Prafulla Dhariwal, Arvind Neelakantan, et al. 2020. "Language Models Are Few-Shot Learners." *Advances in Neural Information Processing Systems* 33.

Brown, Valerie A., John A. Harris, and Jacqueline Y. Russell, eds. 2010. *Tackling Wicked Problems through the Transdisciplinary Imagination*. London: Earthscan.

Brundtland, Gro Harlem. 1987. *Report of the World Commission on Environment and Development: Our Common Future*. Geneva: United Nations.

Brynjarsdottir, Hronn, Maria Håkansson, James Pierce, Eric Baumer, Carl DiSalvo, and Phoebe Sengers. 2012. "Sustainably Unpersuaded: How Persuasion Narrows Our Vision of Sustainability." In *Proceedings of the SIGCHI Conference on Human Factors in Computing Systems*, 947–956. CHI '12. New York: Association for Computing Machinery. https://doi.org/10.1145/2207676.2208539.

Buchanan, Richard. 1992. "Wicked Problems in Design Thinking." *Design Issues* 8 (2): 5–21. https://doi.org/10.2307/1511637.

Bullard, Robert D. 1993. "Anatomy of Environmental Racism." *Toxic Struggles: The Theory and Practice of Environmental Justice* 25: 30–32.

Butler, Octavia E. 1993. *Parable of the Sower*. New York: Open Road Integrated Media.

Calero, Coral, and Mario Piattini, eds. 2015. *Green in Software Engineering*. Electronic resource. Cham, UK: Springer International.

Cambridge Dictionaries. 2011. *Cambridge Business English Dictionary*. Cambridge: Cambridge University Press.

Camerer, Colin, and Martin Weber. 1992. "Recent Developments in Modeling Preferences: Uncertainty and Ambiguity." *Journal of Risk and Uncertainty* 5 (4): 325–370. https://doi.org/10.1007/BF00122575.

Campbell, T. D. 1970. "The Normative Fallacy." *The Philosophical Quarterly* 20 (81): 368–377. https://doi.org/10.2307/2217655.

Carrington, Damian. 2022. "Microplastics Found in Human Blood for First Time." *The Guardian*, March 24, sec. Environment. https://www.theguardian.com/environment/2022/mar/24/microplastics-found-in-human-blood-for-first-time.

Cech, Erin A. 2014. "Culture of Disengagement in Engineering Education?" *Science, Technology, & Human Values* 39 (1): 42–72. https://doi.org/10.1177/0162243913504305.

Center for Humane Technology. 2020. "Ledger of Harms." December 2020. https://ledger.humanetech.com/.

Chambers, Robert. 1997. *Whose Reality Counts*. Vol. 25. London: Intermediate Technology Publications.

Chancel, L., T. Piketty, E. Saez, and G. Zucman. 2021. *World Inequality Report 2022*. World Inequality Lab. https://wir2022.wid.world/.

Checkland, Peter. 1981. *Systems Thinking, Systems Practice*. New York: J. Wiley.

Checkland, Peter. 2000. "Soft Systems Methodology: A Thirty Year Retrospective." *Systems Research and Behavioral Science* 17: S11–58.

Checkland, Peter, and John Poulter. 2007. *Learning for Action: A Short Definitive Account of Soft Systems Methodology, and Its Use for Practitioners, Teachers and Students*. Hoboken, NJ: Wiley.

Chitchyan, Ruzanna, Christoph Becker, Stefanie Betz, Leticia Duboc, Birgit Penzenstadler, Norbert Seyff, and Colin C. Venters. 2016. "Sustainability Design in Requirements Engineering: State of Practice." In *Proceedings of the 38th International Conference on Software Engineering Companion*, 533–542. ICSE '16. New York: Association for Computing Machinery. https://doi.org/10.1145/2889160.2889217.

Chitchyan, Ruzanna, Stefanie Betz, Leticia Duboc, Birgit Penzenstadler, Steve Easterbrook, Christophe Ponsard, and Colin Venters. 2015. "Evidencing Sustainability Design through Examples." In *Proceedings of the Fourth International Workshop on*

Requirements Engineering for Sustainable Systems. Ottawa, Canada. http://ceur-ws.org/Vol-1416.

Choi, Jaz Hee-jeong, and Ann Light. 2020. "'The Co-': Feminisms, Power and Research Cultures: A Dialogue." *Interactions* 27 (6): 26–28.

Churchman, C. West. 1967. "Guest Editorial: Wicked Problems." *Management Science* 14 (4): B141–42.

Churchman, C. West. 1971. *The Design of Inquiring Systems: Basic Concepts of Systems and Organization*. New York: Basic Books.

Churchman, C. West. 1979a. *The Systems Approach*. Rev. and updated. A Delta Book. New York: Dell.

Churchman, C. West. 1979b. *The Systems Approach and Its Enemies*. Basic Books.

Coakes, E., and Tony Elliman. 1999. "Focus Issue on Legacy Information Systems and Business Process Change: The Role of Stakeholders in Managing Change." *Communications of the Association for Information Systems* 2: art. 4. https://aisel.aisnet.org/cais/vol2/iss1/4.

Coalition for Critical Technology. 2020. "Abolish the #TechToPrisonPipeline." *Medium*. June 23. https://medium.com/@CoalitionForCriticalTechnology/abolish-the-techtoprisonpipeline-9b5b14366b16.

Combs, Jennifer, Danielle Kerrigan, and David Wachsmuth. 2020. "Short-Term Rentals in Canada: Uneven Growth, Uneven Impacts." *Canadian Journal of Urban Research* 29 (1): 119–134.

Connolly, Randy. 2020. "Why Computing Belongs within the Social Sciences." *Communications of the ACM* 63 (8): 54–59. https://doi.org/10.1145/3383444.

Constine. Josh. 2019. "Airbnb Invests as Zeus Corporate Housing Raises $55M at $205M." *TechCrunch* (blog). https://social.techcrunch.com/2019/12/09/airbnb-zeus/.

Coroama, Vlad, and Friedemann Mattern. 2019. "Digital Rebound—Why Digitalization Will Not Redeem Us Our Environmental Sins." In *Proceedings of the 6th International Conference on ICT for Sustainability*. Vol. 2382. CEUR Workshop Proceedings. http://ceur-ws.org/Vol-2382/ICT4S2019_paper_31.pdf.

Costa, Arthur L., and Bena Kallick. 1993. "Through the Lens of a Critical Friend." *Educational Leadership* 51 (2): 49–51.

Costanza-Chock, Sasha. 2020. *Design Justice: Community-Led Practices to Build the Worlds We Need*. Cambridge, MA: MIT Press.

CPSR. 2007. "CPSR—Timeline 1981–2001." February 5. https://web.archive.org/web/20070205070532/http://www.cpsr.org/prevsite/cpsr/timeline.html.

Crandall, Beth, Gary Klein, and Robert R. Hoffman. 2006. *Working Minds: A Practitioner's Guide to Cognitive Task Analysis*, first edition. Cambridge, MA: A Bradford Book.

Crawford, Kate. 2021. *The Atlas of AI: Power, Politics, and the Planetary Costs of Artificial Intelligence*. New Haven, CT: Yale University Press. https://doi.org/10.2307/j.ctv1ghv45t.

Crawford, Kate, and Vladan Joler. 2018. "Anatomy of an AI System." http://www.anatomyof.ai.

Crawford, Kate, Meredith Whittaker, Madeleine Clare Elish, Solon Barocas, Aaron Plasek, and Kadija Ferryman. 2016. *The AI Now Report: The Social and Economic Implications of Artificial Intelligence Technologies in the Near-Term*. New York: AI Now Institute.

Crenshaw, Kimberle. 1991. "Mapping the Margins: Intersectionality, Identity Politics, and Violence against Women of Color." *Stanford Law Review* 43 (6): 1241–1299. https://doi.org/10.2307/1229039.

"Critical Friend." 2014. In *The SAGE Encyclopedia of Action Research*, by David Coghlan and Mary Brydon-Miller. Thousand Oaks, CA: SAGE. https://doi.org/10.4135/9781446294406.n91.

"Critical Friend Definition." 2013. The Glossary of Education Reform. May 15. https://www.edglossary.org/critical-friend/.

Cross, Nigel. 1984. *Developments in Design Methodology*. New York: Wiley.

Cross, Nigel. 2006. *Designerly Ways of Knowing*. London: Springer.

Crowther-Heyck, Hunter. 1999. "George A. Miller, Language, and the Computer Metaphor and Mind." *History of Psychology* 2 (1): 37–64. https://doi.org/10.1037/1093-4510.2.1.37.

Crowther-Heyck, Hunter. 2005. *Herbert A. Simon: The Bounds of Reason in Modern America*. Baltimore, MD: Johns Hopkins University Press.

CS2023. 2022. "CS2023 Vision Statement." https://csed.acm.org/vision-statement/.

Demaria, Federico, Giorgos Kallis, and Giacomo D'Alisa, eds. 2015. *Degrowth: A Vocabulary for a New Era*. Abingdon, UK: Routledge.

Demaria, Federico, and Serge Latouche. 2019. "Degrowth." In *Pluriverse: A Post-Development Dictionary*, edited by Ashish Kothari, Ariel Salleh, Arturo Escobar, Federico Demaria, and Alberto Acosta, 148–151. New Delhi: Tulika Books.

Denning, Peter J. 2017. "Remaining Trouble Spots with Computational Thinking." *Communications of the ACM* 60 (6): 33–39. https://doi.org/10.1145/2998438.

Denning, Peter J., and Matti Tedre. 2019. *Computational Thinking*. Electronic resource. MIT Press Essential Knowledge Series. Cambridge, MA: MIT Press.

"Denver Manifesto, The—Values in Computing." 2017. http://www.valuesincomputing.org/background/chi2017-values-in-computing-workshop/the-denver-manifesto/.

Devine, Kyle. 2019. *Decomposed: The Political Ecology of Music*. Cambridge, MA: MIT Press.

Diaper, D., and Neville A. Stanton, eds. 2004. *The Handbook of Task Analysis for Human-Computer Interaction*. Mahwah, NJ: Lawrence Erlbaum.

Díaz, Sandra, Josef Settele, Eduardo S. Brondízio, Hien T. Ngo, John Agard, Almut Arneth, Patricia Balvanera, et al. 2019. "Pervasive Human-Driven Decline of Life on Earth Points to the Need for Transformative Change." *Science* 366 (6471): eaax3100. https://doi.org/10.1126/science.aax3100.

Diekmann, Sven, and Martin Peterson. 2013. "The Role of Non-Epistemic Values in Engineering Models." *Science and Engineering Ethics* 19 (1): 207–218. https://doi.org/10.1007/s11948-011-9300-4.

Digiconomist. 2022. "Bitcoin Energy Consumption Index." Digiconomist. April. https://digiconomist.net/bitcoin-energy-consumption/.

D'Ignazio, Catherine, and Lauren Klein. 2020. *Data Feminism*. Cambridge, MA: MIT Press.

D'Ignazio, Catherine, and Lauren Klein. 2021. "Who Collects the Data? A Tale of Three Maps." *MIT Case Studies in Social and Ethical Responsibilities of Computing*. Winter (February). https://doi.org/10.21428/2c646de5.fc6a97cc.

Dillon, Lindsey, Dawn Walker, Nicholas Shapiro, Vivian Underhill, Megan Martenyi, Sara Wylie, Rebecca Lave, et al. 2017. "Environmental Data Justice and the Trump Administration: Reflections from the Environmental Data and Governance Initiative." *Environmental Justice* 10 (6): 186–192.

DiSalvo, Carl. 2014. "Critical Making as Materializing the Politics of Design." *The Information Society* 30 (2): 96–105. https://doi.org/10.1080/01972243.2014.875770.

DiSalvo, Carl, Phoebe Sengers, and Hrönn Brynjarsdóttir. 2010. "Mapping the Landscape of Sustainable HCI." In *Proceedings of the SIGCHI Conference on Human Factors in Computing Systems*, 1975–1984. Association for Computing Machinery.

Dittrich, Yvonne. 2016. "What Does It Mean to Use a Method? Towards a Practice Theory for Software Engineering." *Information and Software Technology* 70 (February): 220–231. https://doi.org/10.1016/j.infsof.2015.07.001.

Dittrich, Yvonne, Ralf Klischewski, and Christiane Floyd, eds. 2002. *Social Thinking-Software Practice*. Cambridge, MA: MIT Press.

Dodds, R. and R. Venables, eds. *Engineering for Sustainable Development: Guiding Principles*. 2005. London: Royal Academy of Engineering.

Dombrowski, Lynn, Ellie Harmon, and Sarah Fox. 2016. "Social Justice-Oriented Interaction Design: Outlining Key Design Strategies and Commitments." In *Proceedings of the 2016 ACM Conference on Designing Interactive Systems*, 656–671. DIS '16. New York: Association for Computing Machinery. https://doi.org/10.1145/2901790.2901861.

Dorst, Kees. 1995. "Analysing Design Activity: New Directions in Protocol Analysis." *Design Studies*, special issue *Analysing Design Activity*, 16 (2): 139–142. https://doi.org /10.1016/0142-694X(94)00005-X.

Dorst, Kees. 2006. "Design Problems and Design Paradoxes." *Design Issues* 22 (3): 4–17.

Dorst, Kees, and Nigel Cross. 2001. "Creativity in the Design Process: Co-Evolution of Problem–Solution." *Design Studies* 22 (5): 425–437. https://doi.org/10.1016/S0142 -694X(01)00009-6.

Dorst, Kees, and Judith Dijkhuis. 1995. "Comparing Paradigms for Describing Design Activity." *Design Studies*, special issue *Analysing Design Activity*, 16 (2): 261– 274. https://doi.org/10.1016/0142-694X(94)00012-3.

Douglas, Heather. 2011. "Facts, Values, and Objectivity." In *The SAGE Handbook of the Philosophy of Social Sciences*, by Ian Jarvie and Jesús Zamora-Bonilla, 513–529. SAGE Publications. https://doi.org/10.4135/9781473913868.n28.

Dourish, Paul. 2010. "HCI and Environmental Sustainability: The Politics of Design and the Design of Politics." In *Proceedings of the 8th ACM Conference on Designing Interactive Systems*, 1–10. DIS '10. Aarhus, Denmark: Association for Computing Machinery. https://doi.org/10.1145/1858171.1858173.

Dreyfus, Hubert L. 1972. *What Computers Can't Do; a Critique of Artificial Reason*. 1st ed. New York: Harper & Row.

Driver, Henry. 2022. "Secrets of Soil." *Interactions* 29 (1): 8–11. https://doi.org/10.1145 /3505277.

Duboc, L., S. Betz, B. Penzenstadler, S. Akinli Kocak, R. Chitchyan, O. Leifler, J. Porras, N. Seyff, and C. C. Venters. 2019. "Do We Really Know What We Are Building? Raising Awareness of Potential Sustainability Effects of Software Systems in Requirements Engineering." In *2019 IEEE 27th International Requirements Engineering Conference (RE)*, 6–16. https://doi.org/10.1109/RE.2019.00013.

Duboc, Leticia, Curtis McCord, Christoph Becker, and Syed Ishtiaque Ahmed. 2020. "Critical Requirements Engineering in Practice." *IEEE Software* 37 (1): 17–24. https:// doi.org/10.1109/MS.2019.2944784.

Duboc, Leticia, Birgit Penzenstadler, Jari Porras, Sedef Akinli Kocak, Stefanie Betz, Ruzanna Chitchyan, Ola Leifler, Norbert Seyff, and Colin C. Venters. 2020. "Requirements Engineering for Sustainability: An Awareness Framework for Designing Software Systems for a Better Tomorrow." *Requirements Engineering* 25 (4): 469–492. https://doi.org/10.1007/s00766-020-00336-y.

Durdik, Z., B. Klatt, H. Koziolek, K. Krogmann, J. Stammel, and R. Weiss. 2012. "Sustainability Guidelines for Long-Living Software Systems." In *Proc. ICSM*, 517–526. https://doi.org/10.1109/ICSM.2012.6405316.

Easterbrook, Steve. 2014a. "From Computational Thinking to Systems Thinking." In *The 2nd International Conference ICT for Sustainability (ICT4S), Stockholm*. Atlantis Press.

Easterbrook, Steve. 2014b. "From Computational Thinking to Systems Thinking." In *Proceedings of the ICT4Sustainability Conference*. Atlantis Press. http://www.atlantis-press.com/php/download_paper.php?id=13446.

Easterbrook, Steve. 2021. "The Discontinuous Future." *Serendipity* (blog). November 6. https://www.easterbrook.ca/steve/2021/11/the-discontinuous-future/.

Editorial. 2018. "The Guardian View on Google and Toronto: Smart City, Dumb Deal." *The Guardian*. February 5. http://www.theguardian.com/commentisfree/2018/feb/05/the-guardian-view-on-google-and-toronto-smart-city-dumb-deal.

Elkington, John. 2004. "Enter the Triple Bottom Line." In *The Triple Bottom Line: Does It All Add Up?* 11 (12): 1–16.

Ellsberg, Daniel. 1961. "Risk, Ambiguity, and the Savage Axioms." *Quarterly Journal of Economics* 75 (4): 643–669. https://doi.org/10.2307/1884324.

Erickson, Paul, Judy L. Klein, Lorraine Daston, Rebecca Lemov, Thomas Sturm, and Michael D. Gordin. 2013. *How Reason Almost Lost Its Mind: The Strange Career of Cold War Rationality*. University of Chicago Press.

Escobar, Arturo. 2011. *Encountering Development: The Making and Unmaking of the Third World*. Princeton, NJ: Princeton University Press.

Escobar, Arturo. 2018. *Designs for the Pluriverse: Radical Interdependence, Autonomy, and the Making of Worlds*. Durham, NC: Duke University Press.

Eubanks, Virginia. 2018. *Automating Inequality: How High-Tech Tools Profile, Police, and Punish the Poor*. New York: St. Martin's Press.

European Commission. 2020. "8th Environment Action Programme Proposal." Proposal for a Decision of the European Parliament and of the Council COM (2020) 652 final. Brussels: European Commission. https://ec.europa.eu/environment/strategy/environment-action-programme-2030_en.

Fagerholm, Fabian, Christoph Becker, Alexander Chatzigeorgiou, Stefanie Betz, Leticia Duboc, Birgit Penzenstadler, Rahul Mohanani, and Colin C. Venters. 2019. "Temporal Discounting in Software Engineering: A Replication Study." In *2019 ACM/IEEE International Symposium on Empirical Software Engineering and Measurement (ESEM)*, 1–12. https://doi.org/10.1109/ESEM.2019.8870161.

Fagerholm, Fabian, Andres De los Rios, Carol Cardenas Castro, Jenny Gil, Alexander Chatzigeorgiou, Apostolos Ampatzoglou, and Christoph Becker. 2022. "It's about Time: How to Study Intertemporal Choice in Systems Design." Manuscript under review.

Fair Tax Mark. 2019. *The Silicon Six and Their $100 Billion Global Tax Gap*. https://fairtaxmark.net/wp-content/uploads/2019/12/Silicon-Six-Report-5-12-19.pdf.

Fanning, Andrew L., Daniel W. O'Neill, Jason Hickel, and Nicolas Roux. 2022. "The Social Shortfall and Ecological Overshoot of Nations." *Nature Sustainability* 5 (1): 26–36. https://doi.org/10.1038/s41893-021-00799-z.

Feenberg, Andrew. 1996. "Marcuse or Habermas: Two Critiques of Technology." *Inquiry* 39 (1): 45–70. https://doi.org/10.1080/00201749608602407.

Feenberg, Andrew. 1999. *Questioning Technology*. London: Routledge.

Feenberg, Andrew. 2002. *Transforming Technology: A Critical Theory Revisited*. New York: Oxford University Press.

Feenberg, Andrew. 2010. "Ten Paradoxes of Technology." *Techné: Research in Philosophy and Technology* 14 (1): 3–15.

Feenberg, Andrew. 2014. *The Philosophy of Praxis: Marx, Lukács, and the Frankfurt School*, new edition. Brooklyn: Verso.

Feenberg, Andrew. 2016. "A Critical Theory of Technology." In *The Handbook of Science and Technology Studies, Fourth Edition*, edited by Ulrike Felt, Rayvon Fouche, Clark A. Miller, and Laurel Smith-Doerr, 635–663. Cambridge, MA: MIT Press.

Feenberg, Andrew. 2017. *Technosystem: The Social Life of Reason*. Cambridge, MA: Harvard University Press.

Felt, Ulrike, Rayvon Fouche, Clark A. Miller, and Laurel Smith-Doerr, eds. 2016. *The Handbook of Science and Technology Studies*. 4th ed. Cambridge, MA: MIT Press. https://mitpress.mit.edu/9780262035682/the-handbook-of-science-and-technology-studies/.

Fernando, Mayanthi. 2019. "Critique as Care." *Critical Times* 2 (1): 13–22. https://doi.org/10.1215/26410478-7769726.

Ferrario, Maria Angela, Will Simm, Stephen Forshaw, Adrian Gradinar, Marcia Tavares Smith, and Ian Smith. 2016. "Values-First SE: Research Principles in Practice." In *Proceedings of the 38th International Conference on Software Engineering Companion*, 553–562. ICSE '16. New York: Association for Computing Machinery. https://doi.org/10.1145/2889160.2889219.

Filho, Marum Simão, Plácido Rogério Pinheiro, and Adriano Bessa Albuquerque. 2016. "Task Allocation in Distributed Software Development Aided by Verbal Decision Analysis." In *Software Engineering Perspectives and Application in Intelligent Systems*, 127–137. Springer International. https://doi.org/10.1007/978-3-319-33622-0_12.

Finn, Megan, and Quinn DuPont. 2020. "From Closed World Discourse to Digital Utopianism: The Changing Face of Responsible Computing at Computer Professionals for Social Responsibility (1981–1992)." *Internet Histories* 4 (1): 6–31. https://doi.org/10.1080/24701475.2020.1725851.

Flanagan, Mary, Daniel C. Howe, and Helen Nissenbaum. 2008. "Embodying Values in Technology: Theory and Practice." In *Information Technology and Moral Philosophy*, edited by Jeroen van den Hoven and John Weckert, 322–353. Cambridge: Cambridge University Press.

Fleischmann, Kenneth R., Cindy Hui, and William A. Wallace. 2017. "The Societal Responsibilities of Computational Modelers: Human Values and Professional Codes of Ethics." *Journal of the Association for Information Science and Technology* 68 (3): 543–552. https://doi.org/10.1002/asi.23697.

Fleischmann, Kenneth R., and William A. Wallace. 2005. "A Covenant with Transparency: Opening the Black Box of Models." *Commun. ACM* 48 (5): 93–97. https://doi.org/10.1145/1060710.1060715.

Flood, R. 1990. "Liberating Systems Theory: Toward Critical Systems Thinking." *Human Relations* 43 (1): 49–75. https://doi.org/10.1177/001872679004300104.

Flood, Robert L., and Michael C. Jackson. 1991a. *Creative Problem Solving: Total Systems Intervention*. Chichester, UK: Wiley.

Flood, Robert L., and Michael C. Jackson. 1991b. "Critical Systems Heuristics: Application of an Emancipatory Approach for Police Strategy toward the Carrying of Offensive Weapons." *Systems Practice* 4 (4): 283–302.

Flood, Robert L., and Michael C. Jackson. 1991c. *Critical Systems Thinking: Directed Readings*. Chichester, UK: J. Wiley.

Flood, Robert L., and Norma Romm. 1996. *Critical Systems Thinking: Current Research and Practice*. New York: Plenum Press.

Flood, Robert Louis. 2010. "The Relationship of 'Systems Thinking' to Action Research." *Systemic Practice and Action Research* 23 (4): 269–284.

Floridi, Luciano. 2019. "Translating Principles into Practices of Digital Ethics: Five Risks of Being Unethical." *Philosophy & Technology* 32 (2): 185–193. https://doi.org/10.1007/s13347-019-00354-x.

Fowler, H. W., and F. G. Fowler, eds. 1995. *The Concise Oxford Dictionary of Current English*. Oxford: Clarendon Press.

Fox, Sarah, Mariam Asad, Katherine Lo, Jill P. Dimond, Lynn S. Dombrowski, and Shaowen Bardzell. 2016. "Exploring Social Justice, Design, and HCI." In *Proceedings of the 2016 CHI Conference Extended Abstracts on Human Factors in Computing Systems*, 3293–3300. CHI EA '16. San Jose, CA: Association for Computing Machinery. https://doi.org/10.1145/2851581.2856465.

Fox, Sarah, Jill Dimond, Lilly Irani, Tad Hirsch, Michael Muller, and Shaowen Bardzell. 2017. "Social Justice and Design: Power and Oppression in Collaborative Systems." In *Companion of the 2017 ACM Conference on Computer Supported Cooperative Work and Social Computing*, 117–122.

Frankel, Todd C., Michael Robinson Chavez, and Jorge Ribas. 2016. "The Cobalt Pipeline: Tracing the Path from Deadly Hand-Dug Mines in Congo to Consumers' Phones and Laptops." *Washington Post*. September. https://www.washingtonpost.com/graphics/business/batteries/congo-cobalt-mining-for-lithium-ion-battery/.

Franquesa, David, and Leandro Navarro. 2018. "Devices as a Commons: Limits to Premature Recycling." In *Proceedings of the 2018 Workshop on Computing within*

Limits, 1–10. LIMITS '18. New York: Association for Computing Machinery. https://doi.org/10.1145/3232617.3232624.

Frauenberger, Christopher, Judith Good, and Wendy Keay-Bright. 2011. "Designing Technology for Children with Special Needs: Bridging Perspectives through Participatory Design." *CoDesign* 7 (1): 1–28. https://doi.org/10.1080/15710882.2011.587013.

Frauenberger, Christopher, and Peter Purgathofer. 2019. "Ways of Thinking in Informatics." *Communications of the ACM* 62 (7): 58–64. https://doi.org/10.1145/3329674.

Frederick, Shane, George Loewenstein, and Ted O'Donoghue. 2002. "Time Discounting and Time Preference: A Critical Review." *Journal of Economic Literature*, 351–401.

Freeman, Rachel, Mike Yearworth, and Chris Preist. 2016. "Revisiting Jevons' Paradox with System Dynamics: Systemic Causes and Potential Cures." *Journal of Industrial Ecology* 20 (2): 341–353. https://doi.org/10.1111/jiec.12285.

Freire, Paulo. 2000. *Pedagogy of the Oppressed*. 30th anniversary ed. New York: Continuum.

Friedman, Batya. 1996. "Value-Sensitive Design." *Interactions* 3 (6): 16–23.

Friedman, Batya, and David Hendry. 2012. "The Envisioning Cards: A Toolkit for Catalyzing Humanistic and Technical Imaginations." In *Proceedings of the SIGCHI Conference on Human Factors in Computing Systems*, 1145–1148. CHI '12. New York: Association for Computing Machinery. https://doi.org/10.1145/2207676.2208562.

Friedman, Batya, and David G. Hendry. 2019. *Value Sensitive Design: Shaping Technology with Moral Imagination*. Cambridge, MA: MIT Press.

Friedman, Batya, Lisa P. Nathan, and Daisy Yoo. 2017. "Multi-Lifespan Information System Design in Support of Transitional Justice: Evolving Situated Design Principles for the Long(Er) Term." *Interacting with Computers* 29 (1): 80–96. https://doi.org/10.1093/iwc/iwv045.

Fry, Tony. 2005. "The Scenario of Design." *Design Philosophy Papers* 3 (1): 19–27. https://doi.org/10.2752/144871305X13966254124158.

Fuchs, Richard, Calum Brown, and Mark Rounsevell. 2020. "Europe's Green Deal Offshores Environmental Damage to Other Nations." *Nature* 586 (7831): 671–673. https://doi.org/10.1038/d41586-020-02991-1.

Fujita, Kentaro, Yaacov Trope, and Nira Liberman. 2015. "On the Psychology of Near and Far." In *The Wiley Blackwell Handbook of Judgment and Decision Making*, 404–430. John Wiley & Sons. https://doi.org/10.1002/9781118468333.ch14.

Galbraith, John Kenneth. 1975. *Economics and the Public Purpose*. New York: New American Library.

Gardiner, Stephen M. 2014. *A Perfect Moral Storm: The Ethical Tragedy of Climate Change*. Cary, UK: Oxford University Press.

Gardner, Charlie J., Aaron Thierry, William Rowlandson, and Julia K. Steinberger. 2021. "From Publications to Public Actions: The Role of Universities in Facilitating Academic Advocacy and Activism in the Climate and Ecological Emergency." *Frontiers in Sustainability* 2 (679019): 1–6. https://doi.org/10.3389/frsus.2021.679019.

Garfinkel, Harold. 1967. *Studies in Ethnomethodology*. Englewood Cliffs, NJ: Prentice -Hall.

Gigerenzer, Gerd. 1996. "On Narrow Norms and Vague Heuristics: A Reply to Kahneman and Tversky." *Psychological Review* 103 (3): 592–596. https://doi.org/10.1037//0033-295X.103.3.592.

Gigerenzer, Gerd, and Daniel G. Goldstein. 1996. "Mind as Computer: Birth of a Metaphor." *Creativity Research Journal* 9 (2–3): 131–144. https://doi.org/10.1080/10400419.1996.9651168.

Gigerenzer, Gerd, and Reinhard Selten, eds. 2001. *Bounded Rationality: The Adaptive Toolbox*. 1st ed. Cambridge, MA: MIT Press.

Gitelman, Lisa. 2013. *"Raw Data" Is an Oxymoron*. Cambridge, MA: MIT Press.

Goguen, Joseph 1992. "The Dry and the Wet." In *Proceedings of the IFIP TC8/WG8.1 Working Conference on Information System Concepts: Improving the Understanding*, 1–17. Amsterdam, Netherlands: North-Holland Publishing Co. https://doi.org/10.5555/645617.660940. Goguen, Joseph. 1994. "Requirements Engineering as the Reconciliation of Technical and Social Issues." In *Requirements Engineering: Social and Technical Issues*, 165–199. San Diego, CA: Academic Press.

Goguen, Joseph 1996. "Formality and Informality in Requirements Engineering." In *ICRE*, 96:102–108.

Gómez-Baggethun, Erik. 2019. "Sustainable Development." In *Pluriverse: A Post-Development Dictionary*, edited by Ashish Kothari, Ariel Salleh, Arturo Escobar, Federico Demaria, and Alberto Acosta, 71–74. New Delhi: Tulika Books.

Goodkind, Andrew L., Benjamin A. Jones, and Robert P. Berrens. 2020. "Cryptodamages: Monetary Value Estimates of the Air Pollution and Human Health Impacts of Cryptocurrency Mining." *Energy Research & Social Science* 59 (January): 101281. https://doi.org/10.1016/j.erss.2019.101281.

Gore, Tim. 2020. "Confronting Carbon Inequality: Putting Climate Justice at the Heart of the COVID-19 Recovery." September 21. Media Briefing. Nairobi: Oxfam. https://www.oxfam.org/en/research/confronting-carbon-inequality.

Gotterbarn, Don, Amy Bruckman, Catherine Flick, Keith Miller, and Marty J. Wolf. 2017. "ACM Code of Ethics: A Guide for Positive Action." *Communications of the ACM* 61 (1): 121–128. https://doi.org/10.1145/3173016.

Graeber, David. 2011. *Debt: The First 5,000 Years*. Brooklyn: Melville House.

Graeff, Erhardt. 2020. "The Responsibility to Not Design and the Need for Citizen Professionalism." *Computing Professionals for Social Responsibility: The Past, Present*

and Future Values of Participatory Design, May. https://doi.org/10.21428/93b2c832
.c8387014.

Grant, Kristen, Fiona C. Goldizen, Peter D. Sly, Marie-Noel Brune, Maria Neira, Martin
van den Berg, and Rosana E. Norman. 2013. "Health Consequences of Exposure to
E-Waste: A Systematic Review." *The Lancet Global Health* 1 (6): e350–361. https://doi
.org/10.1016/S2214-109X(13)70101-3.

Gregory, Wendy 1996a. "Discordant Pluralism: A New Strategy for Critical Systems
Thinking." *Systems Practice* 9 (6): 605–625. https://doi.org/10.1007/BF02169216.

Gregory, Wendy. 1996b. "Dealing with Diversity." In *Critical Systems Thinking: Current Research and Practice,* edited by Robert L. Flood and Norma R. A. Romm, 37–61.
Boston, MA: Springer US. https://doi.org/10.1007/978-0-585-34651-9_3.

Griggs, David, Mark Stafford-Smith, Owen Gaffney, Johan Rockström, Marcus C.
Öhman, Priya Shyamsundar, Will Steffen, Gisbert Glaser, Norichika Kanie, and Ian
Noble. 2013. "Sustainable Development Goals for People and Planet." *Nature* 495
(7441): 305–307. https://doi.org/10.1038/495305a.

Grünbacher, Paul, and Barry Boehm. 2001. "EasyWinWin: A Groupware-Supported
Methodology for Requirements Negotiation." *ACM SIGSOFT Software Engineering
Notes* 26 (5): 320–321. https://doi.org/10.1145/503271.503265.

Guinée, Jeroen B., Reinout Heijungs, Gjalt Huppes, Alessandra Zamagni, Paolo
Masoni, Roberto Buonamici, Tomas Ekvall, and Tomas Rydberg. 2011. "Life Cycle
Assessment: Past, Present, and Future." *Environmental Science & Technology* 45 (1):
90–96. https://doi.org/10.1021/es101316v.

Guo, Yuepu, Rodrigo Oliveira Spínola, and Carolyn Seaman. 2016. "Exploring the
Costs of Technical Debt Management—A Case Study." *Empirical Software Engineering*
21 (1): 159–182. https://doi.org/10.1007/s10664-014-9351-7.

Guzdial, Mark. 2020a. "Becoming Anti-Racist: Learning about Race in CS Education." *Computing Education Research Blog.* June 8. https://computinged.wordpress
.com/2020/06/08/lets-talk-about-race-in-cs-education-more-resources/.

Guzdial, Mark. 2020b. "Paradigm Shifts in Education and Educational Technology:
Influencing the Students Here and Now." *Computing Education Research Blog.* July 6.
https://computinged.wordpress.com/2020/07/06/paradigm-shifts-in-education-and
-educational-technology-influencing-the-students-here-and-now/.

Haberl, Helmut, Dominik Wiedenhofer, Doris Virág, Gerald Kalt, Barbara Plank,
Paul Brockway, Tomer Fishman, et al. 2020. "A Systematic Review of the Evidence
on Decoupling of GDP, Resource Use and GHG Emissions, Part II: Synthesizing the
Insights." *Environmental Research Letters* 15 (6): 065003. https://doi.org/10.1088
/1748-9326/ab842a.

Habermas, Jürgen. 1968. *Technik Und Wissenschaft Als "Ideologie."* Edition Suhrkamp
287. Frankfurt am Main: Suhrkamp.

Habermas, Jürgen. 1972. *Knowledge and Human Interests*. London: Heinemann Educational.

Habermas, Jürgen. 1985. "Jürgen Habermas, A Philosophico-Political Profile." *New Left Review* I/151 (June). https://newleftreview.org/issues/i151/articles/jurgen-habermas-a -philosophico-political-profile.

Hamraie, Aimi, and Kelly Fritsch. 2019. "Crip Technoscience Manifesto." *Catalyst: Feminism, Theory, Technoscience* 5 (1): 1–33. https://doi.org/10.28968/cftt.v5i1 .29607.

Hansen, Morten T., and Bolko von Oetinger. 2001. "Introducing T-Shaped Managers: Knowledge Management's Next Generation." *Harvard Business Review*, March 1. https://hbr.org/2001/03/introducing-t-shaped-managers-knowledge-managements -next-generation.

Hansson, Lon Åke Erni Johannes, Teresa Cerratto Pargman, and Daniel Sapiens Pargman. 2021. "A Decade of Sustainable HCI: Connecting SHCI to the Sustainable Development Goals." In *Proceedings of the 2021 CHI Conference on Human Factors in Computing Systems*, 1–19. CHI '21. New York: Association for Computing Machinery. https://doi.org/10.1145/3411764.3445069.

Hao, Karen. 2021a. "Big Tech's Guide to Talking about AI Ethics." *MIT Technology Review*. 2021. https://www.technologyreview.com/2021/04/13/1022568/big-tech-ai -ethics-guide/.

Hao, Karen. 2021b. "Inside the Fight to Reclaim AI." *MIT Technology Review* 2021. https://www.technologyreview.com/2021/06/14/1026148/ai-big-tech-timnit-gebru -paper-ethics.

Haraway, Donna. 1988. "Situated Knowledges: The Science Question in Feminism and the Privilege of Partial Perspective." *Feminist Studies* 14 (3): 575–599. https://doi .org/10.2307/3178066.

Hardin, Garrett. 1968. "The Tragedy of the Commons." *Science* 162 (3859): 1243– 1248. https://doi.org/10.1126/science.162.3859.1243.

Harding, Sandra. 1992. "Rethinking Standpoint Epistemology: What Is 'Strong Objectivity'?" *The Centennial Review* 36 (3): 437–470.

Harding, Sandra G. 1986. *The Science Question in Feminism*. Ithaca, NY: Cornell University Press.

Hardisty, David J., Katherine Fox-Glassman, David Krantz, and Elke U. Weber. 2011. "How to Measure Discount Rates? An Experimental Comparison of Three Methods." SSRN Scholarly Paper ID 1961367. Rochester, NY: Social Science Research Network. https://papers.ssrn.com/abstract=1961367.

Harper, Richard, Dave Randall, and W. W. Sharrock. 2016. *Choice: The Sciences of Reason in the 21st Century: A Critical Assessment*. Cambridge: Polity Press.

Hauser, Oliver P., David G. Rand, Alexander Peysakhovich, and Martin A. Nowak. 2014. "Cooperating with the Future." *Nature* 511 (7508): 220–223. https://doi.org/10 .1038/nature13530.

Head, Brian W., and John Alford. 2015. "Wicked Problems: Implications for Public Policy and Management." *Administration & Society* 47 (6): 711–739. https://doi.org /10.1177/0095399713481601.

Hertwig, Ralph, and Gerd Gigerenzer. 1999. "The 'Conjunction Fallacy' Revisited: How Intelligent Inferences Look like Reasoning Errors." *Journal of Behavioral Decision Making* 12 (4): 275–305.

Hickel, Jason. 2020. *Less Is More: How Degrowth Will Save the World.* London: Windmill Books.

Hickel, Jason, and Giorgos Kallis. 2020. "Is Green Growth Possible?" *New Political Economy* 25 (4): 469–486. https://doi.org/10.1080/13563467.2019.1598964.

Hickel, Jason, Daniel W. O'Neill, Andrew L. Fanning, and Huzaifa Zoomkawala. 2022. "National Responsibility for Ecological Breakdown: A Fair-Shares Assessment of Resource Use, 1970–2017." *The Lancet Planetary Health* 6 (4): e342–349. https://doi .org/10.1016/S2542-5196(22)00044-4.

Hicks, Mar. 2017. *Programmed Inequality: How Britain Discarded Women Technologists and Lost Its Edge in Computing.* Cambridge, MA: MIT Press.

Hill Collins, Patricia. 1990. *Black Feminist Thought: Knowledge, Consciousness, and the Politics of Empowerment.* Perspectives on Gender, v. 2. Boston: Unwin Hyman.

Hilty, L. M., A. Köhler, F. Von Schéele, R. Zah, and T. Ruddy. 2006. "Rebound Effects of Progress in Information Technology." *Poiesis & Praxis* 4 (1): 19–38. https://doi.org /10.1007/s10202-005-0011-2.

Hilty, Lorenz M., and Bernard Aebischer. 2015a. "ICT for Sustainability: An Emerging Research Field." In *ICT Innovations for Sustainability*, 3–36. Cham, UK: Springer International.

Hilty, Lorenz M., and Bernard Aebischer. 2015b. *ICT Innovations for Sustainability.* Springer. http://link.springer.com/content/pdf/10.1007/978-3-319-09228-7.pdf.

Hilty, Lorenz M., Peter Arnfalk, Lorenz Erdmann, James Goodman, Martin Lehmann, and Patrick A. Wäger. 2006. "The Relevance of Information and Communication Technologies for Environmental Sustainability—A Prospective Simulation Study." *Environmental Modelling & Software*, Environmental Informatics, 21 (11): 1618–1629. https://doi.org/10.1016/j.envsoft.2006.05.007.

Hoffman, Anna Lauren. 2020. *Data Ethics for Non-Ideal Times.* https://www.anna everyday.com/s/Hoffmann-Data-Ethics-for-Non-Ideal-Times-Lecture-Notes.pdf.

Hoffman, Lily M. 1989. *The Politics of Knowledge: Activist Movements in Medicine and Planning.* SUNY Press.

Holt, J. E., D. F. Radcliffe, and D. Schoorl. 1985. "Design or Problem Solving—A Critical Choice for the Engineering Profession." *Design Studies* 6 (2): 107–110. https://doi .org/10.1016/0142-694X(85)90020-1.

Hölzle, Urs. 2020. "Data Centers Are More Energy Efficient than Ever." Google. February 27. https://blog.google/outreach-initiatives/sustainability/data-centers-energy-efficient/.

Horkheimer, Max. 1972. "Traditional and Critical Theory." *Critical Theory: Selected Essays* 188: 243.

Hsu, Yen-Chia, Paul Dille, Jennifer Cross, Beatrice Dias, Randy Sargent, and Illah Nourbakhsh. 2017. "Community-Empowered Air Quality Monitoring System." In *Proceedings of the 2017 CHI Conference on Human Factors in Computing Systems*, 1607–1619. CHI '17. Denver, CO: Association for Computing Machinery. https://doi.org/10.1145 /3025453.3025853.

Hsu, Yen-Chia, and Illah Nourbakhsh. 2020. "When Human-Computer Interaction Meets Community Citizen Science." *Communications of the ACM* 63 (2): 31–34. https://doi.org/10.1145/3376892.

Huang, G. H., and N. B. Chang. 2003. "The Perspectives of Environmental Informatics and Systems Analysis." *Journal of Environmental Informatics* 1 (1): 1–6. http://www .jeionline.org/index.php?journal=mys&page=article&op=view&path%5B%5D=439.

Hughes, Thomas Parke. 1993. *Networks of Power: Electrification in Western Society, 1880–1930.* JHU Press.

Huppatz, D. J. 2015. "Revisiting Herbert Simon's 'Science of Design.'" *Design Issues* 31 (2): 29–40. https://doi.org/10.1162/DESI_a_00320.

Hutchins, Edwin. 1995. "How a Cockpit Remembers Its Speeds." *Cognitive Science* 19 (3): 265–288. https://doi.org/10.1207/s15516709cog1903_1.

IEA. 2019. "Data Centres and Data Transmission Networks—Tracking Buildings—Analysis." IEA. https://web.archive.org/web/20200311204159/https://www.iea.org /reports/tracking-buildings/data-centres-and-data-transmission-networks.

IEEE. 2021. *IEEE 7000–2021—IEEE Approved Draft Model Process for Addressing Ethical Concerns During System Design.* IEEE Press. https://standards.ieee.org/standard/7000 -2021.html.

Iliadis, Andrew, and Federica Russo. 2016. "Critical Data Studies: An Introduction." *Big Data & Society* 3 (2): 2053951716674238. https://doi.org/10.1177/2053951716674238.

Intergovernmental Panel on Climate Change (IPCC). 2022. *Climate Change 2022: Impacts, Adaptation and Vulnerability,* edited by H.-O. Pörtner, D. C. Roberts, M. Tignor, E. S. Poloczanska, K. Mintenbeck, A. Alegría, M. Craig, et al. Cambridge: Cambridge University Press. https://www.ipcc.ch/report/ar6/wg2/.

International Council for Science. 2017. *A Guide to SDG Interactions: From Science to Implementation,* edited by D. J. Griggs, M. Nilsson, A. Stevance, and D. McCollum.

Paris: International Council for Science. https://council.science/publications/a-guide
-to-sdg-interactions-from-science-to-implementation/.

International Energy Agency. 2013. *Electricity in a Climate-Constrained World: Data
and Analyses*. OECD. https://www.oecd-ilibrary.org/energy/climate-and-electricity
-annual-2012_9789264175556-en.

Isenberg, Daniel. 1984. "How Senior Managers Think." *Harvard Business Review*.
November 1. https://hbr.org/1984/11/how-senior-managers-think.

ISO. 2010. *ISO/IEC/IEEE 24765:2010—Systems and Software Engineering—Vocabulary*.
International Standards Organization. http://www.iso.org/iso/catalogue_detail.htm
?csnumber=50518.

Ivanov, K. 1991. "Critical Systems Thinking and Information Technology." *Journal
of Applied Systems Analysis* 18:39–55.

Iwen, Michelle. 2019. "Letters of Recommendation Reaffirm Entrenched Systems
of Bias and Exclusion." *Inside HigherEd*, April. https://www.insidehighered.com
/advice/2019/04/10/letters-recommendation-reaffirm-entrenched-systems-bias-and
-exclusion-opinion.

Jackson, M. C. 1982. "The Nature of Soft Systems Thinking: The Work of Churchman,
Ackoff and Checkland." *Journal of Applied Systems Analysis* 9 (1): 17–29.

Jackson, M. C. 1985a. "Social Systems Theory and Practice: The Need for a Critical
Approach." *International Journal of General System* 10 (2–3): 135–151.

Jackson, M. C. 1985b. "The Itinerary of a Critical Approach." *Journal of the Operational Research Society* 36 (9): 878–881.

Jackson, M. C. 1991. "The Origins and Nature of Critical Systems Thinking." *Systems
Practice* 4 (2): 131–149. https://doi.org/10.1007/BF01068246.

Jackson, Michael C. 1983. "The Nature of 'Soft' Systems Thinking: Comment on the
Three Replies." *Journal of Applied Systems Analysis* 10:109–113.

Jackson, Michael C. 2003. *Systems Thinking: Creative Holism for Managers*. Chichester,
UK: Wiley.

Jackson, Michael C. 2019. *Critical Systems Thinking and the Management of Complexity*. Newark, UK: John Wiley & Sons.

Jackson, Steven J. 2014. "Rethinking Repair." *Media Technologies: Essays on Communication, Materiality, and Society*, 221–239.

Jackson, Tim. 2009. *Prosperity without Growth: Economics for a Finite Planet*. London:
Earthscan.

Jackson, Tim. 2021. *Post Growth: Life after Capitalism*, 1st edition. Cambridge: Polity.

Jacobs, Abigail Z., and Hanna Wallach. 2021. "Measurement and Fairness." In *Proceedings of the 2021 ACM Conference on Fairness, Accountability, and Transparency*, 375–385.

FAccT '21. New York: Association for Computing Machinery. https://doi.org/10
.1145/3442188.3445901.

JafariNaimi, Nassim, Lisa Nathan, and Ian Hargraves. 2015. "Values as Hypotheses:
Design, Inquiry, and the Service of Values." *Design Issues* 31 (4): 91–104. https://doi
.org/10.1162/DESI_a_00354.

Jarke, Matthias, Pericles Loucopoulos, Kalle Lyytinen, John Mylopoulos, and William Robinson. 2011. "The Brave New World of Design Requirements." *Information
Systems* 36 (7): 992–1008. https://doi.org/10.1016/j.is.2011.04.003.

Jeffries, Stuart. 2016. *Grand Hotel Abyss: The Lives of the Frankfurt School.* London: Verso.

Jenkin, Tracy A., Jane Webster, and Lindsay McShane. 2011. "An Agenda for 'Green'
Information Technology and Systems Research." *Information and Organization* 21 (1):
17–40. https://doi.org/10.1016/j.infoandorg.2010.09.003.

Jenkins, Gwilym Al. 1969. "The Systems Approach." *Journal of Systems Engineering* 1 (1).

Johnson, Deborah G., and Helen Nissenbaum. 1995. *Computers, Ethics & Social
Values.* Prentice-Hall.

Jonassen, David H. 2000. "Toward a Design Theory of Problem Solving." *Educational Technology Research and Development* 48 (4): 63–85. https://doi.org/10.1007
/BF02300500.

Jones, Nicola. 2018. "How to Stop Data Centres from Gobbling up the World's Electricity." *Nature* 561 (7722): 163–166. https://doi.org/10.1038/d41586-018-06610-y.

Kaczmarek, Michelle, Saguna Shankar, Rodrigo dos Santos, Eric M. Meyers, and Lisa P.
Nathan. 2020. "Pushing LIMITS: Envisioning beyond the Artifact." In *Proceedings of the
7th International Conference on ICT for Sustainability,* 255–266. ICT4S2020. New York:
Association for Computing Machinery. https://doi.org/10.1145/3401335.3401367.

Kafer, Alison. 2013. *Feminist, Queer, Crip.* Bloomington: Indiana University Press.

Kahneman, Daniel. 2011. *Thinking, Fast and Slow.* Macmillan.

Kahneman, Daniel, and Gary Klein. 2009. "Conditions for Intuitive Expertise: A
Failure to Disagree." *American Psychologist* 64 (6): 515–526. https://doi.org/10.1037
/a0016755.

Kahneman, Daniel, Paul Slovic, and Amos Tversky, eds. 1982. *Judgment Under Uncertainty: Heuristics and Biases.* Cambridge: Cambridge University Press.

Kahneman, Daniel, and Amos Tversky. 1979. "Prospect Theory: An Analysis of Decision under Risk." *Econometrica: Journal of the Econometric Society,* 263–291.

Kallis, Giorgos. 2015. "The Degrowth Alternative." *Great Transition Initiative,* February. https://greattransition.org/publication/the-degrowth-alternative.

Kallis, Giorgos. 2018. *Degrowth (The Economy: Key Ideas).* Newcastle upon Tyne:
Agenda Publishing.

Kallis, Giorgos, Susan Paulson, Giacomo D'Alisa, and Federico Demaria. 2020. *The Case for Degrowth*, 1st edition. Cambridge, UK: Polity.

Karlskrona Initiative. 2015. "The Karlskrona Manifesto for Sustainability Design." *Sustainability Design* (blog). 30 April 2015. https://www.sustainabilitydesign.org/karlskrona -manifesto/.

Kasser, J., and Y. Zhao. 2016. "Wicked Problems: Wicked Solutions." In *2016 11th System of Systems Engineering Conference (SoSE)*, 1–6. https://doi.org/10.1109/SYSOSE .2016.7542904.

Katz, Eric. 2005. "On the Neutrality of Technology: The Holocaust Death Camps as a Counter-Example." *Journal of Genocide Research* 7 (3): 409–421.

Keeney, Ralph L., and Howard Raiffa. 1993. *Decisions with Multiple Objectives: Preferences and Value Trade-Offs*. Cambridge: Cambridge University Press.

Kelly, Annie. 2019. "Apple and Google Named in US Lawsuit over Congolese Child Cobalt Mining Deaths." *The Guardian*, December 16, 2019, sec. Global development. https://www.theguardian.com/global-development/2019/dec/16/apple-and-google -named-in-us-lawsuit-over-congolese-child-cobalt-mining-deaths.

Kember, David, Tak-Shing Ha, Bick-Har Lam, April Lee, Sandra NG, Louisa Yan, and Jessie C. K. Yum. 1997. "The Diverse Role of the Critical Friend in Supporting Educational Action Research Projects." *Educational Action Research* 5 (3): 463–481. https://doi.org/10.1080/09650799700200036.

Keren, Gideon, and George Wu, eds. 2015. *The Wiley Blackwell Handbook of Judgment and Decision Making*. Chichester, UK: Wiley-Blackwell.

Keyes, Os, Jevan Hutson, and Meredith Durbin. 2019. "A Mulching Proposal: Analysing and Improving an Algorithmic System for Turning the Elderly into High-Nutrient Slurry." In *Extended Abstracts of the 2019 CHI Conference on Human Factors in Computing Systems*, 1–11. CHI EA '19. Glasgow, Scotland: Association for Computing Machinery. https://doi.org/10.1145/3290607.3310433.

Kienzle, Jörg, Gunter Mussbacher, Benoit Combemale, Lucy Bastin, Nelly Bencomo, Jean-Michel Bruel, Christoph Becker, et al. 2020. "Towards Model-Driven Sustainability Evaluation." *Communications of the ACM*. https://doi.org/10.1145/3371906.

Kimmerer, Robin Wall. 2013. *Braiding Sweetgrass: Indigenous Wisdom, Scientific Knowledge, and the Teachings of Plants*. Milkweed Editions.

Kirby, M. W. 2003. "The Intellectual Journey of Russell Ackoff: From OR Apostle to OR Apostate." *The Journal of the Operational Research Society* 54 (11): 1127–1140.

Klein, G., K. G. Ross, B. M. Moon, D. E. Klein, R. R. Hoffman, and E. Hollnagel. 2003. "Macrocognition." *IEEE Intelligent Systems* 18 (3): 81–85. https://doi.org/10.1109/MIS .2003.1200735.

Klein, Gary. 1997. "The Recognition-Primed Decision (RPD) Model: Looking Back, Looking Forward." *Naturalistic Decision Making*, 285–92.

Klein, Gary. 2000. "Cognitive Task Analysis of Teams." *Cognitive Task Analysis* 11:417–429.

Klein, Gary A., ed. 1993. *Decision Making in Action: Models and Methods.* Norwood, NJ.: Ablex.

Klein, Gary A. 1998. *Sources of Power: How People Make Decisions.* Cambridge, MA: MIT Press.

Klein, Gary, and Corinne Wright. 2016. "Macrocognition: From Theory to Toolbox." *Frontiers in Psychology* 7:54. https://doi.org/10.3389/fpsyg.2016.00054.

Klein, Naomi. 2014. *This Changes Everything: Capitalism vs. the Climate.* Toronto: Alfred A. Knopf Canada.

Klein, Naomi. 2020. "Naomi Klein: How Big Tech Plans to Profit from the Pandemic." *The Guardian*, 13 May 2020, sec. News. https://www.theguardian.com/news/2020/may/13/naomi-klein-how-big-tech-plans-to-profit-from-coronavirus-pandemic.

Kleinman, Zoe. 2020. "Firm Defends Algorithm That 'Spots Women's Orgasms.'" *BBC News*, June 12, 2020, sec. Technology. https://www.bbc.com/news/technology-53024123.

Knowles, Bran. 2013. "Cyber-Sustainability: Towards a Sustainable Digital Future." Phd, Lancaster, UK: Lancaster University. http://eprints.lancs.ac.uk/68468/.

Knowles, Bran, Oliver Bates, and Maria Håkansson. 2018. "This Changes Sustainable HCI." In *Proceedings of the 2018 CHI Conference on Human Factors in Computing Systems*, 1–12. CHI '18. New York: Association for Computing Machinery. https://doi.org/10.1145/3173574.3174045.

Ko, Amy J. 2020. "Design Methods." July. http://faculty.washington.edu/ajko/books/design-methods/.

Ko, Amy J., Alannah Oleson, Neil Ryan, Yim Register, Benjamin Xie, Mina Tari, Matthew Davidson, Stefania Druga, and Dastyni Loksa. 2020. "It Is Time for More Critical CS Education." *Communications of the ACM* 63 (11): 31–33. https://doi.org/10.1145/3424000.

Kothari, Ashish, Ariel Salleh, Arturo Escobar, Federico Demaria, and Alberto Acosta, eds. 2019. *Pluriverse: A Post-Development Dictionary.* New Delhi: Tulika Books.

Kotonya, Gerald, and Ian Sommerville. 1998. *Requirements Engineering: Processes and Techniques.* Worldwide Series in Computer Science. New York: John Wiley.

Koziolek, H., D. Domis, T. Goldschmidt, and P. Vorst. 2013. "Measuring Architecture Sustainability." *IEEE Software* 30 (6): 54–62. https://doi.org/10.1109/MS.2013.101.

Kranzberg, Melvin. 1986. "Technology and History: 'Kranzberg's Laws.'" *Technology and Culture* 27 (3): 544–560. https://doi.org/10.2307/3105385.

Krausmann, Fridolin, Dominik Wiedenhofer, Christian Lauk, Willi Haas, Hiroki Tanikawa, Tomer Fishman, Alessio Miatto, Heinz Schandl, and Helmut Haberl. 2017.

"Global Socioeconomic Material Stocks Rise 23-Fold over the 20th Century and Require Half of Annual Resource Use." *Proceedings of the National Academy of Sciences* 114 (8): 1880–1885. https://doi.org/10.1073/pnas.1613773114.

Kruchten, Philippe. 2012. "Technical Debt: From Metaphor to Theory and Practice." *IEEE Software* 29 (6): 18–21.

Krznaric, Roman. 2020. *The Good Ancestor: How to Think Long-Term in a Short-Term World*. New York: The Experiment.

Kuehr, Ruediger. 2014. "Solving the E-Waste Problem (Step) White Paper: One Global Definition of E-Waste." *United Nations University*, June. https://collections.unu.edu /view/UNU:6120#viewMetadata.

Kuhn, Thomas S. 1962. *The Structure of Scientific Revolutions*. International Encyclopedia of Unified Science, vol. 2, no. 2. Chicago: University of Chicago Press.

Labaree, David. 2019. "Pluck and Hard Work, or Luck of Birth? Two Stories, One Man." *Aeon*, December. https://aeon.co/essays/pluck-and-hard-work-or-luck-of-birth -two-stories-one-man.

Lago, Patricia, Sedef Akinli Koçak, Ivica Crnkovic, and Birgit Penzenstadler. 2015. "Framing Sustainability as a Property of Software Quality." *Commun. ACM* 58 (10): 70–78. https://doi.org/10.1145/2714560.

Lakoff, George. 1993. "The Contemporary Theory of Metaphor." In *Metaphor and Thought* edited by Andrew Ortony, 202–251. Cambridge: Cambridge University Press.

Lakoff, George, and Sam Ferguson. 2009. "Crucial Issues Not Addressed in the Immigration Debate: Why Deep Framing Matters." *Cognitive Policy Works*. 14 September 2009. https://web.archive.org/web/20101109112241/http://www.cognitivepolicyworks .com/resource-center/rethinking-immigration/crucial-issues-not-addressed-in-the-im migration-debate/.

Lakoff, George, and Mark Johnson. 1980. *Metaphors We Live By*. Chicago: University of Chicago Press.

Landry, Maurice. 1995. "A Note on the Concept of 'Problem.'" *Organization Studies* 16 (2): 315–343. https://doi.org/10.1177/017084069501600206.

Latour, Bruno. 1987. *Science in Action: How to Follow Scientists and Engineers through Society*. Cambridge, MA: Harvard University Press.

Lehman, M. M. 1979. "On Understanding Laws, Evolution, and Conservation in the Large-Program Life Cycle." *Journal of Systems and Software* 1 (January): 213–221. https://doi.org/10.1016/0164-1212(79)90022-0.

Lehman, M. M. 1996. "Laws of Software Evolution Revisited." In *Software Process Technology*, edited by Carlo Montangero, 108–124. Lecture Notes in Computer Science. Berlin, Heidelberg: Springer. https://doi.org/10.1007/BFb0017737.

Lenberg, Per, Robert Feldt, and Lars Göran Wallgren. 2015. "Behavioral Software Engineering: A Definition and Systematic Literature Review." *Journal of Systems and Software* 107 (September): 15–37. https://doi.org/10.1016/j.jss.2015.04.084.

Le Quéré, Corinne, Robert B. Jackson, Matthew W. Jones, Adam J. P. Smith, Sam Abernethy, Robbie M. Andrew, Anthony J. De-Gol, et al. 2020. "Temporary Reduction in Daily Global CO_2 Emissions during the COVID-19 Forced Confinement." *Nature Climate Change* 10 (7): 647–653. https://doi.org/10.1038/s41558-020-0797-x.

Levert, Dan. 2020. *On Cold Iron: A Story of Hubris and the 1907 Quebec Bridge Collapse.* Victoria, BC: FriesenPress.

Levy, Steven. 2020. *Facebook: The Inside Story.* New York: Blue Rider Press.

Liberman, Nira, and Yaacov Trope. 2014. "Traversing Psychological Distance." *Trends in Cognitive Sciences* 18 (7): 364–369. https://doi.org/10.1016/j.tics.2014.03.001.

Liberman, Nira, Yaacov Trope, and E. Stephan. 2007. "Psychological Distance." In *Social Psychology,* edited by A. W. Kruglanski and E. T. Higgins, 2:353–383. Guilford Press.

Liebowitz, Stan J., and Stephen E. Margolis. 1995. "Path Dependence, Lock-in, and History." *Journal of Law, Economics, and Organization* 11 (1): 205–226.

Light, Ann. 2022. "Ecologies of Subversion: Troubling Interaction Design for Climate Care." *Interactions* 29 (1): 34–38. https://doi.org/10.1145/3501301.

Lindberg, Tilmann, Eva Köppen, Ingo Rauth, and Christoph Meinel. 2012. "On the Perception, Adoption and Implementation of Design Thinking in the IT Industry." In *Design Thinking Research: Studying Co-Creation in Practice. Understanding Innovation,* edited by Hasso Plattner, Christoph Meinel, and Larry Leifer, 229–240. Berlin, Heidelberg: Springer. https://doi.org/10.1007/978-3-642-21643-5_13.

Linebaugh, Peter. 2014. *Stop, Thief!: The Commons, Enclosures, and Resistance.* PM Press.

Lipshitz, Raanan, Gary Klein, Judith Orasanu, and Eduardo Salas. 2001. "Taking Stock of Naturalistic Decision Making." *Journal of Behavioral Decision Making* 14 (5): 331–352. https://doi.org/10.1002/bdm.381.

Liu, Wendy. 2020. *Abolish Silicon Valley: How to Liberate Technology from Capitalism.* London: Repeater.

Loewenstein, George, Daniel Read, and Roy F. Baumeister. 2003. *Time and Decision: Economic and Psychological Perspectives of Intertemporal Choice.* New York: Russell Sage Foundation.

Loewenstein, George, Scott Rick, and Jonathan D. Cohen. 2008. "Neuroeconomics." *Annu. Rev. Psychology.* 59:647–672.

Logic School. 2021. "Logic School." https://school.logicmag.io.

Lowman, Emma Battell, and Adam J. Barker. 2015. *Settler: Identity and Colonialism in 21st Century Canada.* Halifax; Fernwood Publishing.

Lurgio, Jeremy. 2019. "Saving the Whanganui: Can Personhood Rescue a River?" *The Guardian*, November 29. World News. https://www.theguardian.com/world/2019/nov/30/saving-the-whanganui-can-personhood-rescue-a-river.

Maarten, Schraagen Jan, Laura G. Militello, Tom Ormerod, and Lipshitz Raanan. 2017. *Naturalistic Decision Making and Macrocognition*. London: CRC Press. https://doi.org/10.1201/9781315597584.

MacKenzie, Donald, and Yuval Millo. 2003. "Constructing a Market, Performing Theory: The Historical Sociology of a Financial Derivatives Exchange." *American Journal of Sociology* 109 (1): 107–145. https://doi.org/10.1086/374404.

MacKenzie, Donald, and Judy Wajcman. 1999. *The Social Shaping of Technology*. Edited by Donald MacKenzie and Judy Wajcman. Buckingham, UK: Open University Press.

Mahaux, Martin, Patrick Heymans, and Germain Saval. 2011. "Discovering Sustainability Requirements: An Experience Report." In *REFSQ*, edited by Daniel Berry and Xavier Franch, 19–33. LNCS 6606. Springer.

Mahmood, Saba. 2015. *Religious Difference in a Secular Age*. Princeton, NJ: Princeton University Press. https://press.princeton.edu/books/hardcover/9780691153278/religious-difference-in-a-secular-age.

Malazita, James W., and Korryn Resetar. 2019. "Infrastructures of Abstraction: How Computer Science Education Produces Anti-Political Subjects." *Digital Creativity* 0 (0): 1–13. https://doi.org/10.1080/14626268.2019.1682616.

Manders-Huits, Noëmi. 2011. "What Values in Design? The Challenge of Incorporating Moral Values into Design." *Science and Engineering Ethics* 17 (2): 271–287.

Mann, Samuel, Oliver Bates, and Raymond Maher. 2018. "Shifting the Maturity Needle of ICT for Sustainability." In *ICT4S2018. 5th International Conference on Information and Communication Technology for Sustainability*, 209–226. https://doi.org/10.29007/d6g3.

Marcuse, Herbert. 1964. *One-Dimensional Man: The Ideology of Advanced Industrial Society*. Sphere Books.

Margolin, Victor, and Richard Buchanan, eds. 1995. *The Idea of Design*. Design Issues Reader. Cambridge, MA: MIT Press.

Martin, Chris J. 2016. "The Sharing Economy: A Pathway to Sustainability or a Nightmarish Form of Neoliberal Capitalism?" *Ecological Economics* 121 (January): 149–159. https://doi.org/10.1016/j.ecolecon.2015.11.027.

Mattern, Shannon. 2017. "A City Is Not a Computer." *Places Journal*, February. https://doi.org/10.22269/170207.

Maturana, Humberto R. 1980. *Autopoiesis and Cognition: The Realization of the Living*. Boston: D. Reidel.

Maturana, Humberto R., and Francisco J. Varela. 1992. *The Tree of Knowledge: The Biological Roots of Human Understanding*. rev. ed. Boston: Shambhala Publications

Mazzucato, Mariana. 2013. *The Entrepreneurial State: Debunking Public vs. Private Sector Myths*. London: Anthem Press.

McConnell, Steve. 2007. "Managing Technical Debt." http://www.construx.com /10x_Software_Development/Technical_Debt/.

McCord, Curtis, and Christoph Becker. 2019. "Sidewalk and Toronto: Critical Systems Heuristics and the Smart City." In *Proceedings of the 6th International Conference on ICT for Sustainability*. Vol. 2382. Lappeenranta, Finland: CEUR Workshop Proceedings. http://ceur-ws.org/Vol-2382/.

McDonald, Rachel I., Hui Yi Chai, and Ben R. Newell. 2015. "Personal Experience and the 'Psychological Distance' of Climate Change: An Integrative Review." *Journal of Environmental Psychology* 44 (December): 109–118. https://doi.org/10.1016/j.jenvp .2015.10.003.

Mcfedries, Paul. 2014. "The City as System [Technically Speaking]." *IEEE Spectrum* 51 (4): 36-36.

McIntyre, Alison. 2014. "Doctrine of Double Effect." In *The Stanford Encyclopedia of Philosophy*, edited by Edward N. Zalta, Winter 2014. Metaphysics Research Lab, Stanford University. https://plato.stanford.edu/archives/win2014/entries/double-effect/.

McKay, Deirdre, and Sharon George. 2019. "The Environmental Impact of Music: Digital, Records, CDs Analysed." The Conversation. http://theconversation.com/the -environmental-impact-of-music-digital-records-cds-analysed-108942.

Meadows, Dennis L., and Club of Rome. 1972. *The Limits to Growth*. New York: Universe Books.

Meadows, Donella H. 1999. *Leverage Points: Places to Intervene in a System*. Sustainability Institute. https://donellameadows.org/archives/leverage-points-places-to-inter vene-in-a-system/.

Meadows, Donella H. 2008. *Thinking in Systems: A Primer*. White River Junction, VT: Chelsea Green Publishing.

Meadows, Donella H., Dennis L. Meadows, and Jørgen Randers. 2004. *Limits to Growth: The 30-Year Update*. White River Junction, VT: Chelsea Green Publishing Company.

Merriam-Webster. 2020a. "Problem." In *Merriam-Webster's Dictionary*. https://www .merriam-webster.com/dictionary/problem.

Merriam-Webster. 2020b. "Diachronic." In *Merriam-Webster's Dictionary*. https://www .merriam-webster.com/dictionary/diachronic.

Merriam-Webster. 2020c. "Judgment." In *Merriam-Webster's Dictionary*. https://www .merriam-webster.com/dictionary/judgment.

Metcalf, Jacob, Emanuel Moss, and danah boyd. 2019. "Owning Ethics: Corporate Logics, Silicon Valley, and the Institutionalization of Ethics." *Social Research: An International Quarterly* 86 (2): 449–476.

Middleton, Neil, and Philip O'Keefe. 2001. *Redefining Sustainable Development*. London: Pluto Press.

Midgley, G. 2002. "Renewing the Critique of the Theory of Knowledge-Constitutive Interests: A Reply to Martin Reynolds." *Journal of the Operational Research Society* 53 (10): 1165–1168. https://doi.org/10.1057/palgrave.jors.2601429.

Midgley, G., I. Munlo, and M. Brown. 1998. "The Theory and Practice of Boundary Critique: Developing Housing Services for Older People." *Journal of the Operational Research Society* 49 (5): 467–478. https://doi.org/10.2307/3009885.

Midgley, Gerald. 1992. "The Sacred and Profane in Critical Systems Thinking." *Systems Practice* 5 (1): 5–16.

Midgley, Gerald. 1996. "What Is This Thing Called CST?' In *Critical Systems Thinking*, 11–24. Boston, MA: Springer. https://doi.org/10.1007/978-0-585-34651-9_1.

Midgley, Gerald. 1997. "Dealing with Coercion: Critical Systems Heuristics and Beyond." *Systems Practice* 10 (1): 37–57.

Midgley, Gerald. 2000. *Systemic Intervention*. Boston: Springer. https://link.springer.com/book/10.1007/978-1-4615-4201-8.

Midgley, Gerald. 2006. "Systemic Intervention for Public Health." *American Journal of Public Health* 96 (3): 466–472. https://doi.org/10.2105/AJPH.2005.067660.

Mildenberger, Matto. 2019. "The Tragedy of the Tragedy of the Commons." *Scientific American* Blog Network. https://blogs.scientificamerican.com/voices/the-tragedy-of-the-tragedy-of-the-commons/.

Miller, Boaz. 2014. "Catching the WAVE: The Weight-Adjusting Account of Values and Evidence." *Studies in History and Philosophy of Science Part A* 47: 69–80. https://doi.org/10.1016/j.shpsa.2014.02.007.

Miller, Boaz. 2020. "Is Technology Value-Neutral?" *Science, Technology, & Human Values*, January. https://doi.org/10.1177/0162243919900965.

Milman, Oliver, and Dominic Rushe. 2021. "The Latest Must-Have among US Billionaires? A Plan to End the Climate Crisis." *The Guardian*, March 25, sec. US News. https://www.theguardian.com/us-news/2021/mar/25/elon-musk-climate-plan-reward-jeff-bezos-gates-investments.

Milne, Alastair, and Neil Maiden. 2012. "Power and Politics in Requirements Engineering: Embracing the Dark Side?' *Requirements Engineering* 17 (2): 83–98. https://doi.org/10.1007/s00766-012-0151-6.

Mingers, John. 1992. "Recent Developments in Critical Management Science." *Journal of the Operational Research Society* 43 (1): 1–10. https://doi.org/10.1057/jors.1992.1.

MIT News Office. 2018. "MIT Reshapes Itself to Shape the Future." *MIT News*. October 15. http://news.mit.edu/2018/mit-reshapes-itself-stephen-schwarzman-college-of-computing-1015.

Mohai, Paul, David Pellow, and J. Timmons Roberts. 2009. "Environmental Justice." *Annual Review of Environment and Resources* 34:405–430.

Mohanani, Rahul, Paul Ralph, and Ben Shreeve. 2014. "Requirements Fixation." In *Proceedings of the 36th International Conference on Software Engineering*, 895–906. Association for Computing Machinery.

Mohanani, Rahul, Iflaah Salman, Burak Turhan, Pilar Rodriguez, and Paul Ralph. 2018. "Cognitive Biases in Software Engineering: A Systematic Mapping Study." *IEEE Transactions on Software Engineering* 46, no. 12, 1318–1339. https://doi.org/10.1109 /TSE.2018.2877759.

Mora, Camilo, Randi L. Rollins, Katie Taladay, Michael B. Kantar, Mason K. Chock, Mio Shimada, and Erik C. Franklin. 2018. "Bitcoin Emissions Alone Could Push Global Warming above 2°C." *Nature Climate Change* 8 (11): 931–933. https://doi.org /10.1038/s41558-018-0321-8.

Morgan, Gareth. 2006. *Images of Organization*. Updated ed. Thousand Oaks, CA: SAGE Publications.

Morgan, Gareth. 2011. "Reflections on Images of Organization and Its Implications for Organization and Environment." *Organization & Environment* 24 (4): 459–478. https://doi.org/10.1177/1086026611434274.

Morozov, Evgeny. 2014. *To Save Everything, Click Here: The Folly of Technological Solutionism*. First Trade Paper Edition. New York: PublicAffairs.

Mosco, Vincent. 2004. *The Digital Sublime: Myth, Power, and Cyberspace*. Cambridge, MA: MIT Press.

Murugesan, San. 2008. "Harnessing Green IT: Principles and Practices." *IT Professional* 10 (1): 24–33. https://doi.org/10.1109/MITP.2008.10.

Myerson, J., L. Green, and M. Warusawitharana. 2001. "Area under the Curve as a Measure of Discounting." *Journal of the Experimental Analysis of Behavior* 76 (2): 235–243. https://doi.org/10.1901/jeab.2001.76.235.

Nakamoto, Satoshi. 2008. "Bitcoin: A Peer-to-Peer Electronic Cash System." Manubot.

Nardi, Bonnie. 2019. "Design in the Age of Climate Change." *She Ji: The Journal of Design, Economics, and Innovation* 5 (1): 5–14. https://doi.org/10.1016/j.sheji.2019.01.001.

Nardi, Bonnie, Bill Tomlinson, Donald J. Patterson, Jay Chen, Daniel Pargman, Barath Raghavan, and Birgit Penzenstadler. 2018. "Computing within Limits." *Communications of the ACM* 61 (10): 86–93. https://doi.org/10.1145/3183582.

Nathan, Lisa P., Batya Friedman, Predrag Klasnja, Shaun K. Kane, and Jessica K. Miller. 2008. "Envisioning Systemic Effects on Persons and Society throughout Interactive System Design." In *Proceedings of the 7th ACM Conference on Designing Interactive Systems*, 1–10. DIS '08. New York: Association for Computing Machinery. https://doi.org /10.1145/1394445.1394446.

Naur, Peter, and Brian Randell, eds. 1969. *Software Engineering: Report on a Conference Sponsored by the NATO Science Committee.* Brussels: NATO Scientific Affairs Division.

Neeley, Kathryn A., and Bernd Steffensen. 2018. "The T-Shaped Engineer as an Ideal in Technology Entrepreneurship: Its Origins, History, and Significance for Engineering Education." https://peer.asee.org/the-t-shaped-engineer-as-an-ideal-in-technology -entrepreneurship-its-origins-history-and-significance-for-engineering-education.

Neumann, Peter G. 2012. "The Foresight Saga, Redux." *Commun. ACM* 55 (10). https://doi.org/10.1145/2347736.2347746.

Neumayer, E. 2013. *Weak versus Strong Sustainability.* Cheltenham, UK: Edward Elgar Publishing.

Newell, Allen, and Herbert Alexander Simon. 1972. *Human Problem Solving.* Englewood Cliffs, NJ: Prentice-Hall.

Nissenbaum, Helen. 1998. "Values in the Design of Computer Systems." *Computers and Society* 28 (1): 38–39.

Nissenbaum, Helen. 2001. "How Computer Systems Embody Values." *Computer* 34 (3): 119–120.

Noble, David F. 1977. *America by Design: Science, Technology, and the Rise of Corporate Capitalism.* 1st ed. New York: Knopf.

Noble, David F. 1984. *Forces of Production: A Social History of Industrial Automation.* 1st ed. Borzoi Books. New York: Knopf.

Noble, Safiya. 2020. "The Loss of Public Goods to Big Tech." *NOEMA* (blog). https:// www.noemamag.com/the-loss-of-public-goods-to-big-tech/.

Noble, Safiya Umoja. 2018. *Algorithms of Oppression: How Search Engines Reinforce Racism.* New York: NYU Press.

Norton, Juliet, Birgit Penzenstadler, and Bill Tomlinson. 2019. "Implications of Grassroots Sustainable Agriculture Community Values on the Design of Information Systems." *Proc. ACM Human-Computer Interactions.* 3 (CSCW): 34:1–34:22. https://doi .org/10.1145/3359136.

Nost, Eric, and Emma Colven. 2022. "Earth for AI: A Political Ecology of Data-Driven Climate Initiatives." *Geoforum* 130 (March): 23–34. https://doi.org/10.1016/j.geoforum .2022.01.016.

Nuseibeh, Bashar, and Steve Easterbrook. 2000. "Requirements Engineering: A Roadmap." In *Proceedings of the Conference on the Future of Software Engineering,* 35–46. Association for Computing Machinery. http://dl.acm.org/citation.cfm?id=336523.

O'Neil, Cathy. 2016. *Weapons of Math Destruction: How Big Data Increases Inequality and Threatens Democracy.* New York: Crown Publishers.

Ongweso, Edward. 2020. "Gig Workers' Only Chance to Pee Is Apparently an App." *Vice*, October. https://www.vice.com/en/article/93w5by/gig-workers-only-chance-to -pee-is-apparently-an-app.

Oreskes, Naomi. 2010. *Merchants of Doubt: How a Handful of Scientists Obscured the Truth on Issues from Tobacco Smoke to Global Warming*. 1st ed. New York: Bloomsbury Press.

Ornstein, Severo M., Brian C. Smith, and Lucy A. Suchman. 1984. "Strategic Computing." *Bulletin of the Atomic Scientists* 40 (10): 11–15. https://doi.org/10.1080/00963402 .1984.11459292.

Ostrom, Elinor. 2015. *Governing the Commons: The Evolution of Institutions for Collective Action*. Canto Classics edition. Cambridge: Cambridge University Press.

Ostrom, Elinor. 2016. "Tragedy of the Commons." In *The New Palgrave Dictionary of Economics*, 1–5. London: Palgrave Macmillan UK. https://doi.org/10.1057/978-1-349 -95121-5_2047-1.

Oulasvirta, Antti, and Kasper Hornbæk. 2016. "HCI Research as Problem-Solving." In *Proceedings of the 2016 CHI Conference on Human Factors in Computing Systems*, 4956–4967. CHI '16. New York: Association for Computing Machinery. https://doi .org/10.1145/2858036.2858283.

Outhwaite, William. 1994. *Habermas: A Critical Introduction*. Key Contemporary Thinkers. Stanford, CA: Stanford University Press.

Owens, Kellie, and Amanda Lenhart. 2020. "Good Intentions, Bad Inventions." Data & Society Research Institute. October 7. https://datasociety.net/library/good -intentions-bad-inventions/.

Oxford English Dictionary. 2020a. "Choice." Lexico Dictionaries. https://www.lexico .com/definition/choice.

Oxford English Dictionary. 2020b. "Decision." Lexico Dictionaries. https://www.lexico .com/definition/decision.

Oxford English Dictionary. 2020c. "Judgement." Lexico Dictionaries. https://www .lexico.com/definition/judgement.

Oxford English Dictionary. 2020d. "Myth." Lexico Dictionaries. https://www.lexico .com/en/definition/myth.

Oxford English Dictionary. 2020e. "Social Justice." Lexico Dictionaries. https://www .lexico.com/definition/social_justice.

Oxford English Dictionary. 2021. "Sustainability." Lexico Dictionaries. https://www .lexico.com/definition/sustainability.

Ozoma, Ifeoma. 2021. "The Tech Worker Handbook." https://techworkerhandbook .org/.

Pal, Joyojeet. 2017. "CHI4Good or Good4CHI." In *Proceedings of the 2017 CHI Conference Extended Abstracts on Human Factors in Computing Systems*, 709–721. CHI EA '17.

New York: Association for Computing Machinery. https://doi.org/10.1145/3027063 .3052766.

Palacin, Victoria, Matti Nelimarkka, Pedro Reynolds-Cuéllar, and Christoph Becker. 2020. "The Design of Pseudo-Participation." In *Proceedings of the 16th Participatory Design Conference 2020—Participation(s) Otherwise—Volume 2*, 40–44. PDC '20. New York: Association for Computing Machinery. https://doi.org/10.1145/3384772.3385141.

Palmer, Jordan Novet, Annie. 2020. "Amazon Salesperson's Pitch to Oil and Gas: "Remember That We Actually Consume Your Products!"' CNBC. May 20. https://www .cnbc.com/2020/05/20/aws-salesman-pitch-to-oil-and-gas-we-actually-consume-your -products.html.

Pansera, Mario, and Mariano Fressoli. 2021. "Innovation without Growth: Frameworks for Understanding Technological Change in a Post-Growth Era." *Organization* 28 (3): 380–404. https://doi.org/10.1177/1350508420973631.

Parnas, D. L. 1972. "On the Criteria to Be Used in Decomposing Systems into Modules." *Communications of the ACM* 15 (12): 1053–1058. https://doi.org/10.1145/361598 .361623.

Parnas, D. L., and P. C. Clements. 1986. "A Rational Design Process: How and Why to Fake It." *IEEE Transactions on Software Engineering* SE-12 (2): 251–257. https://doi .org/10.1109/TSE.1986.6312940.

Parnas, David Lorge. 1994. "Software Aging." In *Proceedings of the 16th International Conference on Software Engineering*, 279–287. ICSE '94. Los Alamitos, CA: IEEE Computer Society Press. http://dl.acm.org/citation.cfm?id=257734.257788.

Parrique, T., J. Barth, F. Briens, C. Kerschner, A. Kraus-Polk, A. Kuokkanen, and J. H. Spangenberg. 2019. *Decoupling Debunked—Evidence and Arguments against Green Growth as a Sole Strategy for Sustainability*. Brussels: European Environmental Bureau. https://eeb.org/library/decoupling-debunked/.

Passi, Samir, and Solon Barocas. 2019. "Problem Formulation and Fairness." In *Proceedings of the Conference on Fairness, Accountability, and Transparency*, 39–48. FAT* '19. Atlanta, GA: Association for Computing Machinery. https://doi.org/10.1145/3287560 .3287567.

Patitsas, Elizabeth, Jesse Berlin, Michelle Craig, and Steve Easterbrook. 2019. "Evidence hat Computer Science Grades Are Not Bimodal." *Communications of the ACM* 63 (1): 91–98. https://doi.org/10.1145/3372161.

Pellow, David N., and Lisa Sun-Hee Park. 2002. *The Silicon Valley of Dreams: Environmental Injustice, Immigrant Workers, and the High-Tech Global Economy*. Critical America. New York: NYU Press.

Penzenstadler, B., A. Raturi, D. Richardson, and B. Tomlinson. 2014. "Safety, Security, Now Sustainability: The Nonfunctional Requirement for the 21st Century." *IEEE Software* 31 (3): 40–47. https://doi.org/10.1109/MS.2014.22.

Penzenstadler, Birgit, Stefanie Betz, Colin Venters, Ruzanna Chitchyan, Norbert Seyff, Leticia Duboc, Christian U. Becker, and Jari Porras. 2018. "Everything Is Interrelated: Teaching Software Engineering for Sustainability." In *Proceedings of the 40th International Conference on Software Engineering (ICSE 2018) Software Engineering Education and Training Track (SEET)*. IEEE Press. https://doi.org/10.1145/3183377.3183382.

Penzenstadler, Birgit, Ankita Rauturi, Christoph Becker, Juliet Norton, Bill Tomlinson, Six Silberman, and Debra Richardson. 2016. 'Bridging Communities: ICT4Sustainability @iConference 2015'. *Interactions* 23 (1): 64–67. https://doi.org/10.1145/2843584.

Persson, Linn, Bethanie M. Carney Almroth, Christopher D. Collins, Sarah Cornell, Cynthia A. de Wit, Miriam L. Diamond, Peter Fantke, et al. 2022. "Outside the Safe Operating Space of the Planetary Boundary for Novel Entities." *Environmental Science & Technology* 56 (3): 1510–1521. https://doi.org/10.1021/acs.est.1c04158.

Peters, Anne-Kathrin. 2019. "Participation and Learner Trajectories in Computing Education." In *Bridging Research and Practice in Science Education: Selected Papers from the ESERA 2017 Conference*, edited by Eilish McLoughlin, Odilla E. Finlayson, Sibel Erduran, and Peter E. Childs, 139–152. Contributions from Science Education Research. Cham, UK: Springer International Publishing. https://doi.org/10.1007/978-3-030-17219-0_9.

Pitt, Joseph C. 2014. "'Guns Don't Kill, People Kill'; Values in and/or Around Technologies." In *The Moral Status of Technical Artefacts*, edited by Peter Kroes and Peter-Paul Verbeek, 89–101. Philosophy of Engineering and Technology. Dordrecht, Holland: Springer Netherlands. https://doi.org/10.1007/978-94-007-7914-3_6.

Pohl, Klaus. 2010. *Requirements Engineering: Fundamentals, Principles, and Techniques*. Heidelberg, Germany: Springer.

Pomerol, Jean-Charles, and Frédéric Adam. 2006. "On the Legacy of Herbert Simon and His Contribution to Decision-Making Support Systems and Artificial Intelligence." In *Intelligent Decision-Making Support Systems: Foundations, Applications and Challenges (Decision Engineering)*, edited by Jatinder N. D. Gupta, Guisseppi A. Forgionne, and Manuel Mora T., 25–43. London: Springer. https://doi.org/10.1007/1-84628-231-4_2.

Popper, Karl R. 1972. *Objective Knowledge*. Vol. 360. Oxford: Oxford University Press.

Price, Margaret. 2011. *Mad at School: Rhetorics of Mental Disability and Academic Life*. Corporealities. Ann Arbor: University of Michigan Press. https://doi.org/10.3998/mpub.1612837.

Proenca, Diogo, Goncalo Antunes, Jose Borbinha, Artur Caetano, Stefan Biffl, Dietmar Winkler, and Christoph Becker. 2013. "Longevity as an Information Systems Design Concern." In *CAISE Forum*. Valencia.

Pronin, Emily, Christopher Y. Olivola, and Kathleen A. Kennedy. 2008. "Doing unto Future Selves as You Would Do unto Others: Psychological Distance and Decision Making." *Personality and Social Psychology Bulletin* 34 (2): 224–236. https://doi.org/10.1177/0146167207310023.

Raji, Inioluwa Deborah, Morgan Klaus Scheuerman, and Razvan Amironesei. 2021. "You Can't Sit with Us: Exclusionary Pedagogy in AI Ethics Education." In *Proceedings of the 2021 ACM Conference on Fairness, Accountability, and Transparency*, 515–525. FAccT '21. New York: Association for Computing Machinery. https://doi.org/10.1145/3442188.3445914.

Ralph, Paul. 2013. "The Illusion of Requirements in Software Development." *Requirements Engineering* 18 (3): 293–296. https://doi.org/10.1007/s00766-012-0161-4.

Ralph, Paul. 2015. "The Sensemaking-Coevolution-Implementation Theory of Software Design." In *Science of Computer Programming*, special issue, *Towards General Theories of Software Engineering*, 101 (April): 21–41. https://doi.org/10.1016/j.scico.2014.11.007.

Ralph, Paul. 2018. "The Two Paradigms of Software Development Research." *Science of Computer Programming* 156 (May): 68–89. https://doi.org/10.1016/j.scico.2018.01.002.

Ralph, Paul, Mike Chiasson, and Helen Kelley. 2016. "Social Theory for Software Engineering Research." In *Proceedings of the 20th International Conference on Evaluation and Assessment in Software Engineering*, 44:1–44:11. EASE '16. New York: Association for Computing Machinery. https://doi.org/10.1145/2915970.2915998.

Ralph, Paul, and Briony J. Oates. 2018. "The Dangerous Dogmas of Software Engineering." *ArXiv:1802.06321 [Cs]*, February. http://arxiv.org/abs/1802.06321.

Ramage, Magnus, and Karen Shipp. 2009. *Systems Thinkers*. Dordrecht, Holland: Springer. http://books.scholarsportal.info/viewdoc.html?id=/ebooks/ebooks0/springer/2010-04-28/1/9781848825253.

Ratto, Matt. 2011. "Critical Making: Conceptual and Material Studies in Technology and Social Life." *The Information Society* 27 (4): 252–260. https://doi.org/10.1080/01972243.2011.583819.

Raworth, Kate. 2017. *Doughnut Economics: Seven Ways to Think like a 21st Century Economist*. White River Junction, VT: Chelsea Green Publishing.

Raymond Francis. 2006. "The Ritual of the Calling of an Engineer." 2006. https://wolfstu.livejournal.com/277038.html.

Redclift, Michael. 2005. "Sustainable Development (1987–2005): An Oxymoron Comes of Age." *Sustainable Development* 13 (4): 212–227. https://doi.org/10.1002/sd.281.

Reid-Henry, Simon. 2012. "Do Resource Extraction and the Legacy of Colonialism Keep Poor Countries Poor?" *The Guardian*, October 22, sec. Global Development. https://www.theguardian.com/global-development/2012/oct/22/resource-extraction-colonialism-legacy-poor-countries.

Reiss, Julian, and Jan Sprenger. 2017. "Scientific Objectivity." In *The Stanford Encyclopedia of Philosophy*, edited by Edward N. Zalta, Winter. Metaphysics Research Lab, Stanford University. https://plato.stanford.edu/archives/win2017/entries/scientific-objectivity/.

Reynolds, M. 2002. "In Defence of Knowledge Constitutive Interests. A Comment on 'What Is This Thing Called CST?' (Midgley, 1996)." *Journal of the Operational Research Society* 53 (10): 1162–1165. https://doi.org/10.1057/palgrave.jors.2601427.

Reynolds, Martin. 2007. "Evaluation Based on Critical Systems Heuristics." In *Using Systems Concepts in Evaluation: An Expert Anthology*, edited by Williams, B. and I. Imam, I., 101–122. Point Reyes, CA: EdgePress.

Ribes, David. 2019. "How I Learned What a Domain Was." *Proceedings of the ACM on Human-Computer Interaction* 3 (CSCW): 38:1–38:12. https://doi.org/10.1145/3359140.

Ribes, David, Andrew S. Hoffman, Steven C. Slota, and Geoffrey C. Bowker. 2019. "The Logic of Domains." *Social Studies of Science* 49 (3): 281–309. https://doi.org/10.1177/0306312719849709.

Rifat, Mohammad Rashidujjaman, Hasan Mahmud Prottoy, and Syed Ishtiaque Ahmed. 2019. "The Breaking Hand: Skills, Care, and Sufferings of the Hands of an Electronic Waste Worker in Bangladesh." In *Proceedings of the 2019 CHI Conference on Human Factors in Computing Systems*, 1–14. CHI '19. Glasgow, Scotland: Association for Computing Machinery. https://doi.org/10.1145/3290605.3300253.

Rittel, Horst W. J., and Melvin M. Webber. 1973. "Dilemmas in a General Theory of Planning." *Policy Sciences* 4 (2): 155–169.

Roberts, Sarah T., Moira Weigel, Miriam Posner, and Safiya Noble. 2018. "Algorithms of Oppression, Algorithms of Liberation: A Conversation with Logic Magazine." http://annenberg.usc.edu/events/algorithms-oppression-algorithms-liberation-conversation-logic-magazine.

Robertson, Suzanne, and James Robertson. 2012. *Mastering the Requirements Process: Getting Requirements Right*. Addison-Wesley.

Robinson, Brett H. 2009. "E-Waste: An Assessment of Global Production and Environmental Impacts." *Science of The Total Environment* 408 (2): 183–191. https://doi.org/10.1016/j.scitotenv.2009.09.044.

Rockström, Johan, Will Steffen, Kevin Noone, Åsa Persson, F. Stuart Chapin, Eric F. Lambin, Timothy M. Lenton, et al. 2009. "A Safe Operating Space for Humanity." *Nature* 461 (7263): 472–475. https://doi.org/10.1038/461472a.

Rogers, Yvonne, and Paul Marshall. 2017. "Research in the Wild." *Synthesis Lectures on Human-Centered Informatics* 10 (3): i–97. https://doi.org/10.2200/S00764ED1V01Y201703HCI037.

Romm, Norma. 1994. *Continuing Tensions between Soft Systems Methodology and Critical Systems Heuristics*. University of Hull, Centre for Systems Studies.

Romm, Norma. 1995. "Some Anomalies in Ulrich's Critical Inquiry and Problem-Solving Approach." In *Critical Issues in Systems Theory and Practice*, edited by Keith Ellis, Amanda Gregory, Bridget R. Mears-Young, and Gillian Ragsdell, 503–509. Boston, MA: Springer US. https://doi.org/10.1007/978-1-4757-9883-8_76.

Rosner, Daniela. 2018. *Critical Fabulations: Reworking the Methods and Margins of Design*. Design Thinking, Design Theory. Cambridge, MA: MIT Press.

Rosner, Daniela, Alex Taylor, and Mikael Wiberg. 2020. "Feminisms in Design." *Interactions* 27 (6): 5.

Rowlands, Bruce, and Karlheinz Kautz. 2020. "Power Relations Inscribed in the Enactment of Systems Development Methods." *Information Systems Journal*. https://doi.org/10.1111/isj.12322.

Royal Society, The. 2020. *Digital Technology and the Planet: Harnessing Computing to Achieve Net Zero*. https://royalsociety.org/topics-policy/ projects/digital-tech nology-and-the-planet.

Rozanski, Nick, and Eoin Woods. 2012. *Software Systems Architecture: Working with Stakeholders Using Viewpoints and Perspectives*. 2nd ed. Upper Saddle River, NJ: Addison-Wesley.

Samuelson, Paul A. 1937. "A Note on Measurement of Utility." *The Review of Economic Studies* 4 (2): 155–161. https://doi.org/10.2307/2967612.

Saunders, Harry D., and Jeffrey Y. Tsao. 2012. "Rebound Effects for Lighting." *Energy Policy*, Special Section: Fuel Poverty Comes of Age: Commemorating 21 Years of Research and Policy, 49 (October): 477–478. https://doi.org/10.1016/j.enpol.2012.06.050.

Saxena, Devansh, Erhardt Graeff, Shion Guha, EunJeong Cheon, Pedro Reynolds-Cuéllar, Dawn Walker, Christoph Becker, and Kenneth R. Fleischmann. 2020. "Collective Organizing and Social Responsibility at CSCW." In *Conference Companion Publication of the 2020 on Computer Supported Cooperative Work and Social Computing*, 503–509. CSCW '20 Companion. New York: Association for Computing Machinery. https://doi.org/10.1145/3406865.3418593.

Schaake, Marietje. 2020. "There Has Never Been a Better Time to Make Big Tech Pay Its Way." May 20. https://www.ft.com/content/1af55118-99bf-11ea-871b-edeb99a20c6e.

Scheuerman, Morgan Klaus, Emily Denton, and Alex Hanna. 2021. "Do Datasets Have Politics? Disciplinary Values in Computer Vision Dataset Development." August. https://doi.org/10.1145/3476058.

Schön, Donald A. 1979. "Generative Metaphor: A Perspective on Problem-Setting in Social Policy." *Metaphor and Thought* 254:283.

Schön, Donald A. 1983. *The Reflective Practitioner: How Professionals Think in Action*. New York: Basic Books.

Schraagen, Jan Maarten, Susan F. Chipman, and Valerie L. Shalin, eds. 2000. *Cognitive Task Analysis*. Expertise: Research and Applications. Mahwah, NJ: L. Erlbaum Associates.

Schrock, Andrew. 2020. "Silicon Valley Is Not a Place." *Public Books* (blog). April 9. https://www.publicbooks.org/silicon-valley-is-not-a-place/.

Schwartz, Shalom H. 1992. "Universals in the Content and Structure of Values: Theoretical Advances and Empirical Tests in 20 Countries." *Advances in Experimental Social Psychology* 25:1–65. https://doi.org/10.1016/S0065-2601(08)60281-6.

Schwartz, Shalom H. 1994. "Are There Universal Aspects in the Structure and Contents of Human Values?" *Journal of Social Issues* 50 (4): 19–45. https://doi.org/10.1111/j.1540-4560.1994.tb01196.x.

Selbst, Andrew D., Danah Boyd, Sorelle A. Friedler, Suresh Venkatasubramanian, and Janet Vertesi. 2019. "Fairness and Abstraction in Sociotechnical Systems." In *Proceedings of the Conference on Fairness, Accountability, and Transparency*, 59–68. FAT* '19. Atlanta, GA: Association for Computing Machinery. https://doi.org/10.1145/3287560.3287598.

Seligman, Martin E. P., Peter Railton, Roy F. Baumeister, and Chandra Sripada. 2013. "Navigating into the Future or Driven by the Past." *Perspectives on Psychological Science* 8 (2): 119–141. https://doi.org/10.1177/1745691612474317.

Semeniuk, Ivan. 2017. "Massive Review of Federal Science Funding Reveals Risks to Younger Researchers." *The Globe and Mail*, April 10. https://www.theglobeandmail.com/news/national/review-calls-for-new-entity-to-oversee-federal-science-funding/article34650444/.

Sen, Amartya Kumar. 2009. *The Idea of Justice*. Cambridge, MA: Harvard University Press.

Senge, Peter M. 1990. *The Fifth Discipline: The Art and Practice of the Learning Organization*. New York: Doubleday/Currency.

Shafer, Glenn. 1986. "Savage Revisited." *Statistical Science* 1 (4): 463–485.

Shankleman, Jessica, Tom Biesheuvel, Joe Ryan, and Dave Merrill. 2017. "We're Going to Need More Lithium." *Bloomberg Businessweek*, September. https://www.bloomberg.com/graphics/2017-lithium-battery-future/.

Sharf, Samantha. 2019. "Designed to Share: Airbnb-Backed Developer Is Building Condos Made for Home Sharing." *Forbes*. https://www.forbes.com/sites/samanthasharf/2019/06/18/airbnb-backed-real-estate-developer-natiivo-niido-austin-miami/.

Shehabi, Arman, Sarah Smith, Dale Sartor, Richard Brown, Magnus Herrlin, Jonathan Koomey, Eric Masanet, Nathaniel Horner, Inês Azevedo, and William Lintner. 2016. "United States Data Center Energy Usage Report." LBNL-1005775. Lawrence Berkeley National Lab (LBNL), Berkeley, CA. https://doi.org/10.2172/1372902.

Shilton, Katie. 2012. "Values Levers: Building Ethics into Design." *Science, Technology, & Human Values* 38 (3). https://doi.org/10.1177/0162243912436985.

Shilton, Katie. 2018. "Values and Ethics in Human-Computer Interaction." *Foundations and Trends® Human–Computer Interaction* 12 (2): 107–171. https://doi.org/10.1561/1100000073.

Simon, H. A. 1956. "Rational Choice and the Structure of the Environment." *Psychological Review* 63 (2): 129–138. https://doi.org/10.1037/h0042769.

Simon, Herbert. 1962. "The Architecture of Complexity." *Proceedings of the American Philosophical Society* 106 (6): 467–482.

Simon, Herbert A. 1955. "A Behavioral Model of Rational Choice." *The Quarterly Journal of Economics* 69 (1): 99–118.

Simon, Herbert A. 1973. "The Structure of Ill Structured Problems." *Artificial Intelligence* 4 (3): 181–201. https://doi.org/10.1016/0004-3702(73)90011-8.

Simon, Herbert A. 1977. *Models of Discovery: And Other Topics in the Methods of Science.* Boston Studies in the Philosophy of Science v. 54. Dordrecht, Holland; D. Reidel.

Simon, Herbert A. 1996. *The Sciences of the Artificial.* 3rd ed. Cambridge, MA: MIT Press.

Slade, Giles. 2009. *Made to Break: Technology and Obsolescence in America.* Cambridge, MA: Harvard University Press. http://public.ebookcentral.proquest.com/choice/public fullrecord.aspx?p=3300262.

Smith, Brian Cantwell. 1993. "Limits of Correctness in Computers." In *Program Verification: Fundamental Issues in Computer Science*, edited by Timothy R. Colburn, James H. Fetzer, and Terry L. Rankin, 275–293. Studies in Cognitive Systems. Dordrecht, Holland: Springer Netherlands. https://doi.org/10.1007/978-94-011-1793-7_13.

Smith, Brian Cantwell. 1998. *On the Origin of Objects.* Cambridge, MA: MIT Press. https://mitpress.mit.edu/books/origin-objects.

Smith, Brian Cantwell. 2019. *The Promise of Artificial Intelligence: Reckoning and Judgment.* Cambridge, MA: MIT Press.

Smith, Craig S. 2019. "Dealing with Bias in Artificial Intelligence." *New York Times*, November 19, sec. Technology. https://www.nytimes.com/2019/11/19/technology /artificial-intelligence-bias.html.

Snipes, Will, Brian Robinson, Yuepu Guo, and Carolyn Seaman. 2012. "Defining the Decision Factors for Managing Defects: A Technical Debt Perspective." In *2012 Third International Workshop on Managing Technical Debt (MTD)*, 54–60. https://doi.org/10 .1109/MTD.2012.6226001.

Solnit, Rebecca. 2010. *A Paradise Built in Hell: The Extraordinary Communities That Arise in Disaster.* Penguin.

Soman, Dilip, George Ainslie, Shane Frederick, Xiuping Li, John Lynch, Page Moreau, Andrew Mitchell, et al. 2005. "The Psychology of Intertemporal Discounting: Why Are Distant Events Valued Differently from Proximal Ones?" *Marketing Letters* 16 (3): 347–360. https://doi.org/10.1007/s11002-005-5897-x.

Sorrell, Steve. 2009. "Jevons' Paradox Revisited: The Evidence for Backfire from Improved Energy Efficiency." *Energy Policy* 37 (4): 1456–1469. https://doi.org/10.1016 /j.enpol.2008.12.003.

Southern Poverty Law Center. n.d. "Garrett Hardin." Southern Poverty Law Center. Accessed December 16, 2020. https://www.splcenter.org/fighting-hate/extremist -files/individual/garrett-hardin.

Spiekermann, Sarah. 2016. *Ethical IT Innovation: A Value-Based System Design Approach.*

Spiekermann, Sarah. 2017. "IEEE P7000—The First Global Standard Process for Addressing Ethical Concerns in System Design." *Proceedings* 1 (3): 159. https://doi.org/10.3390/IS4SI-2017-04084.

Spiekermann, Sarah, and Till Winkler. 2020. "Value-Based Engineering for Ethics by Design." SSRN Scholarly Paper ID 3598911. Rochester, NY: Social Science Research Network. https://doi.org/10.2139/ssrn.3598911.

Spiel, Katta, Christopher Frauenberger, Os Keyes, and Geraldine Fitzpatrick. 2019. "Agency of Autistic Children in Technology Research—A Critical Literature Review." *ACM Transactions on Computer-Human Interaction* 26 (6): 38:1–38:40. https://doi.org/10.1145/3344919.

Spitzberg, Danny, Kevin Shaw, Colin Angevine, Marissa Wilkins, M. Strickland, Janel Yamashiro, Rhonda Adams, and Leah Lockhart. 2020. "Principles at Work: Applying 'Design Justice' in Professionalized Workplaces." *Tech Otherwise*, October. https://doi.org/10.21428/93b2c832.e3a8d187.

Stark, Luke. 2019. "Facial Recognition Is the Plutonium of AI." *XRDS: Crossroads, The ACM Magazine for Students* 25 (3): 50–55. https://doi.org/10.1145/3313129.

Steel, Brent, D. Lach, and Edward P. Weber, eds. 2017. *New Strategies for Wicked Problems: Science and Solutions in the Twenty-First Century.* Corvallis: Oregon State University Press.

Steffen, Will, Katherine Richardson, Johan Rockström, Sarah E. Cornell, Ingo Fetzer, Elena M. Bennett, Reinette Biggs, et al. 2015. "Planetary Boundaries: Guiding Human Development on a Changing Planet." *Science* 347 (6223). https://doi.org/10.1126/science.1259855.

Stephens, Anne. 2013. *Ecofeminism and Systems Thinking.* 1st ed. Routledge Research in Gender and Society; 36. New York: Routledge, Taylor & Francis Group.

Stephens, Anne, Ann Taket, and Monica Gagliano. 2019. "Ecological Justice for Nature in Critical Systems Thinking." *Systems Research and Behavioral Science* 36 (1): 3–19. https://doi.org/10.1002/sres.2532.

Strengers, Yolande. 2014. "Smart Energy in Everyday Life: Are You Designing for Resource Man?" *Interactions* 21 (4): 24–31. https://doi.org/10.1145/2621931.

Strubell, Emma, Ananya Ganesh, and Andrew McCallum. 2019. "Energy and Policy Considerations for Deep Learning in NLP." *ArXiv:1906.02243 [Cs]*, June. http://arxiv.org/abs/1906.02243.

Sturm, Thomas. 2012. "The "Rationality Wars" in Psychology: Where They Are and Where They Could Go." *Inquiry* 55 (1): 66–81. https://doi.org/10.1080/0020174X.2012.643628.

Subramaniam, Banu, Laura Foster, Sandra Harding, Deboleena Roy, and Kim TallBear. 2016. "Feminism, Postcolonialism, Technoscience." In *The Handbook of Science and*

Technology Studies. 4th ed. Edited by Ulrike Felt, Rayvon Fouche, Clark A. Miller, and Laurel Smith-Doerr, 407–433. Cambridge, MA: MIT Press.

Suchman, Lucy. 1987. *Plans and Situated Actions: The Problem of Human-Machine Communication*. Learning in Doing. Cambridge: Cambridge University Press.

Suchman, Lucy. 1998. "Strengthening Our Collective Resources: A Comment on Morten Kyng's 'A Contextual Approach to the Design of Computer Artifacts.'" *Scandinavian Journal of Information Systems* 10 (1–2): 45–51.

Suchman, Lucy. 2002. "Located Accountabilities in Technology Production." *Scandinavian Journal of Information Systems* 14 (2): 91–105.

Suchman, Lucy. 2006. *Human-Machine Reconfigurations: Plans and Situated Actions*. 2nd ed. Learning in Doing: Social, Cognitive and Computational Perspectives. Cambridge: Cambridge University Press. https://doi.org/10.1017/CBO9780511808418.

Supran, Geoffrey, and Naomi Oreskes. 2021. "Rhetoric and Frame Analysis of ExxonMobil's Climate Change Communications." *One Earth* 4 (5): 696–719. https://doi.org/10.1016/j.oneear.2021.04.014.

"Sustainability: Can Our Society Endure?" n.d. Accessed May 13, 2020. https://web.archive.org/web/20200826214142/https://sustainability.com/sustainability/.

Sweeney, Latanya. 2013. "Discrimination in Online Ad Delivery." *Communications of the ACM* 56 (5): 44–54. https://doi.org/10.1145/2447976.2447990.

Sweeney, Linda Booth, and Dennis Meadows. 2010. *The Systems Thinking Playbook: Exercises to Stretch and Build Learning and Systems Thinking Capabilities*. Har/DVD edition. White River Junction, VT: Chelsea Green Publishing.

Tarnoff, Ben. 2020. *The Making of the Tech Worker Movement*. Logic. https://logicmag.io/the-making-of-the-tech-worker-movement/.

Tarnoff, Ben, and Moira Weigel. 2020. *Voices from the Valley: Tech Workers Talk about What They Do—and How They Do It*. New York: FSG Originals.

Taylor, Dorceta E. 2000. "The Rise of the Environmental Justice Paradigm: Injustice Framing and the Social Construction of Environmental Discourses." *American Behavioral Scientist*, July. https://doi.org/10.1177/0002764200043004003.

Taylor, Keeanga-Yamahtta. 2020. "The Black Plague." *New Yorker*. April 16. https://www.newyorker.com/news/our-columnists/the-black-plague.

TechOtherwise Collective. 2021. "Tech Otherwise." *Tech Otherwise*. https://techotherwise.pubpub.org/.

Termeer, Catrien J. A. M., Art Dewulf, and Robbert Biesbroek. 2019. "A Critical Assessment of the Wicked Problem Concept: Relevance and Usefulness for Policy Science and Practice." *Policy and Society* 38 (2): 167–179. https://doi.org/10.1080/14494035.2019.1617971.

Thaler, Richard H., and Cass R. Sunstein. 2008. *Nudge: Improving Decisions about Health, Wealth and Happiness*. Electronic resource. New Haven: Ipswich: Yale University Press.

Thaler, Richard H., Cass R. Sunstein, and John P. Balz. 2010. "Choice Architecture." SSRN Scholarly Paper ID 1583509. Rochester, NY: Social Science Research Network. https://doi.org/10.2139/ssrn.1583509.

Thew, Sarah, and Alistair Sutcliffe. 2017. "Value-Based Requirements Engineering: Method and Experience." *Requirements Engineering* 23, 443–464, June. https://doi.org/10.1007/s00766-017-0273-y.

Thienen, Julia von, Christoph Meinel, and Claudia Nicolai. 2014. "How Design Thinking Tools Help to Solve Wicked Problems." In *Design Thinking Research: Building Innovation Eco-Systems*, edited by Larry Leifer, Hasso Plattner, and Christoph Meinel, 97–102. Understanding Innovation. Cham, UK: Springer International Publishing. https://doi.org/10.1007/978-3-319-01303-9_7.

Thomas, Alan R., and Martin Lockett. 1991. "Marxism and Systems Research: Values in Practical Action." In *Critical Systems Thinking: Directed Readings*, edited by Robert L. Flood and Michael C. Jackson, 85–102. New York: John Wiley & Sons.

Tomlinson, Bill. 2010. *Greening through IT: Information Technology for Environmental Sustainability*. Cambridge, MA: MIT Press.

Tomlinson, Bill, Eli Blevis, Bonnie Nardi, Donald J. Patterson, M. SIX Silberman, and Yue Pan. 2013. "Collapse Informatics and Practice: Theory, Method, and Design." *ACM Transactions on Computer-Human Interaction* 20 (4): 24:1–24:26. https://doi.org/10.1145/2493431.

Trope, Yaacov, and Nira Liberman. 2010. "Construal-Level Theory of Psychological Distance." *Psychological Review* 117 (2): 440–463. https://doi.org/10.1037/a0018963.

Tsao, J. Y., H. D. Saunders, J. R. Creighton, M. E. Coltrin, and J. A. Simmons. 2010. "Solid-State Lighting: An Energy-Economics Perspective." *Journal of Physics D: Applied Physics* 43 (35): 354001. https://doi.org/10.1088/0022-3727/43/35/354001.

Tuovila, Alicia. 2020. "Insolvency." Investopedia. 2020. https://www.investopedia.com/terms/i/insolvency.asp.

Tversky, Amos, and Daniel Kahneman. 1986. "Rational Choice and the Framing of Decisions." *Journal of Business*, S251–78.

UK Engineering Council. 2014. *UK Standard for Professional Engineering Competence*. http://www.engc.org.uk/ukspec.aspx.

Ulrich, Werner. 1980. "The Metaphysics of Design: A Simon-Churchman 'Debate.'" *INFORMS Journal on Applied Analytics* 10 (2): 35–40. https://doi.org/10.1287/inte.10.2.35.

Ulrich, Werner. 1981. "A Critique of Pure Cybernetic Reason: The Chilean Experience with Cybernetics." *Journal of Applied Systems Analysis* 8 (1): 33–59.

Ulrich, Werner. 1983. *Critical Heuristics of Social Planning: A New Approach to Practical Philosophy*. Schriftenreihe Des Management-Zentrums St. Gallen 3. Bern: P. Hapt.

Ulrich, Werner. 1985. "The Way of Inquiring Systems." *Journal of the Operational Research Society* 36 (9): 873–876. https://doi.org/10.1057/jors.1985.155.

Ulrich, Werner. 1993. "Some Difficulties of Ecological Thinking, Considered from a Critical Systems Perspective: A Plea for Critical Holism." *Systems Practice* 6 (6): 583–611. https://doi.org/10.1007/BF01059480.

Ulrich, Werner. 1998. *Systems Thinking as If People Mattered: Critical Systems Thinking for Citizens and Managers*. Lincoln, UK: University of Lincolnshire and Humberside, Lincoln School of Management.

Ulrich, Werner. 2004. "In Memory of C. West Churchman (1913–2004) Reminiscences, Retrospectives, and Reflections." *Journal of Organisational Transformation & Social Change* 1 (2): 199–219. https://doi.org/10.1386/jots.1.2.199/0.

Ulrich, Werner, and Martin Reynolds. 2010. "Critical Systems Heuristics." In *Systems Approaches to Managing Change: A Practical Guide*. Springer.

United Church of Christ Commission for Racial Justice. 1987. "Toxic Wastes and Race in the United States: A National Report on the Racial and Socio-Economic Characteristics of Communities with Hazardous Waste Sites." New York: United Church of Christ.

United Nations. 2015a. "About the Sustainable Development Goals." *United Nations Sustainable Development* (blog). https://www.un.org/sustainabledevelopment/sustainable-development-goals/.

United Nations. 2015b. "Goal 16." Sustainable Development Knowledge Platform. 2015. https://sustainabledevelopment.un.org/sdg16.

United Nations General Assembly. 2015. *Transforming Our World: The 2030 Agenda for Sustainable Development*. https://www.unfpa.org/resources/transforming-our-world-2030-agenda-sustainable-development.

US GAO (General Accounting Office). 1983. "Siting of Hazardous Waste Landfills and Their Correlation with Racial and Economic Status of Surrounding Communities." Washington, DC: GAO 121648. https://www.gao.gov/assets/rced-83-168.pdf.

Vadén, T., V. Lähde, A. Majava, P. Järvensivu, T. Toivanen, E. Hakala, and J. T. Eronen. 2020. "Decoupling for Ecological Sustainability: A Categorisation and Review of Research Literature." *Environmental Science & Policy* 112 (October): 236–244. https://doi.org/10.1016/j.envsci.2020.06.016.

Van Heddeghem, Ward, Sofie Lambert, Bart Lannoo, Didier Colle, Mario Pickavet, and Piet Demeester. 2014. "Trends in Worldwide ICT Electricity Consumption from 2007 to 2012." *Computer Communications*, Green Networking, 50 (September): 64–76. https://doi.org/10.1016/j.comcom.2014.02.008.

van Vliet, Hans, and Antony Tang. 2016. "Decision Making in Software Architecture." *Journal of Systems and Software* 117 (July): 638–644. https://doi.org/10.1016/j .jss.2016.01.017.

Varela, Francisco J, Evan Thompson, and Eleanor Rosch. 1991. *The Embodied Mind: Cognitive Science and Human Experience.* https://doi.org/10.7551/mitpress/9780262529365 .001.0001.

Venters, Colin C., Rafael Capilla, Stefanie Betz, Birgit Penzenstadler, Tom Crick, Steve Crouch, Elisa Yumi Nakagawa, Christoph Becker, and Carlos Carrillo. 2018. "Software Sustainability: Research and Practice from a Software Architecture Viewpoint." *Journal of Systems and Software* 138 (April): 174–188. https://doi.org/10.1016/j.jss.2017.12.026.

Vera, Lourdes A., Dawn Walker, Michelle Murphy, Becky Mansfield, Ladan Mohamed Siad, Jessica Ogden, and EDGI. 2019. "When Data Justice and Environmental Justice Meet: Formulating a Response to Extractive Logic through Environmental Data Justice." *Information, Communication & Society* 22 (7): 1012–1028. https://doi.org/10 .1080/1369118X.2019.1596293.

Vertesi, Janet, and David Ribes, eds. 2019. *DigitalSTS: A Field Guide for Science & Technology Studies.* Electronic resource. Gale EBooks. Princeton, NJ: Princeton University Press.

Vincent, James. 2019. "Google Employee Who Helped Lead Protests Leaves Company." The Verge. July 16. https://www.theverge.com/2019/7/16/20695964/google -protest-leader-meredith-whittaker-leaves-company.

Vries, Alex de, and Christian Stoll. 2021. "Bitcoin's Growing e-Waste Problem." *Resources, Conservation and Recycling* 175 (December): 105901. https://doi.org/10 .1016/j.resconrec.2021.105901.

Wachsmuth, David, David Chaney, Danielle Kerrigan, Andrea Shillolo, and Robin Basalaev-Binder. 2018. *The High Cost of Short-Term Rentals in New York City.* Vol. 2. Montreal: McGill University.

Wachsmuth, David, Kerrigan Danielle, David Chaney, and Andrea Shillolo. 2017. "Short-Term Cities: Airbnb's Impact on Canadian Housing Markets." Montreal: McGill University. https://upgo.lab.mcgill.ca/publication/short-term-cities/.

Wachsmuth, David, and Alexander Weisler. 2018. "Airbnb and the Rent Gap: Gentrification through the Sharing Economy." *Environment and Planning A: Economy and Space* 50 (6): 1147–1170. https://doi.org/10.1177/0308518X18778038.

Wajcman, Judy. 1991. *Feminism Confronts Technology.* University Park, PA: Pennsylvania State University Press.

Wang, Yan, Andrea Peris, Mohammad Rashidujjaman Rifat, Syed Ishtiaque Ahmed, Nirupam Aich, Linh V. Nguyen, Jakub Urík, et al. 2020. "Measuring Exposure of E-Waste Dismantlers in Dhaka Bangladesh to Organophosphate Esters and Halogenated Flame Retardants Using Silicone Wristbands and T-Shirts." *Science of The Total Environment* 720 (June): 137480. https://doi.org/10.1016/j.scitotenv.2020.137480.

Weber, Elke U. 2006. "Experience-Based and Description-Based Perceptions of Long-Term Risk: Why Global Warming Does Not Scare Us (Yet)". *Climatic Change* 77 (1–2): 103–120. https://doi.org/10.1007/s10584-006-9060-3.

Weizenbaum, Joseph. 1976. *Computer Power and Human Reason: From Judgement to Calculation*. San Francisco: WHFreeman.

Whittaker, Meredith. 2020. "Panel Response to Lilly Irani." Presented at the Collective Organizing and Social Responsibility at CSCW, 23rd ACM Conference on Computer-Supported Cooperative Work and Social Computing, November. https://sites.google.com/view/collectiveorganizing.

Whittaker, Meredith. 2021. "The Steep Cost of Capture." *Interactions* 28 (6): 50–55. https://doi.org/10.1145/3488666.

Whittle, Jon, Maria Angela Ferrario, William Simm, and Waqar Hussain. 2019. "A Case for Human Values in Software Engineering." *IEEE Software*, November. https://doi.org/10.1109/MS.2019.2956701.

Widmer, Rolf, Heidi Oswald-Krapf, Deepali Sinha-Khetriwal, Max Schnellmann, and Heinz Böni. 2005. "Global Perspectives on E-Waste." *Environmental Impact Assessment Review*, Environmental and Social Impacts of Electronic Waste Recycling, 25 (5): 436–458. https://doi.org/10.1016/j.eiar.2005.04.001.

Wiedmann, Thomas, Manfred Lenzen, Lorenz T. Keyßer, and Julia K. Steinberger. 2020. "Scientists' Warning on Affluence." *Nature Communications* 11 (1): 3107. https://doi.org/10.1038/s41467-020-16941-y.

Wiedmann, Thomas O., Heinz Schandl, Manfred Lenzen, Daniel Moran, Sangwon Suh, James West, and Keiichiro Kanemoto. 2015. "The Material Footprint of Nations." *Proceedings of the National Academy of Sciences* 112 (20): 6271–6276. https://doi.org/10.1073/pnas.1220362110.

Wiek, Arnim, Lauren Withycombe, and Charles L. Redman. 2011. "Key Competencies in Sustainability: A Reference Framework for Academic Program Development." *Sustainability Science* 6 (2): 203–218. https://doi.org/10.1007/s11625-011-0132-6.

Willmott, H. 1989. "Or as a Problem Situation: From Soft Systems Methodology to Critical Science." In *Operational Research and the Social Sciences*, edited by M. C. Jackson, P. Keys, and S. A. Cropper, 65–78. Boston, MA: Springer US. https://doi.org/10.1007/978-1-4613-0789-1_8.

Wing, Jeanette W. 2015. "A Systemic Framework for Improving Clients' Understanding of Software Requirements." http://aisel.aisnet.org/ecis2015_rip/10.

Wing, Jeannette M. 2006. "Computational Thinking." *Communications of the ACM* 49 (3): 33–35.

Winner, Langdon. 1980. "Do Artifacts Have Politics?" *Daedalus* 109 (1): 121–136.

Winograd, Terry, and Fernando Flores. 1986. *Understanding Computers and Cognition: A New Foundation for Design*. Language and Being. Norwood, NJ: Ablex.

Wisnioski, Matthew H. 2012. *Engineers for Change: Competing Visions of Technology in 1960s America*. Engineering Studies Series. Cambridge, MA: MIT Press.

Wong, Euphemia. 2020. "What Is a Wicked Problem and How Can You Solve It?" The Interaction Design Foundation. https://www.interaction-design.org/literature /article/wicked-problems-5-steps-to-help-you-tackle-wicked-problems-by-combining -systems-thinking-with-agile-methodology.

Wylie, Bianca. 2020. "In Toronto, Google's Attempt to Privatize Government Fails— For Now." Text. Boston Review. May 12. http://bostonreview.net/politics/bianca -wylie-no-google-yes-democracy-toronto.

Yu, Eric. 2011. *Social Modeling for Requirements Engineering*. Cambridge, MA: MIT Press.

Zannier, Carmen, Mike Chiasson, and Frank Maurer. 2007. "A Model of Design Decision Making Based on Empirical Results of Interviews with Software Designers." *Information and Software Technology*, Qualitative Software Engineering Research, 49 (6): 637–653. https://doi.org/10.1016/j.infsof.2007.02.010.

Zdun, Uwe. 2013. "Sustainable Architectural Design Decisions." *IEEE Software* 30 (6): 46–53.

Zowghi, Didar, and Chad Coulin. 2005. "Requirements Elicitation: A Survey of Techniques, Approaches, and Tools." In *Engineering and Managing Software Requirements*, edited by Aybüke Aurum and Claes Wohlin, 19–46. Berlin, Heidelberg: Springer. https://doi.org/10.1007/3-540-28244-0_2.

Zsambok, Caroline E., and Gary A. Klein, eds. 1997. *Naturalistic Decision Making*. Mahwah, NJ: L. Erlbaum Associates.

INDEX

Note: Figures and tables are indicated by *italicized* page numbers.